MERCURY CONTAMINATION: A HUMAN TRAGEDY
Patricia A. D'Itri and Frank M. D'Itri

POLLUTANTS AND HIGH RISK GROUPS

MENTAL AND

Vitamins

nerals and Macronutrients

RESOURCE RECOVERY AND RECYCLING
A. F. M. Barton

QUANTITATIVE TOXICOLOGY
V. A. Filov, A. A. Golubev, E. I. Liublina, and N.A. Tolokontsev

WASTES IN THE OCEAN

ENVIRONMENTAL SCIENCE AND TECHNOLOGY

A Wiley-Interscience Series of Texts and Monographs

WASTES IN THE OCEAN

Editors: Iver W. Duedall
Dana R. Kester
Bostwick H. Ketchum
P. Kilho Park

Volume 1: INDUSTRIAL AND SEWAGE WASTES IN THE OCEAN
Editors: Iver W. Duedall, Bostwick H. Ketchum, P. Kilho Park, and Dana R. Kester

Volume 2: DREDGED-MATERIAL DISPOSAL IN THE OCEAN
Editors: Dana R. Kester, Bostwick H. Ketchum, Iver W. Duedall, and P. Kilho Park

Volume 3: RADIOACTIVE WASTES AND THE OCEAN
Editors: P. Kilho Park, Dana R. Kester, Iver W. Duedall, and Bostwick H. Ketchum

In Preparation

Volume 4: ENERGY WASTES IN THE OCEAN
Editors: Iver W. Duedall, Dana R. Kester, P. Kilho Park, and Bostwick H. Ketchum

Volume 5: DEEP-SEA WASTE DISPOSAL
Editors: Dana R. Kester, Wayne V. Burt, Judith M. Capuzzo, P. Kilho Park, Bostwick H. Ketchum, and Iver W. Duedall

Volume 6: NEAR-SHORE WASTE DISPOSAL
Editors: Bostwick H. Ketchum, Judith M. Capuzzo, Wayne V. Burt, Iver W. Duedall, P. Kilho Park, and Dana R. Kester

WASTES IN THE OCEAN Volume 1

INDUSTRIAL AND SEWAGE WASTES IN THE OCEAN

Edited by

IVER W. DUEDALL
State University of New York
Stony Brook, New York

BOSTWICK H. KETCHUM
Woods Hole Oceanographic Institution
Woods Hole, Massachusetts

P. KILHO PARK
National Oceanic and Atmospheric Administration
Rockville, Maryland

DANA R. KESTER
University of Rhode Island
Kingston, Rhode Island

A WILEY-INTERSCIENCE PUBLICATION

JOHN WILEY & SONS

New York Chichester Brisbane Toronto Singapore

Library of Congress Cataloging in Publication Data:

Main entry under title:

Industrial and sewage wastes in the ocean.

(Wastes in the ocean; v. 1) (Environmental science and technology, ISSN 0194-0287)
"A Wiley-Interscience publication."
Includes indexes.
1. Waste disposal in the ocean--Addresses, essays, lectures. I. Duedall, Iver W. II. Series. III. Series: Environmental science and technology.

TD763.I53 1983 363.7'28 82-13695
ISBN 0-471-09772-1

Printed in the United States of America

10 9 8 7 6 5 4 3 2 1

We, the remaining editors, honor

DR. BOSTWICK H. KETCHUM

January 21, 1912–July 15, 1982
who, until his death, had worked for and
believed in the immortality of
humanity and science

Iver W. Duedall
P. Kilho Park
Dana R. Kester

CONTRIBUTORS

BAXTER, II, LINCOLN, Woods Hole Oceanographic Institution, Woods Hole, Massachusetts 02543

BODAMMER, SHAWN M., U.S. Department of Commerce, National Marine Fisheries Service, Northeast Fisheries Center, Oxford, Maryland 21654

BODENNEC, GUY, Centre National Pour L'Exploitation Des Oceans, Biote Postale 337, 29243 Brest Cedex, France

BROOKS, JAMES M., Department of Oceanography, Texas A & M University, College Station, Texas 77843

BROWN, JEFFREY F., Bigelow Laboratory for Ocean Sciences, West Boothbay Harbor, Maine 04575

BROWN, MARY F., Graduate School of Oceanography, University of Rhode Island, Kingston, Rhode Island 02881

BUYER, JEFFREY H., University of California San Diego, D-006, La Jolla, California 92093

CARLETON, HERBERT R., College of Engineering and Applied Sciences, State University of New York, Stony Brook, New York, 11794

CAVARI, BENZION, Department of Microbiology, University of Maryland, College Park, Maryland 20742

COLWELL, RITA R., Department of Microbiology, University of Maryland, College Park, Maryland 20742

CSANADY, GABRIEL T., Woods Hole Oceanographic Institution, Woods Hole, Massachusetts 02543

DAYAL, RAMESH, Deparment of Nuclear Energy, Brookhaven National Laboratory, Upton, New York, 11973

DEMING, JODY W., Department of Microbiology, University of Maryland, College Park, Maryland 20742

DOWD, JOYCE M., Graduate School of Oceanography, University of Rhode Island, Kingston, Rhode Island 02881

DUEDALL, IVER W., Department of Oceanography and Ocean Engineering, Florida Institute of Technology, Melbourne, Florida, 32901

FLIERL, GLENN R., Department of Meterology and Physical Oceanography, Massachusetts Institute of Technology, Cambridge, Massachusetts 02139

GUNN, BRUCE, Department of Microbiology, University of Maryland, College Park, Maryland 20742

HAUGEN, ELIN M., Bigelow Laboratory for Ocean Sciences, West Boothbay Harbor, Maine 04575

HEATON, MONTEITH G., Interstate Electronics Corporation, 707 East Vermont Avenue, P.O. Box 3117, Anaheim, California 92803

HERSHEY, J. PETER, Rosenstiel School of Marine and Atmospheric Science, University of Miami, Miami, Florida 33149

KESTER, DANA R., Graduate School of Oceanography, University of Rhode Island, Kingston, Rhode Island 02881

KETCHUM, BOSTWICK H., Senior Scientist Emeritus, Woods Hole Oceanographic Institute, Woods Hole, Massachusetts 02543

KULLENBURG, GUNNAR E. B., Institute of Physical Oceanography, University of Copenhagen, Haralsgade 6, 2200 Copenhagen N., Denmark

LEAR, DONALD W., U.S. Environmental Protection Agency–U.S. National Oceanic and Atmospheric Administration Oceanographic Unit, 839 Bestgate Road, Annapolis, Maryland 21401

LEE, WEN YUH, The University of Texas Marine Science Institute, Port Aransas Marine Laboratory, Port Aransas, Texas 78373

MANHEIM, FRANK T., U.S. Geological Survey, Woods Hole, Massachusetts 02543

MUIR, WILLIAM C., Environmental Impacts Branch, U.S. Environmental Protection Agency (Region III), Sixth and Walnut Streets, Philadelphia, Pennsylvania 19106

MUKHERJI, PRITHVIRAJ, Revere Research, Inc., P.O. Box 1352, Edison, New Jersey 08837

MURPHY, LYNDA S., Bigelow Laboratory for Ocean Sciences, West Boothbay Harbor, Maine 04575

OAKLEY, SUE A., Oceanographic Sciences Division, Brookhaven National Laboratory, Upton, New York 11973

OKUBO, AKIRA, Marine Sciences Research Center, State University of New York, Stony Brook, New York 11794

O'MALLEY, MARRIA L., U.S. Environmental Protection Agency–National Oceanic and Atmospheric Administration Oceanographic Unit, 839 Bestgate Road, Annapolis, Maryland 21401

OOSTDAM, BERNARD L., Department of Earth Sciences, Millersville State College, Millersville, Pennsylvania 17551

ORR, MARSHALL H., Seismology Department, Gulf Research and Developent Company, P.O. Drawer 2038, Pittsburgh, Pennsylvania, 15230

PARK, P. KILHO, Ocean Dumping Program, National Oceanic and Atmospheric Administration, Rockville, Maryland 20852

PARKER, JR., Faust R., The University of Texas Marine Science Institute, Port Aransas, Texas 78373

PEELE, EMILY R., Department of Microbiology, University of Maryland, College Park, Maryland 20742

ROETHEL, FRANK J., Marine Sciences Research Center, State University of New York, Stony Brook, New York 11794

SAUER, JR., THEODOR C., Exxon Production Research Company, P.O. Box 2189, Houston, Texas 77001

SAWYER, THOMAS K., U.S. Department of Commerce, National Marine Fisheries Service, Northeast Fisheries Center, Oxford, Maryland 21654

SINGLETON, FRED L., Department of Microbiology, University of Maryland, College Park, Maryland 20742

TATRO, MARK, Perkin-Elmer Incorporated, Elmwood, New Jersey 07407

WANG, FRANKLIN F. Y., College of Engineering and Applied Sciences, State University of New York, Stony Brook, New York 11794

WILKE, RICHARD J., Oceanographic Sciences Division, Brookhaven National Laboratory, Upton, New York 11973

WIESENBURG, DENIS A., Naval Ocean Research and Development Activity, NSTL Station, Mississippi 39529

WOHLSCHLAG, DONALD E., The University of Texas Marine Science Institute, Port Aransas, Texas 78373

SERIES PREFACE

Environmental Science and Technology

The Environmental Science and Technology Series of Monographs, Textbooks, and Advances is devoted to the study of the quality of the environment and to the technology of its conservation. Environmental science therefore relates to the chemical, physical, and biological changes in the environment through contamination or modification, to the physical nature and biological behavior of air, water, soil, food, and waste as they are affected by man's agricultural, industrial, and social activities, and to the application of science and technology to the control and improvement of environmental quality.

The deterioration of environmental quality, which began when man first collected into villages and utilized fire, has existed as a serious problem under the ever-increasing impacts of exponentially increasing population and of industrializing society. Environmental contamination of air, water, soil, and food has become a threat to the continued existence of many plant and animal communities of the ecosystem and may ultimately threaten the very survival of the human race.

It seems clear that if we are to preserve for future generations some semblance of the biological order of the world of the past and hope to improve on the deteriorating standards of urban public health, environmental science and technology must quickly come to play a dominant role in designing our social and industrial structure for tomorrow. Scientifically rigorous criteria of environmental quality must be developed. Based in part on these criteria, realistic standards must be established and our technological progress must be tailored to meet them. It is obvious that civilization will continue to require increasing amounts of fuel, transportation, industrial chemicals, fertilizers, pesticides, and countless other products; and that it will continue to produce waste products of all descriptions. What is urgently needed is a total systems approach to modern civilization through which the pooled talents of scientists and engineers, in cooperation with social scientists and the medical profession, can be focused on the development of order and equilibrium in the presently disparate segments of the human environment. Most of the skills and tools that are needed are already in existence. We surely have a right to hope a technology that has created such manifold environmental problems is also capable of solving them. It is our hope that this series in Environmental

Sciences and Technology will not only serve to make this challenge more explicit to the established professionals, but that it also will help to stimulate the student toward the career opportunities in this vital area.

Robert L. Metcalf
Werner Stumm

PREFACE TO *WASTES IN THE OCEAN*

This is the first of six volumes considering the problems of *Wastes in the Ocean* in which we consider the following subjects:

Volume 1, "Industrial and Sewage Wastes in the Ocean,"
Volume 2, "Dredged-Material Disposal in the Ocean,"
Volume 3, "Radioactive Wastes and the Ocean,"
Volume 4, "Energy Wastes in the Ocean,"
Volume 5, "Deep-Sea Waste Disposal," and
Volume 6, "Near-Shore Waste Disposal."

The objectives are to present a comprehensive overview of the state of our knowledge concerning the disposal of waste in the ocean and to present new and original contributions to the evaluation of the impact of the disposal of waste materials on human life and well-being, on the marine biota, on amenities, and on legitimate uses of the ocean. The chapters included in this and succeeding volumes of this series have been subjected to both external and editorial reviews. We are especially grateful to the reviewers of these chapters for the time and effort they devoted to the development of the final manuscripts.

The burgeoning human population on earth and the continuing development of complex industrial technology have inevitably led to enormous increases in both the quantity and the kind of waste material that must be disposed of in ways that do not cause an intolerable degradation of our environment. The optimum solution to the problem is to recycle the waste material in ways that produce a beneficial effect. This is not always possible and various types of treatment can be employed to minimize the quantity of waste and to make the product of the treatment less damaging to our environment. Even after treatment, there will be some residue that must be contained or discharged to the environment.

The options for environmental discharge are limited: on land, into the atmosphere, or into the hydrosphere. Before any one of these particular environments is selected for any specific waste material there should be a careful scientific analysis of the possible impacts of such a disposal operation. It is hoped that this series will provide the framework for the evaluation of the impact of specific types of waste in the ocean.

From time immemorial people have been disposing of waste materials into the marine environment or into the rivers and streams which ultimately lead to the sea. For millenia it was assumed that the oceans are so vast that our puny efforts would have no measurable or damaging impact. Within the last century it has become clear that some semiconfined bodies of water were being degraded seriously and that the disposal of waste into the hydrosphere must be managed and controlled in order to preserve the integrity of the oceans. We now know that some pollutants are distributed worldwide and can be identified and measured in the waters of the open sea far from the source. The problems associated with ocean disposal of waste material require careful and critical evaluation so that we may assure that the valuable resources of the sea are preserved and protected for future generations. This will require the most careful evaluation of the impact of waste disposal at sea so that we may use the oceans wisely.

We are very thankful to Mr. Treville Leger, Editor, John Wiley & Sons, for his constant encouragement and for being very helpful in the preparation of these volumes.

The Editors

PREFACE

Every year millions of metric tons of industrial wastes and sewage sludges are dumped into the ocean. Scientific research and public debate about the behavior, fate, and effects of these wastes in the sea have increased greatly since 1970. On the global scale, ocean dumping of wastes will probably increase with time. Future dumping will be controlled more rigorously by national laws and international conventions. Application of these laws and conventions will require the understanding, which can be attained through scientific research, of wastes in the ocean.

The decision to dump a particular waste at sea will depend on the economics of disposal, the public perception of its impact, and consideration of environmental effects. In this book, *Industrial and Sewage Wastes in the Ocean*, we are concerned with the latter consideration. We believe that the findings presented here can provide a basis for better informed public perceptions.

This book contains 20 chapters most of which are technical considerations written for scientists. The chapters were organized in seven parts along lines of a theme or discipline, where possible, in order to provide continuity.

The authors provide information on global ocean dumping, the role of U.S. Federal agencies in ocean dumping research, physical oceanographic aspects and chemical effects of ocean dumping, toxic effects of pharmaceutical and other industrial wastes, ocean dumping at the U.S. Mid-Atlantic dumpsites, the role of marine amoebae in sediment, and physical and chemical properties of stabilized coal wastes. The ocean dumping of coal wastes may become an important disposal alternative for populated coastal cities where land is scarce. Scientific strategy on industrial and sewage wastes disposal in the ocean is the main topic of the concluding chapter.

Although this book is not intended to provide a systematic presentation or treatise on all the scientific aspects of ocean dumping, it does reflect the strong and continuing interest in both the theoretical and descriptive studies on the dumping of industrial and sewage wastes. It provides a better understanding of the problems related to the behavior and the effects of these wastes in the sea.

We are grateful to many people who assisted us in the preparation of this book. We express our appreciation to members of the Second International Ocean Disposal Symposium Executive Committee who, in addition to the editors, included Professor Michael A. Champ, Dr. Thomas P. O'Connor, and Dr. Marshall H. Orr.

We thank the reviewers who accepted the arduous task of refereeing the manuscripts. Several people at the Marine Sciences Research Center of the State University of New York at Stony Brook helped significantly in the preparation of the manuscript. We thank Laura Antonacci who patiently retyped most of the chapter manuscripts and maintained organization in the more or less constant communication between authors and editors; Jennifer Jesty who was meticulous in providing editorial assistance and technical comment on each chapter manuscript; Jacqueline Restivo and Mary Ann Lau for technical assistance on a word processor; Marie Gladwish, Marie Eisel, and Vivian Abolins for their care in the touch-up and redrafting of many figures to ensure consistency in illustrations throughout the book; and Anne Lannak, Yong C. Park, and Tom Edwards for doing paste-up and mechanicals, and also for completing countless other tasks. For the photographs on the part title pages we thank W. N. Adams, David Ball, J. M. Brooks, James W. Brown, G. M. Capriulo, Dennis K. Clark, R. R. Colwell, C. E. Firstenberg, G. R. Flierl, Howard R. Gordon, J. C. Hathaway, P. P. Lapennas, W. Y. Lee, J. H. Parker, and F. J. Roethel. This work was supported in part by U.S. National Oceanic and Atmospheric Administration grant 04-8-M01-192.

IVER W. DUEDALL
BOSTWICK H. KETCHUM
P. KILHO PARK
DANA R. KESTER

Stony Brook, New York
December 1982

CONTENTS

PART III CHEMICAL EFFECTS: Acid-Iron and Pharmaceutical Wastes

PART IV BIOLOGICAL AND TOXIC EFFECTS: Pharmaceutical Wastes

PART V THE MID-ATLANTIC AND NEW YORK BIGHT DUMPSITES

GLOSSARY OF ACRONYMS

AAS	Atomic absorption spectrophotometry
AODC	Acridine orange direct count
ASTM	American Society for Testing Materials
ATP	Adenosine triphosphate
BLM	U.S. Bureau of Land Management
CEQ	U.S. Council for Environmental Quality
CFA	Continuous flow analyzer
CFU	Colony-forming units
CTD	Conductivity-temperature-depth
DDW	Distilled-deionized water
DMA	*N,N*-Dimethylaniline
DMRP	U.S. Dredged Material Research Program
DNA	Deoxyribose nucleic acid
DOD	U.S. Department of Defense
DOE	U.S. Department of Energy
DOI	U.S. Department of Interior
DVC	Direct viable counts
EDTA	Ethylenediamine tetraacetic acid
EPA	U.S. Environmental Protection Agency
GC	Gas chromatography
GCMS	Gas chromotography-mass spectrometry
GESAMP	IMCO/FAO/UNESCO/WMO/WHO/IAEA/UN/UNEP Joint Group of Experts on the Scientific Aspects of Marine Pollution
GEOSECS	Geochemical Ocean Sections Study
HEW	U.S. Department of Health, Education, and Welfare

ICOP	Interagency Committee on Ocean Pollution
IMCO	Inter-Governmental Maritime Consultative Organization of United Nations
IUCS	IU Conversion Systems, Inc.
JOIDES	Joint Oceanographic Institutes' Deep Earth Sampling Program
LDC	London Dumping Convention
LORAN	Long Range Navigation
MESA	Marine EcoSystems Analysis
MIBK	Methylisobutylketone
MPRSA	U.S. Marine Protection, Research, and Sanctuaries Act
MS	Mass Spectrometry
NACOA	U.S. National Advisory Committee on Oceans and Atmosphere
NASA	U.S. National Aeronautics and Space Administration
NOAA	U.S. National Oceanic and Atmospheric Administration
NODC	U.S. National Oceanographic Data Center
NSF	U.S. National Science Foundation
NTIS	U.S. National Technical Information Service
OCS	Outer continental shelf
PCB	Polychlorinated biphenyl
PVC	Polyvinyl chloride
SIC	Standard Industrial Classification
SSIE	Smithsonian Science Information Exchange
WHOI	Woods Hole Oceanographic Institution
XBT	Expendable bathythermograph

WASTES IN THE OCEAN

PART I: INTRODUCTION

1

GLOBAL INPUTS, CHARACTERISTICS, AND FATES OF OCEAN-DUMPED INDUSTRIAL AND SEWAGE WASTES: AN OVERVIEW

Iver W. Duedall

Marine Sciences Research Center
State University of New York
Stony Brook, New York

Bostwick H. Ketchum

Senior Scientist Emeritus
Woods Hole Oceanographic Institution
Woods Hole, Massachusetts

P. Kilho Park

Ocean Dumping Program
National Oceanic and Atmospheric Administration
Rockville, Maryland

Dana R. Kester

Graduate School of Oceanography
University of Rhode Island
Kingston, Rhode Island

Dr. Duedall's present address: Department of Oceanography and Ocean Engineering, Florida Institute of Technology, Melbourne, Florida, 32901.

ABSTRACT

During the 1976–1979 period, the Inter-Governmental Maritime Consultative Organization (IMCO) had been notified of over 2000 permits which were issued mainly for the disposal of dredged material, industrial wastes, and sewage sludge into the ocean. For these wastes, total annual tonnages ranged from 35×10^6 to 231×10^6 metric tons for 1976 and 1978, respectively. Estimated tonnages of industrial wastes ranged from 10×10^6 in 1978 to 18×10^6 t y^{-1} in 1979, with the Unites States, France, and the United Kingdom leading all other countries. Annual tonnages of sewage sludge, essentially all from the United States, United Kingdom, and Federal Republic of Germany during the 1976 through 1979 period remained nearly constant at about 16×10^6 t y^{-1}.

Physical and chemical properties of industrial and sewage wastes are characterized. Abundances of toxic metallic elements in these wastes varied by orders of magnitude when compared on an element-to-element basis. Fly ash and sewage sludge have the highest concentrations of elements of environmental concern. The distribution and fate of an ocean-dumped waste in the sea are complicated, depending on (1) the physical processes of dispersion, advection, and sedimentation; (2) chemical processes such as volatilization, neutralization, precipitation, flocculation, adsorption, desorption, dissolution, oxidation, and reduction; and (3) biological processes involving responses of marine organisms to waste materials, incorporation of these materials within the organism, and modification of waste substances by organisms.

1.1. INTRODUCTION

The ease of disposal of wastes in the ocean must have been recognized by early civilizations living near the sea. But environmentally degrading pollution through activities by man in rivers and streams leading to the ocean probably began during the industrial revolution when primitive sewage systems of great cities like London became overburdened by industrial and sewage wastes (Topping, 1976; Ketchum, 1980). Now late in the twentieth century millions of metric tons of industrial wastes and sewage sludges enter the ocean worldwide every year via scheduled ocean dumping by many of the most technologically advanced countries (Fig. 1.1).

The practice of ocean dumping has created, within those countries, a conflict over appropriate activities and uses of the sea. Some would give the ocean preferential status viewing it as the last frontier, to be left unspoiled by man's activities. Others believe the ocean can be used wisely in the assimilation of the wastes and pollutants that enter it (Goldberg, 1981; Osterberg, 1981). In either case, the need for marine disposal sites for wastes will continue to require progressively accurate scientific understanding of choices as other options are exhausted.

The effects of ocean disposal of wastes on the marine environment, including the biota, depend on complicated interactions among many factors or processes, such as the quantities and kinds of waste, chemical and physical properties of the waste and its rate of discharge into the ocean, the interaction of waste with seawater, its toxicity to organisms, oceanic depth, and mixing and other physical processes (Johnston, 1976).

The main purpose of this volume is to increase our scientific understanding of the effects on the sea of ocean disposal of wastes. How widespread is ocean dumping and what are the quantities and types of waste? Except for a recently published article on global ocean dumping of sewage sludge by Champ and Park (1981), the magnitude of ocean disposal on a global scale has not been examined. Thus, in this introductory chapter we present, on a worldwide basis, the number of ocean dumping permits (excluding pipeline discharges), the magnitude of industrial and sewage wastes dumped in the ocean, some physical and chemical properties of these wastes, and a brief discussion on physical, chemical, and biological processes that affect the behavior of wastes in the ocean.

1.2. SOURCE OF INFORMATION: LONDON DUMPING CONVENTION

Since 1976 the Inter-Governmental Maritime Consultative Organization (IMCO) has been receiving reports of dumping permits issued by the Contracting Parties to the London Dumping Convention (LDC). The LDC originated in June 1972 from a recommendation of the United Nations Conference on the Human Environment, Stockholm (Timagenis, 1980). The LDC, which covers all marine waters other than the internal waters of Contracting Parties, entered into force on 30 August 1975, is known formally as the *Convention on the Prevention of Marine Pollution by Dumping of Wastes and Other Matter* (IMCO, 1976).

Each Contracting Party to the LDC is required to establish a national system to control ocean dumping, to maintain records of ocean dumping activities through a

Figure 1.1. The disposal of wastes in the open ocean. (a) Pharmaceutical wastes discharged in the Puerto Rico dumpsite; (b) acid-iron wastes at Deepwater Dumpsite-106 (DWD-106). In (b), the wake behind the barge is bright yellow due to precipitating iron oxide particles.

permit program, to monitor the conditions of the sea for the purposes of the LDC, and to report records of permits and information on monitoring activities to IMCO (Moore, 1976; Timagenis, 1980). IMCO prepares summary reports annually for circulation to all Contracting Parties, providing information on the number of permits, quantities and types of wastes to be discharged at sea, location of the dumpsite, and other related information. The Fourth Consultative Meeting of Contracting Parties to LDC in October 1979 adopted a procedure for the notification of types and quantities of wastes which have been actually dumped per calendar year. The first report of actually dumped wastes is being prepared by IMCO and will be circulated soon (M. Nauke of IMCO, personal communication).

The LDC defines ocean dumping as follows (IMCO, 1976):

1. Any deliberate disposal at sea of wastes or other matter from vessels, aircraft, platforms or other man-made structures at sea.
2. Any deliberate disposal at sea of vessels, aircraft, platforms or other man-made structures at sea.

The LDC excludes from the definition of ocean dumping:

1. The disposal at sea of wastes or other matter incidental to, or derived from the normal operations of vessels, aircraft, platforms, or other man-made structures at sea and their equipment, other than wastes or other matter transported by or to vessels, aircraft, platforms, or other man-made structures at sea, operating for the purpose of disposal of such matter or derived from the treatment of such wastes or other matter on such vessels, aircraft, platforms or structures.
2. Placement of matter for a purpose other than the mere disposal thereof, provided that such placement is not contrary to the aims of the Convention.
3. The disposal of wastes or other matter directly arising from, or related to the exploration, exploitation and associated off-shore processing of sea-bed mineral resources.

1.3. OCEAN DUMPING PERMITS

1.3.1. Number of Permits and Categories of Wastes

In the period 1976–1979, IMCO (1979a, 1979b, 1980, 1981) reported that over 2000 permits were issued for the disposal of wastes in the sea by the LDC states (Fig. 1.2). Several countries have indicated that they issue additional permits, for example, for the disposal of wastes in marine waters that are internal waters (bays and estuaries) and therefore do not fall within the scope of the LDC.

In the 1976–1979 time period, which is too short to draw many conclusions from, the number of permits issued per year has increased at an average of about 70 per year. In some cases, permits valid for periods of more than one year for the disposal of wastes at sea have been issued for validity periods of more than one year. The number of permits therefore may not necessarily reflect the dumping activity of the Contracting Parties.

Permitted annual tonnages of industrial wastes are increasing ($\simeq 11 \times 10^6$ in 1976 to $\simeq 18 \times 10^6$ wet tons in 1979), but at a variable rate (Fig. 1.2). The permit tonnages of sewage sludge from 1977 to 1979 has remained the same at $\simeq 16 \times 10^6$ wet tons per year; in 1976, the permit tonnage of sewage sludge was 13.5×10^6 tons (Fig. 1.2). We make clear, however, that the permit tonnages (Fig. 1.2) of

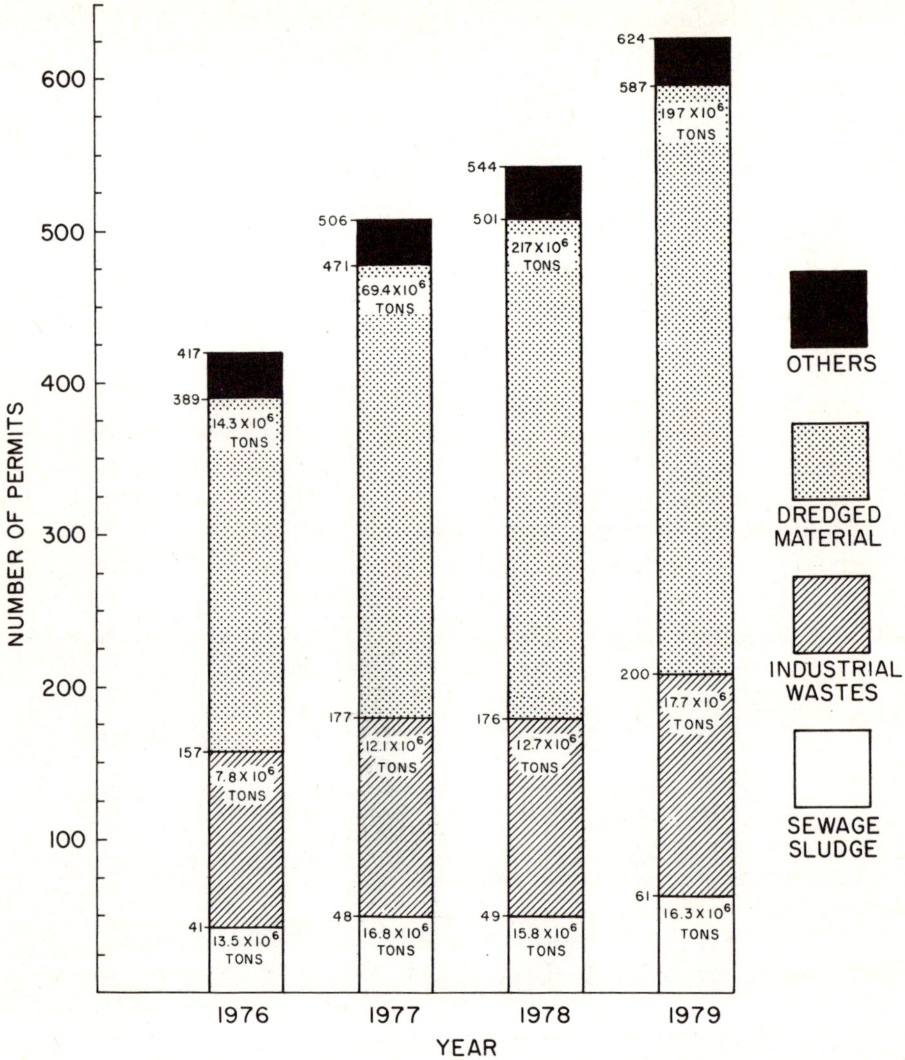

Figure 1.2. Total number of permits issued and estimated tonnages (metric tons), on a global basis, for the disposal of wastes in the sea (IMCO, 1979a, 1979b, 1980, 1981). The tonnages for "Others" are not given because the quantities were either not reported or the data in the IMCO reports were considered not applicable in estimating total tonnages. Organohalogens, for example, are incinerated at sea rather than dumped in the ocean; their tonnages are not included in the industrial wastes.

waste published in the IMCO summary reports are anticipated or permitted quantities for disposal as reported in the permits; actual tonnages dumped at sea are not given in the IMCO reports and may be quite different from those in the permits. It can, however, be assumed that 80% of the amounts permitted for disposal have been actually dumped at sea (M. Nauke and M. A. Champ, personal communication).

The principal waste categories listed in the IMCO reports in order of both decreasing numbers of permits and tonnages are dredged material, industrial wastes, and sewage sludge. Industrial wastes and sewage sludge, the two wastes of main interest in this volume, together accounted for 33-37% of the permits issued per year during the 1976-1979 period. The majority of permits issued were for dredged material; the tonnages (Fig. 1.2) in 1977-1979 were several times those of either the industrial wastes or sewage sludge [see Kester et al. (1983) for further discussion of dredged material].

Relatively fewer permits were issued in 1976-1979 for a rather diverse range of other wastes such as ammunition, engine parts, excavation material, handcuffs, organohalogens and driftwood (both for incineration at sea), radioactive material, seal carcasses, old ships, and weapons and firearms (see "Others" in Fig. 1.2). Radioactive waste disposal at sea requires special permits (see Hagen, 1983). The tonnages of these other wastes are difficult to determine accurately and are small compared to the total of sewage sludge, industrial wastes, and dredged material.

IMCO also reports the cases of *force majeure*, the term used to describe the dumping of cargo at sea in order to prevent danger to human life or to vessels, aircraft, platforms, or other man-made structures caused by the stress of weather. Between 1976 and 1979 two cases—a fishing vessel in 1978 and a tanker in 1979, both Canadian—of *force majeure* were reported to IMCO (1980, 1981).

A more detailed account of who dumps what is given in Table 1.1, which provides a breakdown of the types of waste authorized to be dumped at sea in 1979 by each Contracting Party to the LDC. During 1979, the United States, Canada, and the United Kingdom led by far all countries in the issuing of the largest numbers of permits for ocean dumping. As we will see later, however, Canada's permitted tonnages of wastes in the sea was only 1% of the total global industrial and sewage wastes for 1979. Thus, a direct relationship between the number of permits and tonnages is not always observed.

A majority of the Contracting Parties to the LDC either issued no reports or provided no information on dumping at sea (Table 1.1). Many of the countries providing no information and an undetermined number of non-LDC member countries may conduct some form of ocean dumping. Thus the anticipated tonnage for each year shown in Fig. 1.2 represents the lower limit for global input of wastes during the 1976-1979 period.

Table 1.1 also gives the number of permits for the different subcategories of industrial and sewage wastes. These include aqueous solutions, dilute acids, industrial sludges (sometimes stored in drums which are dumped into the ocean), coal wastes, caustic materials, and fish wastes. The IMCO summary reports contain no information on the chemical composition (quantitative analysis) of the various wastes for which permits were issued.

The number of Contracting Parties to the LDC is increasing (M. Nauke, personal communication). As of 1 October 1981, 47 countries have ratified or acceded to the convention. With Japan, which became a Contracting Party in late 1980, and the anticipated ratification of the convention by Ireland, Italy, and Australia in 1982 (M. Nauke, personal communication), most industrialized countries will be Contracting Parties to the LDC. At present ocean disposal activities of emerg-

Table 1.1. Members of the London Dumping Convention and Record of Permit Activity for 1979 (IMCO, 1981)

Country and Wastes	Number of Permits Issued	No Information
Afghanistan	–	X
Argentina	–	X
Byelorussian SSR	Nil	
Belgium[a]		
Industrial wastes		
Aqueous solutions	2	
Sludges	1	
Diluted acids	2	
Dredged material	8	
Radioactive wastes	1[b]	
Canada		
Industrial wastes		
Fish offal	17	
Solids (some to break up ice)	6	
Ammunition	1	
Dredged materials	183	
Scientific experiment (crude oil)	1	
Ship	1	
Force majeure[c]	1	
Cape Verde	Nil	
Chile	–	X
Cuba	–	X
Denmark		
Industrial waste	1	
Dredged material	7	
Dominican Republic	–	X
Finland		
Incineration of organohalogens	1	
France		
Industrial waste	1	
Dredged material	2	
Germany, Democratic Republic of	–	X
Germany, Federal Republic of		
Industrial waste		
Diluted acid containing copper	1	
Sewage sludge		
Organohalogens (for incineration)	4	
Guatemala	–	X
Haiti	–	X
Hungary	–	X
Iceland	Nil	
Japan[d]	–	X
Jordan	–	X
Kenya	–	X

Table 1.1. *Continued*

Country and Wastes	Number of Permits Issued	No Information
Libyan Arab Jamahiriya	–	X
Mexico	Nil	
Monaco	–	X
Morocco	–	X
Netherlands		
Industrial waste		
Aqueous solutions	1	
Diluted acids	3	
Polluted water	1	
Dredged material	2	
Radioactive waste[b]	1	
Organohalogens (for incineration)	3	
New Zealand		
Industrial wastes		
Liquid	1	
Dredged material	17	
Nigeria	–	X
Norway		
Organohalogens (for incineration)	1	
Panama	–	X
Philippines	–	X
Poland	–	X
Portugal	Nil	
South Africa	Nil	
Spain		
Industrial waste		
Diluted acid with coppers	1	
Coal ashes	1	
Sweden		
Organohalogens (for incineration)	3	
Switzerland[b]	Nil	
Tunisia	–	X
Ukrainian SSR	Nil	
USSR	Nil	
United Arab Emirates	–	X
United Kingdom		
Industrial wastes		
Liquids	24	
Sludges (drums containing sludge)	43	
Diluted acids	5	
Caustic materials	18	
Colliery waste and coal ash	3	
Drums, containers, sludges, and solids	4	
Sewage sludge	39	

Table 1.1. *Continued*

Country and Wastes	Number of Permits Issued	No Information
Dredged material	67	
Radioactive waste	1	
United Kingdom (Hong Kong Government)		
Mutilated weapons	1	
Excavation material	7	
Dredged material	15	
United States		
Industrial wastes		
Aqueous and liquid wastes	12	
Diluted acids	3	
Solids (construction debris)	1	
Sewage sludge	26	
Dredged material	86	
Driftwood	2	
Experiment (oil and dispersants)	1	
Vessel	1	
Yugoslavia	–	X
Zaire	–	X

[a] Belgium is not yet a contracting party to the London Dumping Convention.
[b] Joint disposal operation by Belgium, Netherlands, and Switzerland.
[c] Bow section of a tanker.
[d] Japan recently became a Contracting Party of the London Dumping Convention.

ing industrialized countries is small. It is assumed that disposal activities not covered by the issuing of permits is roughly 25% of those activities covered by the LDC permit program (M. Nauke, personal communication).

1.3.2. The Dumpsites

The IMCO summary reports also give the coordinates and location maps for each dumpsite. Based on the information available locations of major dumpsites have changed little during 1976–1979. Figure 1.3 is a composite map based on permit information for the 1979 reporting period (IMCO, 1981). Nearly all of the dumping permits for industrial and sewage wastes are for disposal operations carried out in marine waters. The English Channel, those coastal areas on the southern side of the North Sea, and the apex of the New York Bight (United States), contain as of 1979 the greatest concentration of industrial wastes and sewage sludge dumpsites.

1.3.3. Tonnages and Percentages of Wastes Dumped into the Ocean by Different Countries

Relative quantities of industrial wastes and sewage sludge dumped in the ocean by those Contracting Parties to the LDC that contribute more than 0.2% (by total

Figure 1.3. Ocean dumpsites in 1979 used by Contracting Parties to the London Dumping Convention. Maps and the location of the dumpsites were adapted from maps given in IMCO (1981).

weight) of the total permit tonnages published in the IMCO summary reports are given in Fig. 1.4. As previously mentioned dredged material accounts for the greatest quantity of wastes to be dumped in the ocean (Fig. 1.2), but these tonnages are not included in Fig. 1.4. Discussion of the global tonnages of dredged material to the sea is given by Kester et al. (1983). Except for Canada and to a lesser degree for United Kingdom, percentages of the total annual permit quantities of industrial wastes dumped at sea by various countries have remained more or less the same from year to year. In 1976 Canada issued dumping permits for the ocean disposal of about 2×10^6 tons of these wastes or about 20% of the total permit tonnage for that year; most of Canada's permits in 1976 were for "liquids" to be dumped at a site in the Pacific. During 1977–1979, Canada's annual input to the ocean was reduced to 0.1×10^6 tons per year, or less than 1% of the total annual tonnage (Fig. 1.4). The abrupt decrease after 1976 in Canada's tonnage of waste was probably due to its recently enacted Canadian Ocean Control Act (Canada Gazette, 1975), which came into force in 1977 (M. Ludwig, personal communication).

The sudden appearance of Hong Kong in 1979 (Fig. 1.4) as a dumper of industrial wastes is due to its discharge of relatively large quantities of demolition and excavation rubble which were classified in the IMCO reports as industrial wastes.

For countries such as France, Spain, the Federal Republic of Germany, and the United States, the annual tonnage of waste, on an absolute basis, for ocean dumping was more or less constant during 1976–1979, although the percentages of the total waste for 1979 showed some increase and variation due mainly to the large variable quantities of industrial waste associated with permit information for United Kingdom. It should be pointed out that France issued permits for ocean disposal in late December 1975 for the disposal of 2.3×10^6 tons of waste in 1976 (M. Nauke, personal communication). To avoid similar confusions, the LDC permit notification format was amended by the Sixth Consultative meeting of the LDC in October 1981. Thus, in future, reference to permits will be made to the date from which a permit is valid and when it ends, rather than the date of issue of the permit (M. Nauke, personal communication).

For sewage sludge, the United Kingdom, United States, and the Federal Republic of Germany are the only three countries that issue dumping permits for the ocean disposal of large quantities of sewage sludge. As with France above, a valid permit for 1976 and 1977 by the Federal Republic of Germany for the disposal of sewage sludge had been issued earlier and was therefore not actually reported to IMCO for these two years (IMCO, 1979a, 1979b); the quantities of sludge per year for ocean disposal by the Federal Republic of Germany for 1976 and 1977 were the same as that indicated in the 1978 permit (M. Nauke, personal communication).

From 1977 to 1979 (Fig. 1.4), the estimated total annual tonnages of sewage sludge are of the order of 16×10^6 tons. In the United Kingdom the sewage sludge dumpsites can be found at several locations in the coastal waters around that nation (Fig. 1.3).

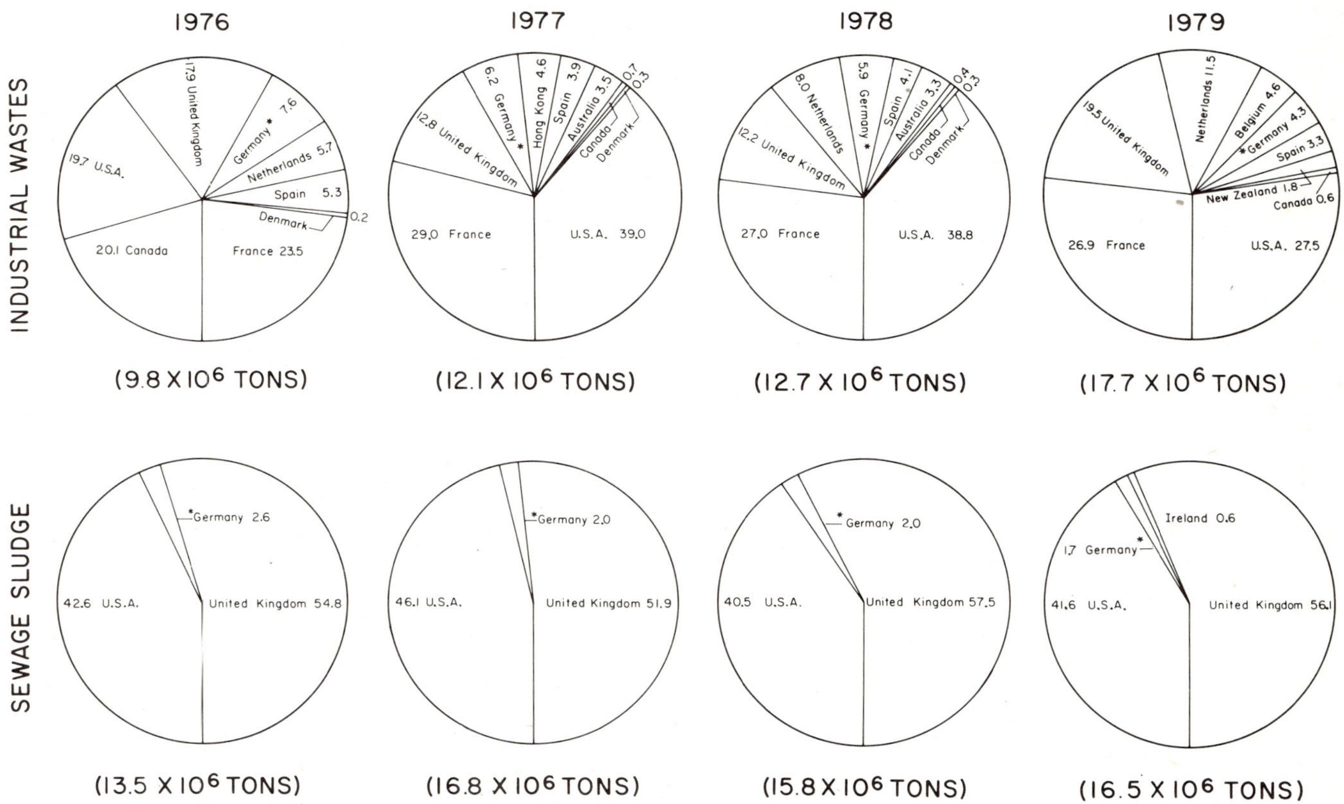

Figure 1.4. Percentages of the total estimated tonnage of waste (belonging to each country) for ocean disposal. Calculated from data in IMCO (1979a, 1979b, 1980, 1981). Germany* is the Federal Republic of Germany.

In the United States and with the recent closing of the Philadelphia dumpsite in the Mid-Atlantic Bight (Chapter 14), all shipboard disposal of sewage sludge now takes place at a single dumpsite in the apex of the New York Bight, located 22 km from the entrance to New York Harbor (see Chapters 2 and 17; Ketchum et al., 1981). This dumpsite, which receives sludges from New York metropolitan region, was scheduled for closure by 31 December 1981. The Congress of the United States mandated through an act to amend the Marine Protection, Research, and Sanctuaries Act of 1972 (U.S. Congress, 1972, 1977) that ocean dumping of all "harmful sewage sludge" shall end after 31 December 1981 (see also U.S. Environmental Protection Agency, 1977; Manheim, Chapter 2). "Harmful sewage sludge" is defined by the U.S. Congress (U.S. Congress, 1977) as that material "which may unreasonably degrade or endanger human health, welfare, amenities, or the marine environment, ecological systems, or economic potentialities." Recently, however, a Federal Court decision in New York in *The City of New York* versus *U.S. Environmental Protection Agency* ruled that the U.S. Environmental Protection Agency (EPA), which established the criteria for disposal of wastes, has not proved that New York City sewage sludges "unreasonably degrade" the marine environment. Also, the U.S. National Advisory Committee on Oceans and Atmospheres (NACOA) has recommended recently that the United States reexamine its current waste management practices (NACOA, 1981) with the ultimate aim of adopting a multimedia approach (air, land, and sea) for the disposal of its wastes. Thus, at present in the United States it does not appear that sewage dumping at sea will cease, since the Federal court case with New York City could serve as a precedent to be followed in the future by other municipalities and waste dumpers (Kamlet, 1981).

1.4. WASTE CHARACTERISTICS

Several types of waste are dumped into the ocean (Table 1.1). For each type, there are large variations or ranges in physical and chemical properties; industrial and municipal authorities requesting ocean dumping permits may have little control on the composition of the wastes they produce.

IMCO summary reports do not contain information on, or from the scientific community about, the characteristics of the waste material.

In this section on waste characteristics, we present a brief overview of the physical and chemical properties of the industrial and sewage wastes considered in this book. Often, in the discussion, a waste's density will be compared to seawater's density which is close to 1.02 g cm^{-3} over the range of temperature and salinity at coastal and oceanic dumpsites.

1.4.1. Physical Properties

Dispersion, advection, and settling of a waste in the ocean will depend, in part, on several of the waste's physical properties, such as bulk density, solid density, suspended solid concentration (for a liquid), and particle size (Chapter 4). Chemical composition through the flocculation process may also affect the behavior of a waste in the water column. Table 1.2 shows some values for these properties for the wastes examined in this book. More information is available on the properties of fly ash and sewage sludge than for acid-iron or pharmaceutical wastes. Although it is not comprehensive, due to lack of data for some wastes, Table 1.2 does give an idea of the range in values of selected physical properties and shows also where gaps in knowledge exist.

If a waste's density were the only factor to determine its settling behavior in the ocean, it is evident from Table 1.2 that each waste considered here would be distributed differently in a column of seawater after dumping. Consider fly ash, Fig. 1.5, a solid produced in large quantities at a power plant from the combustion of coal (Chapters 18 and 19). Its density is between 2.1 and 2.8 g cm^{-3}, depending on many factors. Thus fly ash dumped at sea from a barge should sink fairly rapidly to the seabed since its density, although quite variable, is at least twice as great as that of seawater (Table 1.2). Microscopic examination of fly ash shows, however, that it consists mostly of small, solid spheres (Fig. 1.5b and c) whose diameters, when different ashes are compared, are in the range 0.5 μm to $\simeq$200 μm (Table 1.2; see also Hulett et al., 1981; Bamber, 1980). However, for any given power plant burning a specific type of coal the range in diameter will be small for the majority of fly ash particles produced (Bamber, 1980). Small-sized particles, $\simeq$10 μm, have very low settling rates of $\simeq 10^{-3}$ cm s^{-1} in seawater (Schubel, 1971). Thus, in the absence of flocculation or turbulent diffusion the settlement of small fly ash particles on the seabed may be considerably slower than for larger particles.

In addition to those fly ash particles that will sink, regardless of the settling rate, in the ocean, there is another group of particles in the ash that will float on seawater. These are the cenospheres, glassy spheres that range in diameter from 20 to 200 μm and are reportedly filled with nitrogen and carbon dioxide gasses (Torrey, 1978). Cenospheres may comprise up to 20% by volume of fly ash (Torrey, 1978).

Thus, even though fly ash is more dense than seawater, its physical characteristics are complex and would preclude any simple description of settling behavior. The wide variations in physical properties, such as density, particle size, of fly ash are due to a number of factors, the most important being origin of the coal, degree of pulverization of the coal before it is burned, and the type of operation of the boiler unit (Torrey, 1978; Elliott, 1981; Bahor et al., 1981). A wealth of new data on the properties of fly ash is rapidly accumulating as coal is increasingly substituted for oil in the production of electricity (Bahor et al., 1981; Elliott, 1981; Hulett et al., 1981).

In a new approach to waste disposal, fly ash and scrubber sludge [principally gypsum and calcium sulfite hemihydrate (Chapters 18 and 19)] can be mixed together with lime to form solid blocks of consolidated coal waste. These blocks can be manufactured at conventional concrete block plants (Fig. 1.6) and have densities in the range 1.5–1.7 g cm^{3} (Table 1.2), depending on mix design, density of starting materials, and machine settings at the factory (Parker et al., 1981). A

Table 1.2. Physical Properties of Industrial and Sewage Wastes[a]

Wastes	Suspended Solids (g liter^{-1})	Specific Gravity (g cm^{-3})	Solids Density (g cm^{-3})	Particle Size (μm)
Industrial				
Acid-Iron[b]				
(1) H_2SO_4-iron[c]	0.010	1.084	–	–
(2) H_2SO_4-iron[d]	10	1.168	–	–
(3) H_2SO_4 (90%)-HCl (10%)-iron[e]	–	1.18	–	–
(4) H_2SO_4 (25%)-HCl (75%)-iron[e]	–	1.22	–	–
(5) HCl-iron[e]	–	1.18–1.35	–	–
(6) HCl-iron[e]	1.5	1.15	–	–
Pharmaceutical[f]	0.5	1.025	–	"Fine"
Fly ash[g]	NA	NA	2.1–2.8[h]	22–177[h] 10[i], 200[j]
Consolidated fly ash and scrubber sludge	NA	NA	1.5–1.7[k]	NA
Sewage sludge[l]	27 ± 12	1.0090 ± 0.0048	1.51 ± 0.18	Variable

[a]The dash within the table denotes no information available; NA denotes not applicable.

[b]The "acid" in acid-iron wastes is either sulfuric acid (H_2SO_4) or hydrochloric acid (HCl) or a combination of the two (Chapter 14).

[c]NL Industries, Inc., Sagreville, New Jersey, United States; data taken from Anderson and Dewling (1981).

[d]Allied Chemical Corp., Elizabeth, New Jersey, United States; data taken from Anderson and Dewling (1981).

[e]E. I. Du Pont de Nemours & Co., Edge Moor, Delaware, United States; data taken from Chapter 14 (Table 14.5) for waste (3), (4), and (5); see Chapter 8 (Table 8.1) for waste (6).

[f]Properties of a composite waste sample containing wastes from six pharmaceutical companies in Puerto Rico. These companies contracted as a group, PCI International, Inc., to ocean dump the wastes. Wastes data for suspended solids and density taken from Anderson and Dewling (1981); description of particle size as "fine" (see text) is from Presley et al. (1981).

[g]The properties of fly ash can be extremely variable (see text); the values reported here demonstrate the range in properties.

[h]Fly ash from Blyth power station, Northumberland, United Kingdom; data taken from Bamber (1980).

[i] Fly ash from Conesville power station, Coshocton, Ohio, United States; data from Parker et al. (1981).

[j] Fly ash samples taken from four Tennessee Valley Authority (United States) power station. Data taken from Hulett et al. (1981). Only about 1.8–4.5% of the fly ashes belonging to the four stations had particle diameters 200 μm; the majority of the particles had diameters in the range 37–100 μm.

[k]Taken from Chapter 18.

[l] These properties for City of New York sewage sludge; data taken from Callaway et al. (1976).

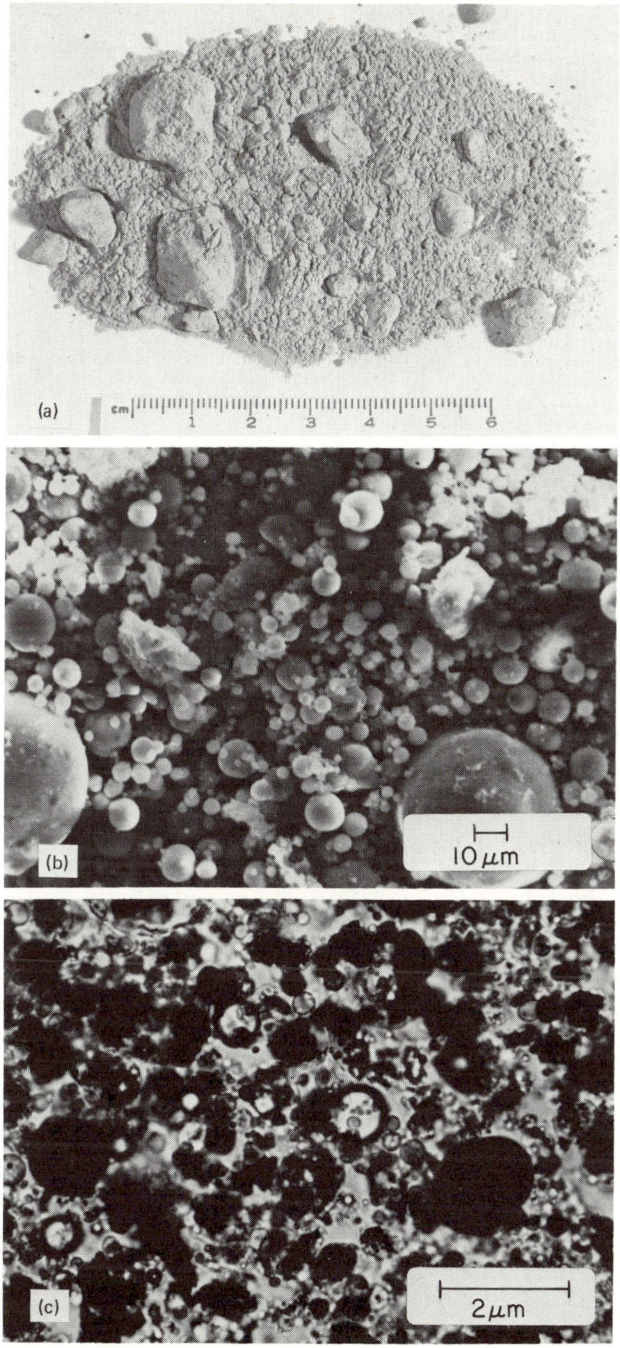

Figure 1.5. (a) Fly ash; (b) scanning electron and (c) optical photomicrographs of fly ash.

coal waste block is thus a form of waste self-containment as opposed to dispersion and dilution in the case of liquid wastes, and, as such, dissolution or leaching of major and minor components from the blocks is minimized (Chapter 19). In the sea these blocks provide habitat for marine organisms. The blocks are now being tested as a substrate for artificial reefs in the ocean (Fig. 1.6) (Duedall et al., 1981; Parker et al., 1981; Roethel, 1981; see also Chapters 18 and 19).

Sewage sludge is a principal waste produced at municipal waste-water treatment plants. Sewage sludge is a liquid in which up to 10% of its weight is solids (Fig. 1.7). It has a bulk density of about 1.01 g cm^{-3}, a value slightly less than the density of seawater and a solid density of $\simeq$1.5 g cm^{-3} (Table 1.2).

The solid fraction in sewage sludge is a heterogeneous mixture of a variety of solids including microorganisms, organic detritus or aggregates, fibers (including hair), mineral grains, and food residues (Fig. 1.7c). Accordingly, sewage sludge has no well-defined morphology; the size of its particulate matter literally varies from readily visible to microscopic to colloidal.

Besides the liquid and solid phases, sewage sludge also contains a film(s) (or film fragments) floating on the surface (Fig. 1.7b). This material has not been characterized physically or chemically.

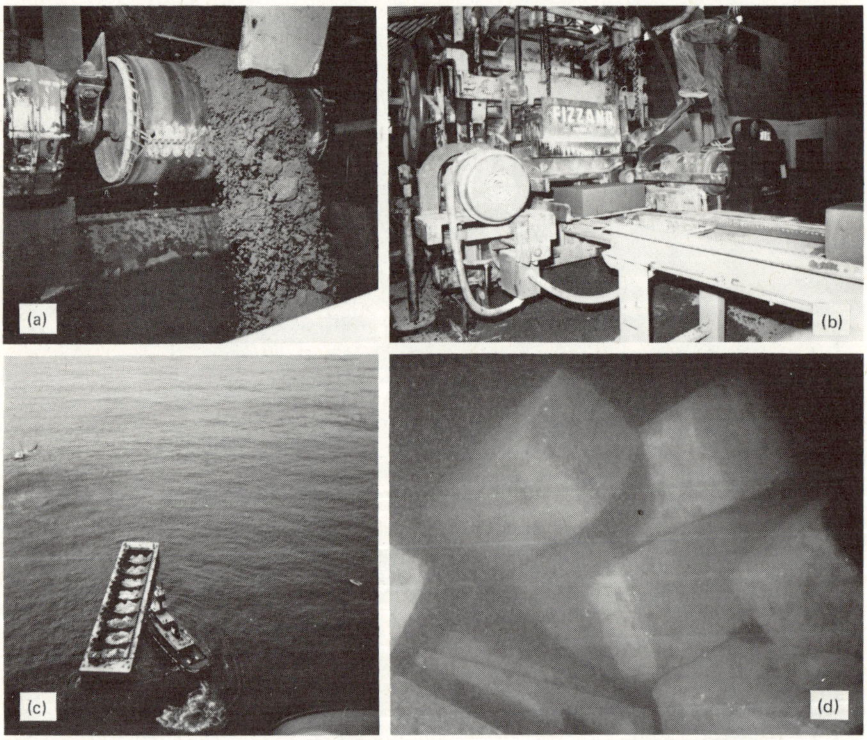

Figure 1.6. Coal waste blocks (Parker et al., 1981): (a) and (b) block fabrication at a concrete block factory; (c) ocean disposal of coal waste blocks; (d) coal waste blocks on the sea bed.

(c)

Figure 1.7. Sewage sludge. (a) Left bottle contains sewage sludge; the right bottle, also containing the same sewage sludge, shows the separation after centrifugation of liquid and solids in sludge. (b) Flakes of films taken from the surface of a sample of sewage sludge. (c) Sludges (after freeze-drying) from two different treatment plants; note the differences between textures of the two different sludges.

Sewage sludge would be expected to fractionate itself in the water column after disposal in the ocean, since it is a mixture containing both a liquid phase less dense than seawater and a solid fraction with densities which may be more or less than that of seawater (Table 1.2; Fig. 1.7a). Observations of fractionation of sewage sludge in the water column have been made during controlled dumping experiments and will be discussed later in this chapter. Variability of different sewage sludges is large, and is due to many factors, such as origin of the influent entering a treatment plant, efficiency of the treatment, and the technique used to process the sludge.

Physical properties or characteristics of both acid-iron and pharmaceutical wastes are not well known, or at best they have not been reported in the literature. Acid-iron wastes are liquids with densities that range from about 1.1 to 1.5 g cm^{-3} (Table 1.1; see also Chapter 14; Anderson and Dewling, 1981). In the United States, a majority of such wastes have been produced from the processing of titanium dioxide ores for the production of paint pigment; the amount of particulate matter in an acid-iron waste will depend on whether the acid used is sulfuric acid (H_2SO_4) or hydrochloric acid (HCl) (Chapter 14). Suspended solids are much greater in abundance in H_2SO_4-iron waste than in HCl-iron waste (Chapter 14; W. Muir, personal communication). The precipitate formed by reaction of acid-iron waste with seawater is flocculent; it can remain suspended in seawater and is clearly visible from the air for several hours after disposal (Fig. 1.1b).

Few data are available on the physical properties of pharmaceutical wastes. These materials have been dumped at a deepwater dumpsite off the coast of Puerto Rico (Fig. 1a; see also Chapters 9–13; Anderson and Dewling, 1981). They have a bulk density of 1.025 g cm^{-3} (Table 1.2), a value in the range of oceanic seawater densities, and a concentration of suspended solids of 0.5 g $liter^{-1}$ (Table 1.2); the density of the solids fraction has not been reported. Presley et al. (1981) described the particulates in pharmaceutical wastes as "very fine" grained particles which do not settle upon prolonged standing.

1.4.2. Elemental Composition of Wastes

Except for fly ash and the consolidated coal wastes, the wastes considered in subsequent chapters of this volume are liquids containing variable amounts and kinds of solids (Table 1.2). Ideally each phase in a waste should be characterized, since the distribution and ecological effects of each phase in the ocean may be quite different. Unfortunately, however, data for elemental content of wastes are reported on a dry weight basis, that is a waste usually being evaporated to dryness before the chemical analysis is performed. In Table 1.3 we compare the elemental composition of *total* solids present in each of four different wastes. Also, for comparison purposes, the elemental composition of rock and soil are examined (Table 1.3). We acknowledge that the absence of analytical data for each of the phases in the wastes limits, in a fundamental sense, the interpretation on total concentration, vis-à-vis chemical behavior or processes.

The most complete data sets are for coal wastes and sewage sludges, followed by pharmaceutical and acid-iron wastes. Similarities and differences in the major elements in these wastes are evident from the results in Table 1.3: Aluminum, silicon, calcium, iron, magnesium, manganese, and titanium are major elements in both fly ash and sewage sludge; calcium and magnesium are also abundant in pharmaceutical wastes; and carbon and nitrogen are major elements in sewage sludge. All these major elements, with exceptions of carbon and nitrogen (which are both derived from undigested organic matter in sewage sludge), are major con-

Table 1.3. Elemental Composition (mg g^{-1}, dry weight) of Wastes and Geological Material[a]

	Industrial Waste				Sewage Sludge			Geological Material[g]		
Element (Symbol)	Acid–Iron[b]	Pharmaceutical[c]	Fly Ash[d]	Consolidated Fly Ash and Scrubber Sludge[e]	Camden[b]	New York[f]	Philadelphia[f]	Rock	Soil	Sea Salt
Aluminum (Al)	0.001	0.70	127[h]	54–82	23	12	23	81	71	1.4×10^{-5}
Arsenic (AS)	0.001	0.0035	0.058[h]	0.044–0.068	0.001	0.008	0.016[e]	0.005	0.006	4.2×10^{-5}
Antimony (Sb)	–	0.002	0.069[h]	–	–	0.011	0.044	(0.001)	(0.002–0.010)	6.7×10^{-6}
Barium (Ba)	–	–	2.7[h]	0.34–0.68	–	0.42	0.53	0.43	0.50	5.5×10^{-4}
Beryllium (Be)	–	–	0.012	–	0.001	0.012	0.007	0.006 (sed)	0.006	–
Bismuth (Bi)	–	–	–	–	–	0.0076	0.010	0.0002	–	–
Boron (B)	–	–	0.50[i]	–	–	0.050	0.044	0.010	0.010	0.12
Bromine (Br)	–	–	0.012[h]	–	–	0.12	0.035	0.0025	0.005	1.9
Cadmium	0.001	1×10^{-5}	0.0014[i]	0.00078–0.0015	0.072	0.030	0.041–0.054	0.00018	0.00006	8×10^{-7}
Calcium (Ca)	–	6.8	47[h]	100–191	–	16	20	36	14	11.5
Carbon (C)	–	–	–	–	–	–	–	0.32	20	0.8
Cerium (Ce)	–	–	0.146[h]	–	–	0.071	0.030	0.042 (sed)	0.050	8×19^{-8}
Cesium (Cs)	–	–	0.0086[h]	–	–	0.00085	0.0016	0.0032	0.006	1×10^{-5}
Chlorine (Cl)	–	–	0.89[h]	–	–	8.5	1.1	0.48	0.10	553
Chromium (Cr)	0.037–0.41		0.13[h]	0.071–0.096	0.058	0.65	2.3	0.20	0.10	8×10^{-6}
Cobalt (Co)	–		0.041[h]	0.026–0.027	–	0.0062	0.016	0.040	0.008	1.4×10^{-6}
Copper (Cu)	4–8	0.002	0.13[i]	0.060–0.067	0.068	1.9	2.7	0.070	0.020	3×10^{-6}
Dysprosium (Dy)	–	–	–	–	–	0.0017	0.0012	0.0045 (sed)	–	–
Erbium (Er)	–	–	–	–	–	0.0005	0.0006	0.0025 (sed)	–	2×10^{-8}
Europium (Eu)	–	–	0.0025[h]	–	–	0.002	0.0007	0.0011 (sed)	–	3×10^{-9}
Fluorine (F)	–	–	0.020[i]	–	–	0.0051	0.28	0.80	0.20	3×10^{-2}
Gadolinium (Gd)	–	–	–	–	–	0.0065	0.0015	0.0064 (sed)	–	2×10^{-9}
Gallium (Ga)	–	–	–	–	–	0.0123	0.012	0.015	0.030	8×10^{-9}
Germanium (Ge)	–	–	–	–	–	0.0043	0.0015	0.007	0.001	–
Gold	–	–	–	–	–	0.0015	0.00048	0.000001	–	1×10^{-7}
Hafnium (Hf)	–	–	0.0079[h]	–	–	0.0015	0.0029	0.0045	0.006	2×10^{-7}
Holmium (Ho)	–	–	–	–	–	0.00053	0.00018	0.0012 (sed)	–	6×10^{-5}
Indium (In)	–	–	0.0027[i]	–	–	0.00016	0.00058	0.0001	–	–
Iodine (I)	–	–	0.0029[h]	–	–	0.012	0.011	0.0003	–	–
Iridium (Ir)	–	–	–	–	–	0.00005	0.00007	0.000001	–	–

Table 1.3. (*Continued*)

	Industrial Waste				Sewage Sludge			Geological Material[g]		
Element (Symbol)	Acid–Iron[b]	Pharmaceutical[c]	Fly Ash[d]	Consolidated Fly Ash and Scrubber Sludge[e]	Camden[b]	New York[f]	Philadelphia[f]	Rock	Soil	Sea Salt
Iron (Fe)	50–133	4.0	62[h]	60–153[d]	–	21	20	50	38	6×10^{-9}
Lanthanum (La)	–	–	0.082[h]	–	–	0.014	0.013	0.018 (sed)	0.030	8×10^{-8}
Lead (Pb)	0.013–0.10	0.002	0.075[h]	0.025–0.033	–	2.0	7.6	0.016	0.010	8×10^{-8}
Lithium (Li)	–	0.01	–	0.060–0.12	–	–	–	0.065	0.030	0.01
Lutetium (Lu)	–	–	0.00098[i]	–	–	0.00024	0.00014	0.00075 (sed)	–	6×10^{-9}
Magnesium (Mg)	–	0.78	18[h]	6.70–7.3	–	6.0	6.1	21	6.3	35.9
Manganese (Mn)	1.1	0.12	0.500[h]	0.17–0.21	–	0.15	0.095	1	0.85	6×10^{-6}
Mercury (Hg)	0.001	9×10^{-5}	0.00014[i]	–	–	0.015	0.0047	0.0005 (sed)	0.03	–
Molybdenum (Mo)	–	0.023	0.031[i]	–	–	0.005	0.0081	0.0023	0.002	3×10^{-4}
Neodymium (Nd)	–	–	0.064[i]	–	–	0.0011	0.0015	0.024 (sed)	–	8×10^{-8}
Nickel (Ni)	0.010–0.050	0.007	0.098[h]	0.081–0.085	–	0.014	0.43	0.040	0.040	6×10^{-6}
Nitrogen (N)	–	–	–	–	–	–	23	–	1.0	4×10^{-3}
Osmium (Os)	–	–	–	–	–	0.00008	0.00009	–	–	–
Palladium (Pd)	–	–	–	–	–	0.0011	0.0008	0.00001	–	–
Phosphorus (P)	–	–	0.88[i]	–	–	–	19	1.2	0.65	2×10^{-3}
Platinum (Pt)	–	–	–	–	–	0.00011	0.00074	0.000005	–	–
Palladium (Pd)	–	–	–	–	–	0.0011	0.0008	0.00001	–	–
Potassium (K)	–	8	16[h]	11–13	–	3.3	14	26	14	10.6
Praseodymium (Pr)	–	–	–	–	–	0.0042	0.0014	0.0055 (sed)	–	2×10^{-8}
Rhenium (Re)	–	–	–	–	–	0.00005	0.00004	0.000001	–	–
Rubidium (Rb)	–	–	0.12[h]	–	–	0.033	0.031	0.28	0.10	3.3×10^{-3}
Ruthenium (Ru)	–	–	–	–	–	0.0003	0.001	–	–	–
Samarium (Sm)	–	–	0.012[h]	–	–	0.0067	0.0026	0.0065 (sed)	–	1×10^{-8}
Scandium (Sc)	–	–	0.027[h]	–	–	0.0031	0.0028	0.005 (sed)	0.007	1.7×10^{-7}
Selenium (Se)	–	–	0.010[h]	–	–	0.0027	0.0034	0.00009	0.0002	–
Silicon (Si)	–	–	210[h]	11–16	–	–	–	280	330	0.06
Silver (Ag)	0.001	0.0005	–	–	–	–	–	0.00002	0.0001	1×10^{-6}
Sodium (Na)		233	3.2[h]	1.7–2.0	–	15	2.7	28	6.3	299
Strontium (Sr)	–	0.12	1.7[h]	0.39–0.50	–	0.33	0.032	0.15	0.30	0.22
Sulfur (S)	–	–	5.0[h]	–	–	–	–	0.52	0.70	25.8
Tantalum (Ta)	–	–	0.0018	–	–	0.00049	0.00066	0.0021	–	5×10^{-8}

Tellurium (Te)	–	–	–	–	–	0.00010	0.00025	(0.0000018)	–	–
Terbium (Tb)	–	–	0.0019[h]	–	–	0.00050	0.00034	0.00091	–	3×10^{-9}
Thallium (Tl)	–	–	0.0038[i]	–	–	–	–	0.0003	0.0001	–
Thorium (Th)	–	–	0.0025[h]	–	–	0.0078	0.0065	0.0115 (sed)	0.005	3×10^{-7}
Thulium (Tm)	–	–	–	–	–	0.00011	0.00037	0.00020 (sed)	–	6×10^{-9}
Tin (Sn)	–	–	0.011[i]	–	–	0.20	0.49	0.040	0.010	–
Titanium (Ti)	–	–	7.4[h]	2.9–4.3	–	2.0	2.7	4.4	5.0	3×10^{-5}
Tungsten (W)	–	–	0.0046[h]	–	–	0.0105	0.032	0.001	0.001	–
Uranium (U)	–	–	0.0120[h]	–	–	0.0010	0.0009	0.004	0.001	9×10^{-5}
Vanadium (V)	–	–	0.24[h]	–	–	0.054	0.0384	0.15	0.10	6.9×10^{-5}
Ytterbium (Yb)	–	–	0.007[h]	–	–	0.00074	0.00064	0.0027 (sed)	–	2×10^{-8}
Yttrium (Y)	–	–	0.062[h]	–	–	0.0042	0.0021	0.028 (sed)	0.050	4×10^{-8}
Zinc (Zn)	–	–	0.22[h]	0.16–0.22	–	1.3	6.9	0.080	0.050	3×10^{-6}
Zirconium (Zr)	–	–	0.30[h]	–	–	0.0054	0.011	0.22	0.30	–

[a] Data taken from the literature rounded to two significant figures; the uncertainty in the concentrations in the wastes is estimated to be about ±10%. Absence of data, indicated by a dash in table, denotes no information available.

[b] Taken from Chapter 14.

[c] Values taken from Presley et al. (1981); solids content 42.5 g liter^{-1} from the work of Presley et al.

[d] National Bureau of Standards Fly Ash (SRM-1633); See footnotes *h* and *i* for reference.

[e] Unpublished results from P. Harder (personal communication), Marine Sciences Research Center, State University of New York, Stony Brook.

[f] Furr et al. (1976); see Chapter 14 for additional data for city of Philadelphia sewage sludge. At present, the City of New York is carrying out a review of its existing data at the city's 12 municipal sewage treatment plants (J. T. Miller, personal communication).

[g] Values for rock are from Goldschmidt (1958) and the values for soil from Bowen (1966) as reported by Rahn (1976). The notation (sed) in the table represents concentration in sediment; the concentrations in parentheses are uncertain. A more recent compilation of the elemental composition of rock and soil, as well as that for dissolved and particulate matter in rivers and oceans, can be found in Martin and Whitfield (1982); their values show some deviations, although not large, from the results given in this table. Composition of sea salt taken from the compilation in Martin and Whitfield (1982); based on salinity equal 35‰. Values for nitrogen, chlorine, and sulfur not given in Martin and Whitfield but taken from formula for seawater with nitrogen equal to 10 μM.

[h] Ondov et al. (1975).

[i] Nadkarni (1980).

stituents in rock, soil, and sea salt (Table 1.3). In coal wastes the silicon is present as glass or quartz and the aluminum as alumina (Al_2O_3) or mullite ($2Al_2O_3 \cdot SiO_2$); both elements are very abundant in fly ash, since the predecessor to fly ash is mineral grains (that is, weathered rock) that impregnate coal seams during past geological periods.

Many of the so called "trace elements" present in trace amounts in natural materials (viz., rock, soil) are very enriched in the wastes (Table 1.3). Sewage sludge generally has the highest concentrations of trace elements. Chromium, copper, lead, nickel, and zinc, elements of concern from the ecological point of view (Johnston, 1976), are very abundant ($\simeq$hundred to several thousand $\mu g\ g^{-1}$). These concentrations are orders of magnitude greater than those found in the natural materials. The composition of sewage sludge may be expected to vary (Table 1.3) depending on the input of industrial wastes to a city's sewage system (Chapter 14).

Fly ash also contains abundant amounts of trace elements, but its trace element concentrations are only about twice that found in natural materials (Table 1.3). The rank in trace element concentrations for a given fly ash will depend on the origin of the parent coal (Elliott, 1981) and also on factors at the electric power station, such as the fly ash collecting device, the combustion process used, and boiler configuration (Bahor et al., 1981). The latter two factors affect formation of high temperature fly ash phases such as glass, mullite-quartz, and a magnetic phase. These phases are selective in their ability to concentrate specific trace elements (Hulett et al., 1981).

Although the concentrations listed in Table 1.3 show an interesting comparison among different wastes, they provide little, if any, in the way of an understanding of the chemical reactivity of a particular waste in seawater. Fate and effects of waste in the ocean are dependent on chemical processes, which in turn are dependent on specific types or kinds of compounds and species present in a waste. In theory it may be possible to predict (based on chemical principles) the availability or form (speciation) of a waste-derived element based on a waste's chemical characteristics (Stumm and Morgan, 1981). However, qualitative and quantitative analyses for both compounds and dissolved species in a waste can present great difficulties because the waste being complex and its chemical analyses difficult. Sewage sludge is probably the typical example of a complex mixture. We have already noted sewage sludge's enormous heterogeneity in particle size. Also, sewage sludge contains abundant quantities of organic carbon and nitrogen (Table 1.3). On trace organic compounds present in sewage sludge, polychlorinated hydrocarbons and other halogenated hydrocarbons are present in the 1 to 10 $\mu g\ g^{-1}$ (dry wt) range (Furr et al., 1976). The major compounds are believed to be mainly carbohydrates, proteins, and fats (Hunter and Heukelekien, 1965). Unfortunately, quantitative data are lacking and varying. The liquid phase no doubt contains a wide range of these materials both in true solution and in colloidal suspension, but there have been few data reported on the form of either dissolved or particulate material in sewage sludge. Also, an immiscible material, which has never been characterized, can often be seen floating on the surface of sewage sludge (Fig. 1.7b); such material may also be attached directly to the particles in the solids fraction of sewage sludge.

For fly ash the problem of chemical characterization is tractable, since fly ash is an inorganic solid whose principal matrices are fairly well defined and can be separated and then analyzed both chemically and mineralogically. As previously mentioned, in fly ash the glass phase is primarily a glassy silica (SiO_2), the crystalline phase mainly a mullite-quartz mixture ($3Al_2O_3 \cdot 2SiO_2$), and iron phase mainly a magnetic constituent with a formula similar to $Fe_2O_3 \cdot Al_{0.7}O_4$ (Hulett et al., 1981). These materials (for a variety of fly ashes) are high temperature phases and have been analyzed for selected trace elements in order to determine elemental distributions between the different matrices (Hulett et al., 1981). Results show that arsenic, lead, and cadmium are concentrated in the glass; vanadium, chromium, gallium, zirconium, iron, and tin are present in minor amounts in the mullite-quartz phase; and vanadium, chromium, manganese, iron, cobalt, nickel, copper, and zinc are very enriched in the magnetic component. The release of these elements from fly ash to an aqueous system, such as seawater, will depend largely on the solubilities of the fly ash matrices in seawater.

1.5. PROCESSES IN THE OCEAN

After a waste enters the water column from ocean disposal it will be dispersed, advected, and changed by a number of physical, chemical, and biological processes (Fig. 1.8). Some processes may occur simultaneously, others may not occur until much later. Chemical and biological processes occurring in the water column are for the most part short-term and will be very dependent on physical processes which are at all times acting to dilute wastes. Geochemical and biological processes taking place on the seafloor occur on much longer time scales. Once on the seafloor, concentration rather than dilution of a waste will occur due to settling, sedimentation, and bioaccumulation processes. But the mobility of some waste components will be dependent upon geochemical factors or processes. Thus, in considering both short-term and long-term effects of ocean disposal, the fate of a dumped waste (or its components) will be very difficult to determine. Put in another way, the fate of a waste results from the integration of biological, chemical, physical, and sedimentological processes occurring both in the water column and in the benthic region.

In this section we describe briefly some physical, chemical, and biological processes that affect the distribution and behavior of waste dumped into the sea. Later chapters in this volume provide more complete treatments of specific processes and other factors related to the fate of wastes in the ocean. There are available a number of other recent publications dealing specifically with processes, fates, and effects of wastes in the ocean (Goldberg, 1976; Johnston, 1976; The Royal Society of London, 1979; Morel and Schiff, 1980; Ketchum et al., 1981).

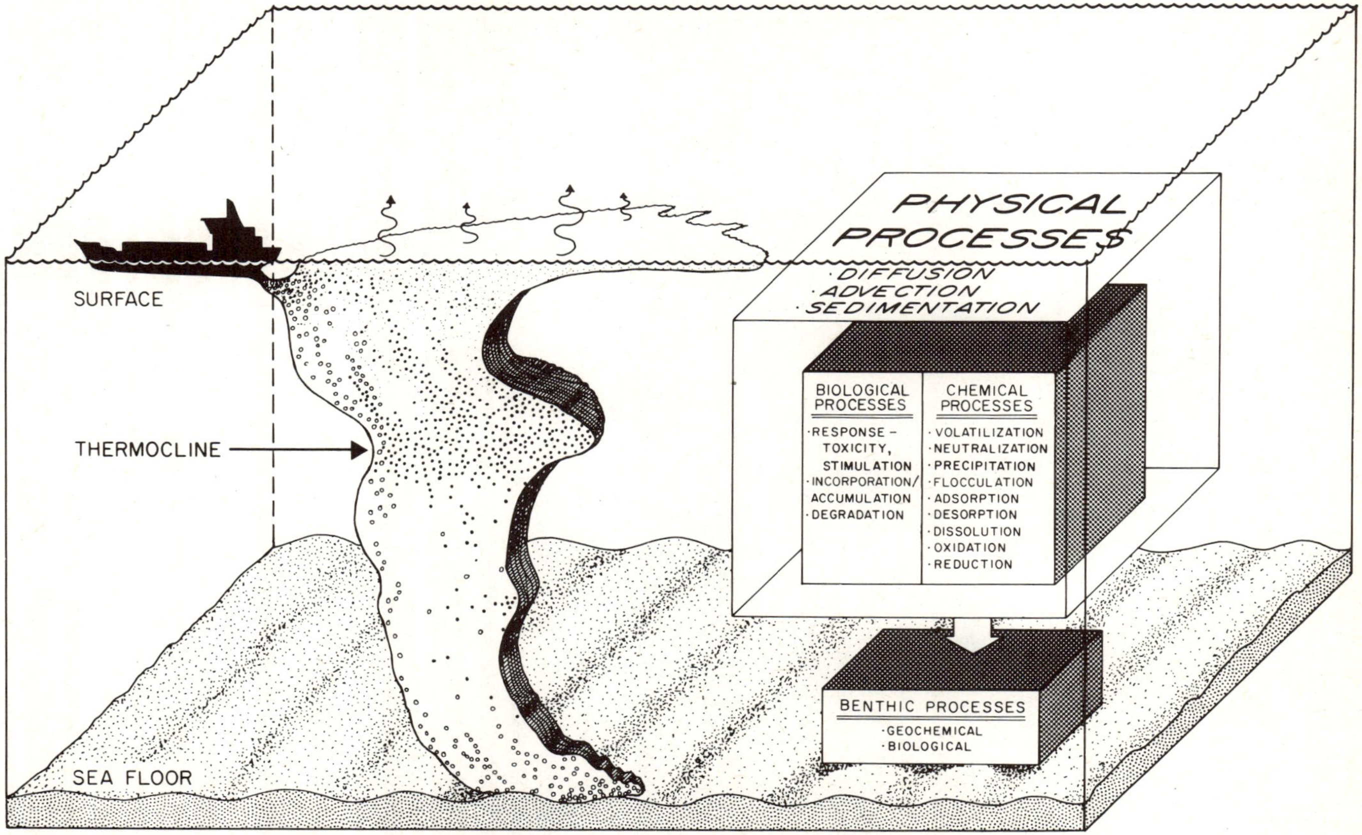

Figure 1.8. Schematic of the distribution of waste in the water column after a dumping event, and some processes that will affect distribution and fate of the waste.

1.5.1. Physical Processes

We begin with the dumping operation (Fig. 1.1). Current dumping practice provides the initial, rapid dispersion of waste. For instance the dilution of various waste dumped at sea from a moving transport ship are in the order of 10^3–10^5 by the ship's wake (Csanady, 1981; Hatcher et al., 1981; Kester et al., 1981). Data for acid-iron waste dumped at sea show that the mixing coefficient is proportional to the rate of waste discharge from the transport vessel (Ketchum and Ford, 1952).

As the wake generated by the discharge vessel subsides, usually in tens of minutes to hours, naturally occurring turbulence and currents at a dumpsite and surrounding waters will further disperse the dumped waste to nondetectable or background levels; this may occur within a few hours or take up to several days after a dumping event (Chapter 7). Dilution of a waste will depend on the waste, the disposal site, and the time scales of physical processes. Actual replacement of the entire water mass containing the diluted waste will depend on the extent of large-scale circulation processes and may take months to years (Csanady, Chapter 5).

The depth of penetration in the water column of a waste plume will depend on the type or kind of waste and its physical and chemical properties, water-column stratification, and the currents present at the dumpsite (Chapters 4 and 6; Fischer et al., 1979). Some wastes may sink rapidly to the bottom. Others will sink slowly or remain in the surface water for several hours (during calm weather) after a dumping event (Fig. 1.9).

Some wastes may fractionate. Figure 1.10, for example, shows an aerial view of the surface plumes created by the spot and line dumps of sewage sludge in the New York Bight. Figure 1.11 gives time-history plots of the concentrations of ammonium and suspended matter along with vertical profiles of the temperature and salinity for the spot dump shown in Fig. 1.10b. Differences are clearly seen in the sinking and dispersion behavior of the dissolved and particulate phases in the dumped sewage sludge. The near-neutral density phase (mostly liquid, see Fig. 1.7a) of the sewage sludge remains (as a plume or cloud) in surface waters for about two hours after dumping (Fig. 1.11a). The solids phase (Fig. 1.11b) in the dumped sludge appears as a fairly well-defined plume descending to the top of the thermocline (Fig. 1.11c) at about 17 m. Clearly, stratification in the water column is an important barrier to *some* of the settling sludge particles (Duedall et al., 1977).

The observations shown in Fig. 1.11 require some additional interpretation. Apparently, a more dense and rapidly sinking fraction of sewage sludge can entrain, during its descent to the bottom, some of the dissolved components in sewage sludge. This would explain the elevated bottom ammonium concentrations. The rapid descent to the bottom ($\simeq$24 m within a few minutes) of some sewage sludge components has been confirmed by Proni and Hansen (1981). The bottom concentrations of ammonium and suspended solids shown in Fig. 1.11 appear displaced from those above the thermocline and thus may not be necessarily representative

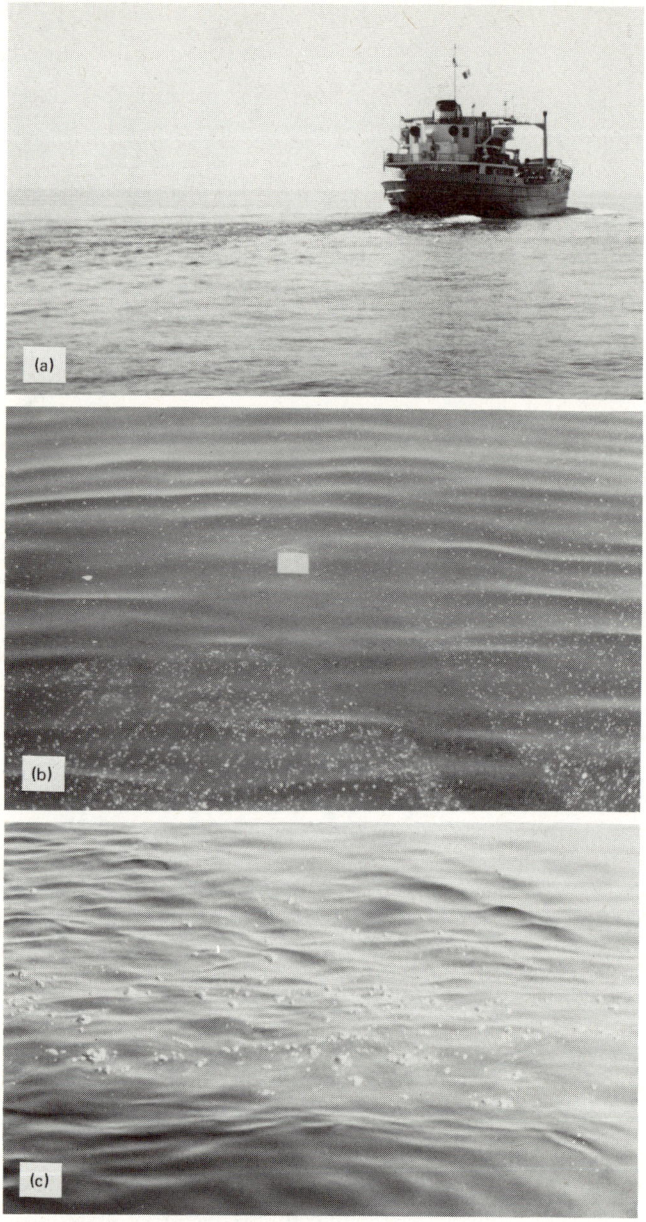

Figure 1.9. (a) Discharge of sewage sludge in a calm sea. (b) and (c) Sewage sludge remaining in surface waters after dumping event. In (b) white sheet of paper is 22 × 28 cm.

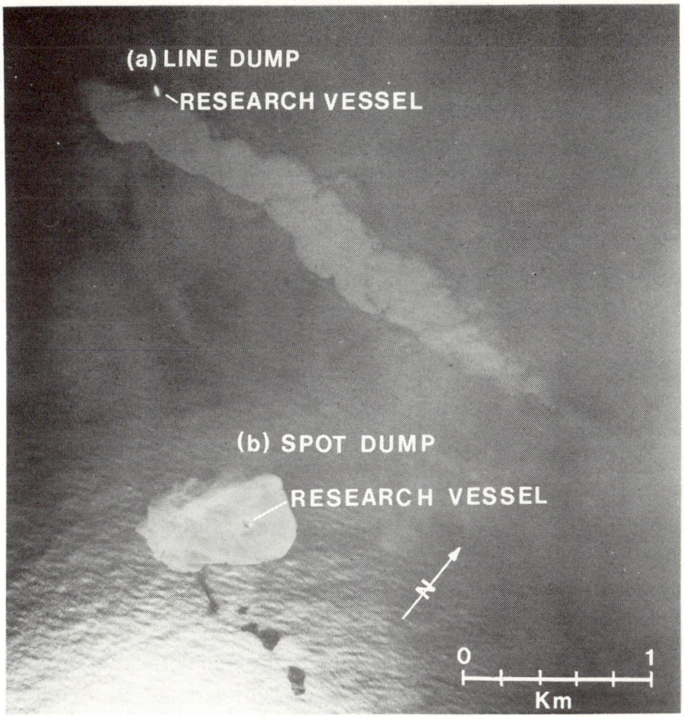

Figure 1.10. Aerial view of surface plumes resulting from the disposal of sewage sludge (adapted from Johnson and Ohlhorst, 1981; illustration modified slightly and with permission of Plenum Publishing Corp.). (a) Line dump; (b) spot dump.

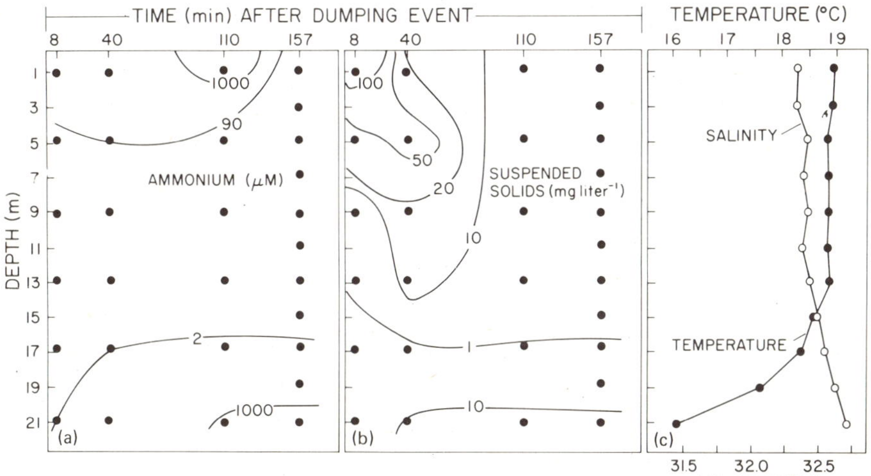

Figure 1.11. Time history distributions of (a) ammonium and (b) suspended solids after a spot dump of sewage sludge in the New York Bight. (c) Vertical profiles of temperature and salinity. (Taken from Duedall et al., 1977.)

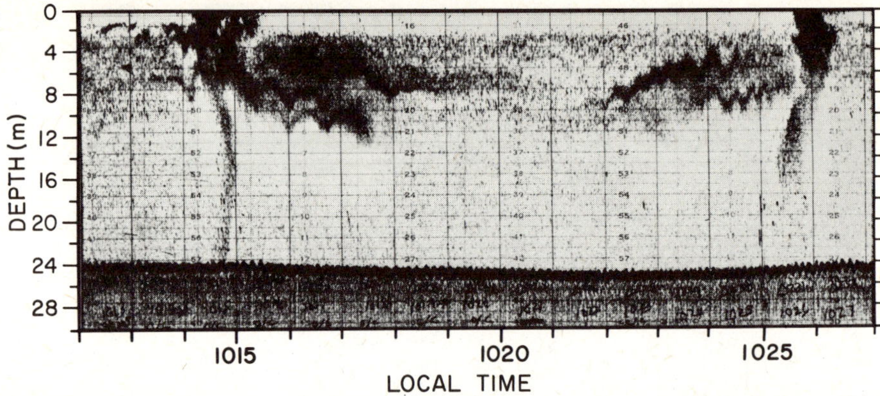

Figure 1.12. Acoustic record of sewage plume showing two successive ship passages over a spot dump in the New York Bight (Proni and Hansen, 1981; with permission of Plenum Publishing Corp.). Note the particulate material spreading along the thermocline in the 5–10 m depth range.

of the center of mass of the sludge plume. This is probably due to differences in surface and bottom currents.

While the thermocline is a surface where some waste particles from dumping may accumulate (Chapter 6), it is also that portion of the water column in which particles can be spread and mixed horizontally. Proni and Hansen (1981) using an acoustic method, similar to that described by Orr and Baxter (Chapter 6), have detected sludge particles spreading horizontally on the thermocline from the apex of a plume after a dumping event (Fig. 1.12); the rate of horizontal movement is about 50 cm s^{-1} (Proni and Hansen, 1981). Thus, waste particles may move long distances along density surfaces in the thermocline.

1.5.2. Chemical Processes

The physical and chemical characteristics of a waste can be substantially modified by a variety of processes that can occur upon mixing with seawater. These processes include acid-base neutralization, dissolution or precipitation and flocculation of solid phases, surface exchange reactions between particles and seawater (adsorption, desorption), volatilization at the sea surface, and changes in oxidation state (Fig. 1.8). These processes, some of which are discussed in Chapters 7, 9, 16, and 19 and also by Stumm and Morgan (1981), generally determine the chemical speciation of constituents in the waste and in seawater; they, therefore, are important factors controlling the chemical reactivity of these constituents.

Studies of acid-iron dumping (Chapters 6 through 8) provide clear examples of some chemical processes that occur in the water column after a dumping event. The high acid concentration of these wastes is neutralized rapidly by the buffering system in seawater and hydrous iron oxide precipitates forming a turbid plume in the wake of the discharged vessel (Fig. 1.1). This conversion of iron from a dissolved to a particulate phase upon mixing with seawater also includes an oxidation reaction of Fe^{2+} to Fe^{3+} for those wastes in which iron is initially in the reduced state. These processes have been examined by Ketchum and Ford (1948) and in Chapters 7, 8, and 14. The adsorption process is demonstrated in the scavenging ability of iron oxide particles for trace metals such as copper, lead, and cadmium (Kester et al., 1981). Flocculation, the agglomeration of small particles to form larger particles, may also occur in acid-iron-seawater mixtures (Ketchum and Ford, 1948).

The form of an element in a liquid waste, rather than an element's total concentration (Table 1.3), will determine a waste's chemical and biological behavior in the ocean. Chemical speciation of an element includes consideration of its oxidation state, its formation of ion-pairs and organic complexes, and its incorporation into solid phases. Figure 1.13 provides an illustration of speciation classification using size as a general criterion. In recent years more and more values for equilibrium and stability constants for reactions involving ion-pair or metal organic complex formation are becoming available (Stumm and Morgan, 1981). Thus, it is now possible to calculate the distribution for many dissolved species in solutions.

Elements concentrated in wastes, like iron and manganese, may also change chemical form through a change in valency as a result of oxidation-reduction reactions. Thus, the occurrence or distribution of the different forms of iron and manganese will depend on the redox potential (Eh) and pH of the system as illustrated in Fig. 1.14a. Under normal water-column conditions in the ocean, the dominant iron species is $Fe(OH)_3$ (Fig. 1.14b). But for manganese, the boundaries separating several species are situated in the normal pH range of seawater (Fig. 1.14c). Hence, several forms of manganese potentially will be present in seawater depending upon the water properties. The manganese and iron speciation is more complicated, however, due to kinetic factors. The rates of oxidation of Fe^{2+} and Mn^{2+} are pH dependent. At pH greater than 4, Fe^{2+} is oxidized only very slowly, and in the pH range of marine waters the time scale for Fe^{2+} oxidation ranges from minutes to days depending on pH, salinity, and temperature (Kester et al., 1975a). The rate of this reaction is 100 times slower than would be inferred from studies in fresh water systems. Kinetic factors are important in the redox speciation of manganese also (Trudinger and Swaine, 1979).

Morel et al. (1975) calculated trace metal speciation for a hypothetical primary sewage effluent (Table 1.4), with and without organic molecules or adsorbing particles. Their calculations show that for the trace metals it is the insoluble metal sulfides that make up the bulk of the speciation in this hypothetical waste. Even with the addition of dissolved organics and adsorbing surfaces (Table 1.4), only chromium and nickel will change their speciation significantly. For sewage sludge, which is enriched in particulate organic carbon, one may expect the trace elements to be associated more with the particulate organic matter, especially the smaller-sized particles. The equilibrium model for sewage (Table 1.4) should be recalculated based on oxic conditions.

When the settling fraction of waste reaches the seafloor it will be further changed. Here there will likely be strong coupling among various geochemical and biological processes. Organisms may ingest some waste particles over time, converting them to biomass or fecal material; other organisms may feed on the fecal material. Thus, any particular element may be recycled many times before it becomes part of the sedimentary record.

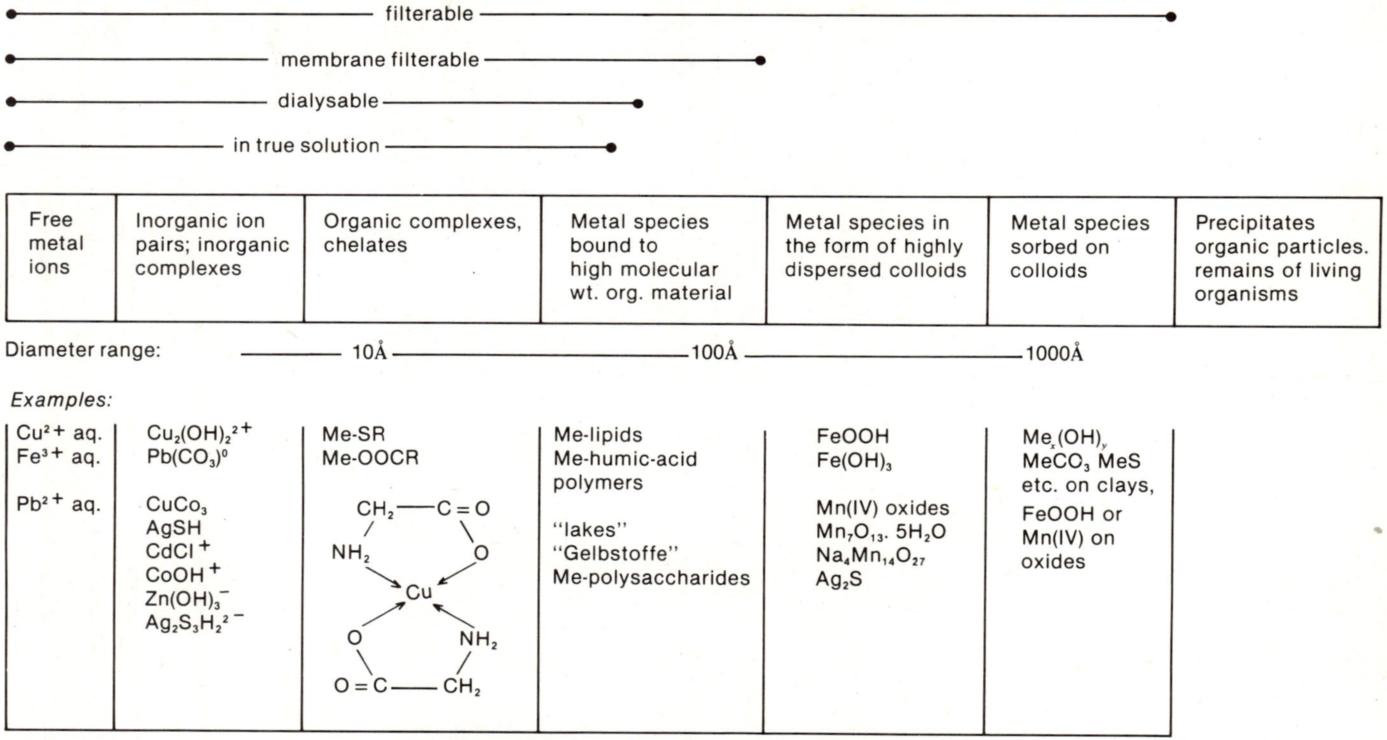

Figure 1.13. Examples of metals species in seawater (Stumm and Morgan, 1981).

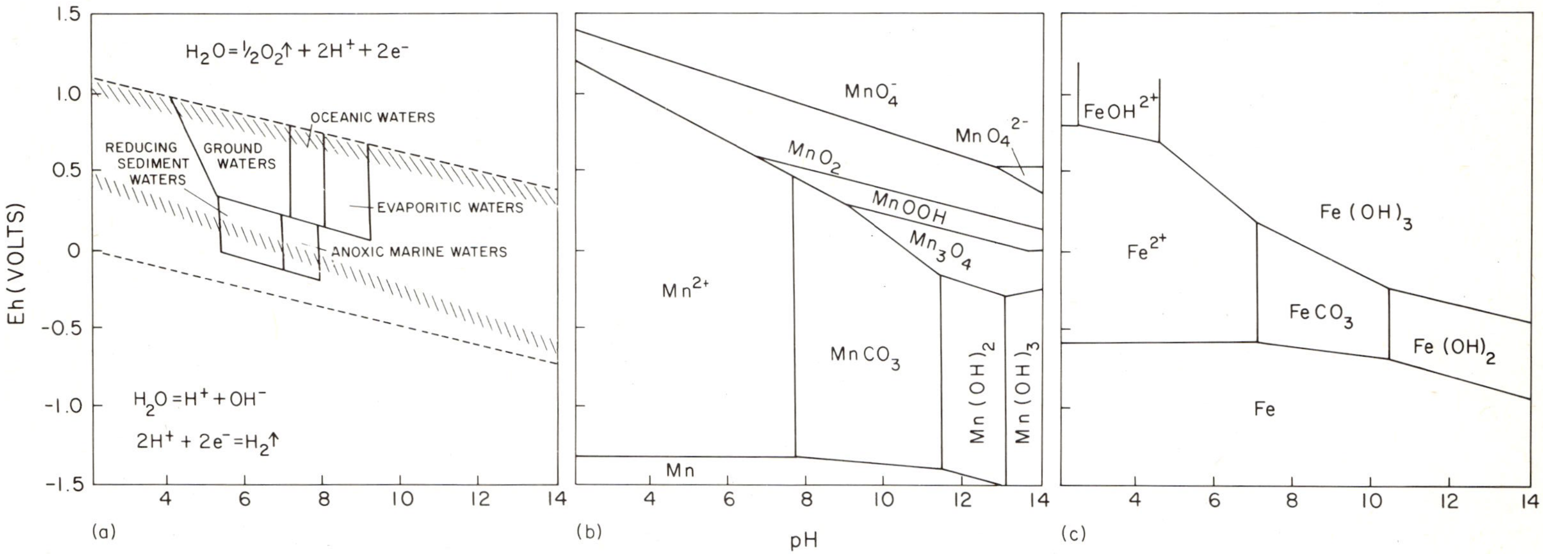

Figure 1.14. Eh–pH stability diagrams. (a) Eh–pH ranges for several natural systems (adapted from Kester et al., 1975b; with permission of *Thalassia Jugoslavica*, Rudjer Boskovic Institute, Yugoslavia.); (b) iron; and (c) manganese (adapted from Stumm and Morgan, 1981). The shaded area in (a) is the stability band for Eh control by dissolved oxygen concentrations of 0.01–6.00 ml liter^{-1}.

Table 1.4. Equilibrium Models of Sewage[a]

−Log Total Concentration	Inorganic Model (%)[b]	Additions of Organics (%)[c]	Addition of Adsorbing Surface (%)[d]
Fe, 3.7	Fe_3O_4(s), 100	Fe_3O_4(s), 100	Fe_3O_4(s), 100
Cr, 4.8	$Cr(OH)_3$(s), 97 $Cr(OH)_4^-$, 3	$Cr(OH)_3$(s), 97 $Cr(OH)_4^-$, 3	$Cr(OH)_3$, 52 $Cr(OH)_4^-$, 3 ADS, 45
Cu, 5.0	CuS(s), 100	CuS(s), 100	CuS(s), 100
Cd, 6.5	CdS(s), 100	CdS(s), 100	CdS(s), 100
Pb, 6.0	PbS(s), 100	PbS(s), 100	PbS(s), 100
Zn, 4.5	ZnS(s), 100	ZnS(s), 99	ZnS(s), 99
Ag, 6.7	Ag_2S(s), 100	Ag_2S(s), 100	Ag_2S(s), 100
Hg, 8.3	HgS(s), 100	HgS(s), 100	HgS(s), 100
Ni, 5.4	NiS(s), 42 Ni^{2+}, 2 $Ni(CN)_4{}^{2-}$, 56	NiS(s), 22 Ni^{2+}, 2 $Ni(CN)_4{}^{2-}$, 56 GLY, 10 GLU, 10	NiS(s), 14 Ni^{2+}, 2 $Ni(CN)_4{}^{2-}$, 56 GLY, 10 GLU, 10 ADS, 8
Co, 6.8	CoS(s), 97 Co^{2+}, 2	CoS(s), 95 Co^{2+}, 2 GLU, 2	CoS(s), 93 Co^{2+}, 2 GLU, 2 ADS, 2
Mn, 5.6	Mn^{2+}, 50 $MnHCO_3{}^+$, 24 $MnSO_4$, 19 $MnCl^+$, 6	Mn^{2+}, 50 $MnHCO_3{}^+$, 24 $MnSO_4$, 19 $MnCl^+$, 6	Mn^{2+}, 45 $MnHCO_3{}^+$, 21 $MnSO_4$, 14 $MnCl^+$, 6 ADS, 9

[a] From Morel et al. (1975).
[b] Inputs to the model: pH = 7.7, pCa = 2.75, pMg = 3.0, pBa = 5.0, pAl = 4.5, pCO_3 = 2.0, pSO_4 = 2.3, $pS(-II)_t$ = 4.7, pCl = 1.8, pF = 4.0, pNH_3 = 2.2, pPO_4 = 3.3, pCn = 5.0. Ionic strength = 0.01, T = 25°C; see also Stumm and Morgan (1981).
[c] Organic ligands (L) added pXL: pAcetate = 3.3, pGlycine = 3.8, pTartrate = 3.6, pGlutamate = 3.7, pSalicylate = 3.8, pPhthalate = 3.9.
[d] Adsorbing surface (ADS): 10 m^2 $liter^{-1}$.

Certain waste-derived trace elements may become mobile if the sediment is enriched in organic matter. If, on the other hand, the sediment is also rich in sulfide ions, a fraction of the trace elements may become immobilized through precipitation as insoluble metal sulfides.

An organic waste, such as sewage sludge, situated on the seabed will undergo further decomposition by bottom dwelling microorganisms; some particulates are transformed to their dissolved form and may release important plant nutrients. If these nutrients are mixed upward into the photic zone of the water column, they can be utilized by the plants.

Waste particles sitting on the seabed may also be mixed deeper in the sediment column by the burrowing organisms (bioturbation). The subsequent chemical mobility of these particles in the sediment will depend on the physiochemical properties of the sediment (Aller, 1977).

Thus, behavior of a waste on the seafloor will be quite different from that in the water column. Generally, the processes acting on the waste in the benthic zone (Fig. 1.8) will occur over a longer time scale.

1.5.3. Biological Processes

One objective of the regulation and control of the disposal of wastes into the ocean is to prevent damage to marine organisms and adverse modificaion of the functioning of the marine ecosystem. Human health, which might be endangered by the ingestion of contaminated seafood, is generally assumed to be adequately protected if the marine organisms are not adversely affected.

Three categories of biological processes should be considered when evaluating the impact of waste disposal:

1. *Responses of marine organisms* Waste materials may include toxic substances and the impact of these must be evaluated. Other substances in waste materials may stimulate biological activity which might be advantageous under certain circumstances but disastrous under others.
2. *Incorporation of substances of the waste within the living organism* Living organisms can incorporate substances in their body and the concentrations within the organisms may be orders of magnitude greater than that in the surrounding environment.
3. *Modification of waste substances by biological processes* Organic materials in the waste can be metabolized by the marine organisms resulting either in a change of the character of the compound or complete oxidation to carbon dioxide and water.

1.5.3a. Response of Marine Organisms to Waste Substances

Many of the substances in a waste material are toxic. They include many heavy metals and numerous organic compounds. The London Dumping Convention (IMCO, 1976) prohibits the dumping of mercury and cadmium and their compounds, organohalogen compounds, and various petroleum products. These are known toxic substances and their concentration in the marine environment should not be increased, at least by deliberate waste disposal activities. Numerous other substances require special care for their disposal and special permits are required for these.

The response of an organism to a toxic substance is a function both of the concentration of the toxic and of the time of exposure of the organism (see Chapters 10 through 13). In laboratory bioassays of toxicity, it is customary to use a notation such as "96-hour LC_{50}." This identifies the concentration of the toxic substance killing half of the population of the test organism in a period of 96 hours. It is clear that this is not a safe concentration, and it is customary to specify that the concentration in the environment should not exceed one-tenth or one one-hundredth of the lethal concentration identified in the bioassay (U.S. Environmental Protection Agency, 1976).

The chronic, sublethal effects of toxic substances are more difficult to evaluate under laboratory conditions. With organisms that can be raised and bred in the laboratory, long-term tests can be performed including the complete life cycle of the organism. In such tests, a flow-through environmental setup is required so that the concentration of a toxicant in the environment will not change with time (Chapter 13). Careful studies of polluted environments in nature can also, of course, be used to evaluate these long-term sublethal effects (The Royal Society of London, 1979). Without killing the organism involved, these modifications of essential biological processes may prevent the maintenance of the population, including, for example, changes in migration routes, in feeding behavior, in breeding activities, in escape from predators, or in the metamorphosis or survival of various critical stages of the life cycle. These processes may be mediated by chemical signals which can be altered by waste constituents.

Stimulation of various biological processes is also an impact of waste disposal. The most common stimulants include the essential nutrients for photosynthesis and growth of the phytoplankton populations, particularly compounds of nitrogen and phosphorus. These compounds are always present in sewage effluents and sewage sludge, and when released to the environment result in cultural eutrophication of the receiving body of water. The resulting enhanced photosynthesis can increase the total productivity of the ecosystem, including those species harvested for human consumption. If the fertilization is excessive, however, the normal phytoplankton assemblage can be destroyed and replaced by nuisance species which may not be as suitable as food or as nutritious for the herbivores in the system with a resulting collapse of the entire natural ecosystem in a heavily polluted environment.

1.5.3b. Incorporation of Substances into the Organism

Seawater is a complex solution and it contains all the elements essential for life. Some of the essential elements are present in low concentrations in seawater and these can be highly concentrated by the living organism in the formation of its structure. It is, perhaps, not surprising that the same organisms can concentrate nonessential constituents, some of which are detrimental. If the marine organism concentrates these substances sufficiently, they may constitute a hazard to humans when eaten as seafood.

NACOA (1981) has defined three main processes to describe the uptake of these substances from the environment by the organism:

BIOCONCENTRATION is a process whereby substances enter aquatic organisms through the gills or integument directly from solution in the water. Most aquatic organisms move large quantities of water across their gill surfaces and through their digestive tracts. This allows an equilibrium to be established between their body tissues and the surrounding water. The equilibrium concentration within the organism may, however, be less than the concentration in the water or more than that concentration by several orders of magnitude.

BIOACCUMULATION is the process whereby the marine organism stores substances in various tissues or organs of the body. The source of the contaminant may either be through direct exchange with substances dissolved in the water or in the food consumed. Different substances will be accumulated to different degrees in different tissues or organisms, for example in the liver, in muscle, in the digestive tract, in bone or in the fatty tissue.

BIOMAGNIFICATION is the process by which the concentration of bioaccumulated contaminants increase as they pass up the food chain through two or more trophic levels.

Since there are numerous organisms in the sea, each one of which may respond differently at different life stages to the various substances in a waste material, the number of possible permutations is extremely large. The best strategy in evaluating this process is probably to select the most critical substance in a waste for evaluation with a suitable organism for laboratory maintenance. Careful studies of natural ecosystems can be valuable; but unfortunately, the most clear-cut answers are found after the damage has been done.

1.5.3c. Modification of Waste by Metabolic Processes

Waste materials contain a variety of organic substances, most of which can be metabolized by marine organisms. Most of the natural organic material in waste are easily metabolized and the half-life of these substances is generally on the order of days or, at the most, of weeks. Some natural organic substances are more resistant to decomposition, however; cellulose and the lignins and tannins of woody plants are examples of these more resistant natural products.

Crude oil is a natural product which has been formed under anoxic conditions over periods of millions of years. This has probably been released to the marine environment in natural seeps for many centuries, but the rate of addition to the oceans has been increased vastly in the last few decades by man's activities. These substances are slowly oxidized in the marine environment under aerobic conditions, and the half-life for different constituents probably ranges from the order of years to many decades. When incorporated in anoxic sediments, these crude oils can persist indefinitely. The deliberate dumping of crude oils and of various petroleum products is prohibited by the London Dumping Convention (IMCO, 1976).

Synthetic organic compounds are even more highly resistant to decomposition in the marine environment. The half-life of some of these synthetic organic substances may range from decades to centuries. The London Dumping Convention prohibits the deliberate disposal of organohalogen compounds and of other persistent plastics (IMCO, 1976).

If the organic material in a waste is completely decomposed, it no longer presents a hazard to the environment. There is, however, an indirect effect which is frequently critical. The decomposition of the organic material removes oxygen from the water and anoxic conditions may be produced with the subsequent release of hydrogen sulfide, which itself is a toxic substance. The amount of oxygen which can be dissolved in seawater at normal atmospheric pressure is a function of the salinity and temperature of the water; the saturation value for oxygen decreases as either the salinity or temperature increases. Oxygen removed by the decomposition of organic material can be replaced by exchanges with the atmosphere and by the photosynthetic processes of plants. Both of these are effective only at or near the surface and decomposition of organic materials in waters can frequently remove all the dissolved oxygen. Since the amount of oxygen the water can contain is well known, the quantity of organic material that can be safely added is readily determined. The acceptable rate of addition is frequently far less than is required to dispose of the effluent and the sewage sludge from a sewage treatment plant. To ensure acceptable environmental conditions, therefore, the natural processes of reaeration, circulation, and mixing to dilute the waste and to introduce additional oxygen into the system must be evaluated. In the past, this has been done poorly, in part because the increase of the pollutant due to population growth was not properly evaluated and in part because the processes involved in maintaining an environment of suitable quality were poorly understood. We are now able to assess the problem more realistically, but economic constraints will always make it difficult to ensure that we maintain a marine environment of desirable integrity.

1.6. CONCLUSIONS

This overview was concerned mainly with three aspects of the problem of dumping wastes at sea: (1) the magnitude of the tonnages, based on ocean dumping permit records, published by IMCO; (2) physical and chemical characteristics of ocean dumped industrial and sewage wastes, with particular reference to acid-iron, pharmaceutical, sewage sludge, and coal wastes, which are the four main wastes to be considered in later chapters of this volume; and (3) physical and chemical processes that influence fates and pathways of the wastes, and the biological processes that influence effects of wastes on marine organisms.

Our study shows that ocean dumping activity is increasing. This conclusion is based on an analysis of both the total number of permits issued and the combined tonnages of industrial wastes and sewage sludges dumped by member states to the London Dumping Convention. This trend is likely to continue into the future as more dumping records become available from new members who join the Convention. With regard to the availability of ocean dumping records, we acknowledge the important role played by the IMCO in providing reports on the extent of dumping activities carried out by the members of the Convention.

Ocean-dumped wastes are very complex mixtures as evidenced by their physical properties and the wide range in composition and concentrations. An utmost effort must be devoted to the comprehensive characterization of these wastes and to the determination of the range in variability in properties within a particular waste type. This conclusion is based on our finding that few scientific reports provide comprehensive and systematic characterization of wastes for ocean dumping. The Crystal Mountain workshop d.scussed the importance for more and better characterization of wastes (Goldberg, 1979). In this regard, industry or municipalities producing wastes for ocean disposal could, through cooperative endeavors with the scientific community, provide both the scientific community and themselves a service by implementing a program of comprehensive waste characterization. The results would be useful to scientists conducting ocean dumping research; without adequate source characterization, subsequent researches are ineffective. Prediction on the consequence of pollution is our goal for future researches.

Knowledge of all the important physical, chemical, and biological processes affecting ocean-dumped wastes is required to determine scientifically the pathways of wastes in the ocean. These processes do not act independently; thus, an investigation of fates of wastes requires a multidisciplinary approach to the problem by the working together of chemists, ecologists, mathematicians, and physicists. Biological processes associated with effects of wastes is a particularly difficult problem as one must consider not only toxic impacts but also sublethal effects and the transfer of contaminated seafood to humans. The latter may occur as the result of oceanic pollutant transports and of bioconcentration, bioaccumulation, and biomagnification processes. Finally, we continue to advocate that the fundamental researches in oceanography must go on concurrently with the urgent marine pollution studies; they must not be separated.

ACKNOWLEDGMENTS

We sincerely thank M. A. Champ, K. S. Kamlet, M. Ludwig, and M. K. Nauke for their helpful comments during the preparation of the manuscript. We are grateful to V. Abolins, L. F. Antonacci, M. T. Eisel, M. H. Gladwish, A. M. Lannak, M. A. Lau, and J. M. Restivo for assisting in the artistic and technical preparation of the manuscript. Support for this work was provided by the New York State Energy Research and Development Authority, Electric Power Research Institute, U.S. Department of Energy, U.S. National Oceanic and Atmospheric Administration, and the Power Authority of the State of New York.

This is Contribution 322 of the Marine Sciences Research Center of the State University of New York at Stony Brook.

REFERENCES

Aller, R. C. 1977. The Influence of Macrobenthos on Chemical Diagenesis of Marine Sediments. Ph.D. Thesis, Yale University, New Haven, Connecticut, 600 pp.

Anderson, P. W., and R. T. Dewling. 1981. Industrial ocean dumping in EPA Region II–Regulatory aspects. *In*: Ocean Dumping of Industrial Wastes, B. H. Ketchum, D. R. Kester, and P. K. Park (Eds). Plenum Press, New York, pp. 25–37.

Bahor, M. P., R. J. McLaren, J. E. Niece, and H. C. Pedersen. 1981. Coal Ash Disposal Manual, 2nd ed. EPRI Report CS-2049, Electric Power Research Institute, Palo Alto, California, 9 sections (paginated separately), 2 appendices.

Bamber, R. N. 1980. Properties of fly ash as a marine sediment. *Marine Pollution Bulletin*, **11**, 323–326.

Bowen, H. J. M. 1966. Trace Elements in Biochemistry. Academic Press, London, 211 pp.

Callaway, R. J., A. M. Teeter, D. W. Browne, and G. R. Ditsworth. 1976. Preliminary analysis of the dispersion of sewage sludge discharged from vessels to New York Bight waters. *In*: Middle Atlantic Continental Shelf and the New York Bight, Special Symposia, Vol. 2, M. G. Gross (Ed.). American Society of Limnology and Oceanography, Lawrence, Kansas, pp. 199–211.

Canada Gazette. 1975. Ocean Dumping Control Act. (19 June 1975). Senate and House of Commons of Canada. Canada Gazette, Volume I, No. 9, Part III, Ottawa, 25 pp. See also Ocean Dumping Control Regulations (2 October 1975). Canada Gazette, Volume 109, No. 20, Part II, pp. 2786–2803.

Champ, M. A., and P. K. Park. 1981. Ocean Dumping of sewage sludge: a global review. *Sea Technology*, **22**, 18–24.

Csanady, G. T. 1981. An analysis of dumpsite diffusion experiments. *In*: Ocean Dumping of Industrial Wastes, B. H. Ketchum, D. R. Kester, and P. K. Park (Eds.). Plenum Press, New York, pp. 109–129.

Duedall, I. W., H. B. O'Connors, S. A. Oakley, and H. M. Stanford. 1977. Short-term water column perturbations caused by wastewater sludge dumping in the New York Bight apex. *Journal Water Pollution Control Federation*, **49**, 2074–2080.

Duedall, I. W., F. J. Roethel, J. D. Seligman, H. B. O'Connors, J. H. Parker, P. M. J. Woodhead, R. Dayal, B. Chezar, B. K. Roberts, and H. Mullen. 1981. Stabilized power plant scrubber sludge and fly ash in the marine environment. *In*: Ocean Dumping of Industrial Wastes, B. H. Ketchum, D. R. Kester, and P. K. Park (Eds.). Plenum Press, New York, pp. 315–346.

Elliott, M. A. (Ed.). 1981. Chemistry of Coal Utilization. John Wiley and Sons, New York, 2374 pp.

Fischer, H. B., E. J. List, R. C. Y. Koh, J. Imberger, and N. H. Brooks. 1979. Mixing in Inland and Coastal Waters. Academic Press, New York, 483 pp.

Furr, A. K., A. W. Lawrence, S. C. Tong, M. C. Grandolfo, R. A. Hofstader, C. B. Bache, W. H. Gutenmann, and D. J. Lisk. 1976. Multielement and chlorinated hydrocarbon analysis of municipal sewage sludges of American cities. *Environmental Science and Technology*, **10**, 683–687.

Goldberg, E. D. 1976. The Health of the Oceans. The Unesco Press, Paris, 172 pp.

Goldberg, E. D. (Ed.). 1979. Proceedings of a Workshop on Assimilative Capacity of U.S. Coastal Waters for Pollutants, Crystal Mountain, Washington. Special Report, National Oceanic and Atmospheric Administration–Environmental Research Laboratories, Boulder, Colorado, 284 pp.

Goldberg, E. D. 1981. The oceans as waste space: The argument. *Oceanus*, **24**, 2–9.

Goldschmidt, V. M. 1958. Geochemistry. Oxford University Press, Oxford, 730 pp.

Hagen, A. A. 1983. History of radioactive waste disposal into the ocean. *In*: Radioactive Wastes and the Ocean, P. K. Park, D. R. Kester, I. W. Duedall, and B. H. Ketchum (Eds.). John Wiley and Sons, New York, in press.

Hatcher, P. G., G. A. Berberian, A. Y. Cantillo, P. A. McGillivary, P. Hanson, and R. H. West. 1981. Chemical and physical processes in a dispersing sewage sludge plume. *In*: Ocean Dumping of Industrial Wastes, B. H. Ketchum, D. R. Kester, and P. K. Park (Eds.). Plenum Press, New York, pp. 347–378.

Hulett, L. D., A. J. Weinbeger, N. M. Ferguson, K. J. Northcutt, and W. S. Lyon. 1981. Trace Element and Phase Relations in Fly Ash. EPRI Report EA-1822. Electric Power Research Institute, Palo Alto, California, 63 pp.

Hunter, J. V., and H. Heukelekian. 1965. The composition of domestic sewage fractions. *Journal Water Pollution Control Federation*, **37**, 1142–1163.

Inter-Governmental Maritime Consultative Organization (IMCO). 1976. Inter-Governmental Conference on the Convention on the Dumping of Wastes at Sea. (London, 30 October–13 November 1971). Final Act of the Conference with Technical Memorandum and Resolution adopted by the Conference and Convention on the Prevention of Marine Pollution by Dumping of Wastes and Other Matter. Inter-Governmental Maritime Consultative Organization, London, 36 pp. plus Amendments (Publication No. 76.14 E), 6 pp.

IMCO. 1979a. Convention on the prevention of Marine Pollution by Dumping of Wastes and Other Matter, 1972. Reports of permits issued for dumping in 1976. LDC.2/Circ. 31 (19 March 1979), Inter-Governmental Maritime Consultative Organization, London, 25 pp.

IMCO. 1979b. Convention on the Prevention of Marine Pollution by Dumping of Wastes and Other Matter, 1972. Reports of permits issued for dumping in 1977. LDC.2/Circ. 33 (10 April 1979), Inter-Governmental Maritime Consultative Organization, London, 23 pp.

IMCO. 1980. Convention on the Prevention of Marine Pollution by Dumping of Wastes and Other Matter, 1971. Reports of permits issued for dumping in 1978. LDC.2/Circ. 47 (12 February 1980), Inter-Governmental Maritime Consultative Organization, London, 24 pp.

IMCO. 1981. Convention on the Prevention of Marine Pollution by Dumping of Wastes and Other Matter, 1972. Reports of permits issued for dumping in 1979. LDC.2/Circ. 64 (30 April 1981), Inter-Governmental Maritime Consultative Organization, London, 26 pp.

Johnson, R. W., and C. W. Ohlhorst. 1981. Application of remote sensing to monitoring and studying dispersion in ocean dumping. *In*: Ocean Dumping of Industrial Wastes, B. H. Ketchum, D. R. Kester, and P. K. Park (Eds.). Plenum Press, New York, pp. 175–191.

Johnston, R. (Ed.). 1976. Marine Pollution. Academic Press, London, 729 pp.

Kamlet, K. S. 1981. Court decision makes future of sewage sludge ocean dumping uncertain. *Coastal Ocean Pollution Assessment News*, **1**(3), 46–47.

Kester, D. R., R. H. Byrne, Jr., and Y.-J. Liang. 1975a. Redox reactions and solution complexes of iron in marine systems. *In*: Marine Chemistry in the Coastal Environment, T. M. Church (Ed.). ACS Symposium Series, American Chemical Society, Washington, D.C., pp. 56–79.

Kester, D. R., T. P. O'Connor, and R. H. Byrne, Jr. 1975b. Solution Chemistry, solubility, and adsorption equilibria of iron, cobalt, and copper in marine systems. *Thalassia Jugoslavica*, **11**, 121–134.

Kester, D. R., R. C. Hittinger, and P. Mukherji. 1981. Transition and heavy metals associated with acid-iron waste disposal at Deep Water Dumpsite 106. *In*: Ocean Dumping of Industrial Wastes, B. H. Ketchum, D. R. Kester, and P. K. Park (Eds.). Plenum Press, New York, pp. 215–232.

Kester, D. R., B. H. Ketchum, I. W. Duedall, and P. K. Park (Eds.). 1983. Dredged Material Disposal in the Ocean. John Wiley and Sons, New York, in press.

Ketchum, B. H., and W. L. Ford. 1948. Waste Disposal at Sea, Preliminary Report on Acid-Iron Waste Disposal. Unpublished Report, Woods Hole Oceanographic Institution, Woods Hole, Massachusetts, 18 pp.

Ketchum, B. H., and W. L. Ford. 1952. Rate of dispersion in the wake of a barge at sea. *Transactions, American Geophysical Union*, **33**, 680–684.

Ketchum, B. H. 1980. Marine Industrial Pollution. *In*: Oceanography: The Past, M. Sears,and D. Merriman (Eds.). Springer-Verlag, New York, pp. 397–413.

Ketchum, B. H., D. R. Kester, and P. K. Park (Eds.). 1981. Ocean Dumping of Industrial Wastes. Plenum Press, New York, 525 pp.

Martin, J.-M., and M. Whitfield. 1982. The significance of the river input of chemical elements to the ocean. *In*: Trace Metals in Sea Water, C. S. Wong, E. A. Boyle, J. D. Burton, K. W. Bruland, and E. D. Goldberg (Eds.). Plenum Press, New York. In press.

Moore, G. 1976. Legal aspects of marine pollution control. *In*: Marine Pollution, R. Johnston (Ed.). Academic Press, London, pp. 589–697.

Morel, F. M. M., J. C. Westall, C. R. O'Melia, and J. J. Morgan. 1975. Fate of trace metals in Los Angeles county wastewater discharge. *Environmental Science and Technology*, **9**, 756–761.

Morel, F. M. M., and S. L. Schiff. 1980. Geochemistry of Municipal Waste in Coastal Waters. Report No. 259, Department of Civil Engineering, Massachusetts Institute of Technology, Cambridge, Massachussetts, 205 pp.

Nadkarni, R. A. 1980. Multitechnique multielemental analysis of coal and fly ash. *Analytical Chemistry*, **52**, 929–935.

National Advisory Committee on Oceans and Atmosphere (NACOA). 1981. The Role of the Ocean in a Waste Management Strategy. National Advisory Committee on Oceans and Atmospheres. U.S. Government Printing Office, Washington, D.C., 103 pp. and Appendices.

Ondov, J. M., W. H. Zoller, I. Olmez, N. K. Aras, G. E. Gordon, L. A. Rancitelli, K. H. Abel, R. H. Filby, K. R. Shah, and R. C. Ragaini. 1975. Elemental concentrations in the National Bureau of Standards Environmental Coal and fly ash standard reference material. *Analytical Chemistry*, **47**, 1102–1109.

Osterberg, C. 1981. Seas: To Waste or Not. (Editorial) *The New York Times*, 9 August.

Parker, J. H., P. M. J. Woodhead, and I. W. Duedall. 1981. Coal Waste Artificial Reef Program, Phase 3. Volume 2: Comprehensive Report. EPRI Report

CS-2009, Electric Power Research Institute, Palo Alto, California, 4 sections (paginated separately) + 1 appendix.

Presley, B. J., J. S. Schofield, and J. Trefry. 1981. Waste material behavior and inorganic geochemistry at the Puerto Rico waste dumpsite. *In*: Ocean Dumping of Industrial Wastes, B. H. Ketchum, D. R. Kester, and P. K. Park (Eds.). Plenum Press, New York, pp. 233–246.

Proni, J. R., and D. V. Hansen. 1981. Dispersion of particulates in the ocean studied acoustically: The importance of gradient surfaces in the ocean. *In*: Ocean Dumping of Industrial Wastes, B. H. Ketchum, D. R. Kester, and P. K. Park (Eds.). Plenum Press, New York, pp. 161–173.

Rahn, K. A. 1976. The Chemical Composition of the Atmospheric Aerosol. Technical Report (1 July), Graduate School of Oceanography, University of Rhode Island, Kingston, 265 pp.

Roethel, F. J. 1981. The Interactions of Stabilized Power Plant Coal Waste with the Marine Environment. Ph.D. Thesis, State University of New York, Stony Brook, 349 pp.

Schubel, J. R. 1971. Tidal variation of the size distribution of suspended sediment at a station in the Chesapeake Bay turbidity maximum. *Netherlands Journal of Sea Research*, 5, 252–266.

Stumm, W., and J. J. Morgan. 1981. Aquatic Chemistry, 2nd ed. Wiley-Interscience, New York, 780 pp.

The Royal Society of London. 1979. The Assessment of Sublethal Effects of Pollutants in the Sea. The Royal Society, London, 235 pp. (First published in *Philosophical Transactions* of the *Royal Society of London, Series B*, **284**, 397–633.)

Timagenis, G. J. 1980. International Control of Marine Pollution, Volume 1. Oceana Publications, Dobbs Ferry, New York, 404 pp.

Topping, G. 1976. Sewage and the sea. *In*: Marine Pollution, R. Johnston (Ed.). Academic Press, London, pp. 303–351.

Torrey, S. 1978. Coal Ash Utilization. Noyes Data Corporation, Park Ridge, New Jersey, 370 pp.

Trudinger, P. A., and D. J. Swaine (Eds.). 1979. Biogeochemical Cycling of Mineral-Forming Elements. Elsevier, Amsterdam, 612 pp.

U.S. Congress. 1972. Marine Protection, Research, and Sanctuaries Act of 1972. Public Law 92–532, 86 Stat. 1052.

U.S. Congress. 1977. Authorization, Appropriation–Marine Protection, Research, and Sanctuaries Act of 1972. Public Law 95–153, 91 Stat. 1255.

U.S. Environmental Protection Agency. 1976. Quality Criteria for Water (July 1976). U.S. Environmental Protection Agency, Washington, D.C., 256 pp.

U.S. Environmental Protection Agency. 1977. Ocean Dumping, Final Revision of Regulations and Criteria. *Federal Register*, **42**(No. 7), 2462–2490.

2

WHO IS DOING WHAT IN MARINE DUMPING?

Frank T. Manheim

U.S. Geological Survey
Woods Hole, Massachusetts

ABSTRACT

In 1978, major categories of United States waste products being discharged to the oceans by vessels were dredged material, 65 million tons (metric); solid sewage waste, 5.0 million tons; industrial and construction wastes, 2.4 million tons. The amount of treated (liquid) sewage waste discharged to the ocean and estuaries from U.S. coastal communities has been estimated at somewhat less than 20 million tons per day.

The principal Federal agencies engaged in in-house or contracted research on waste disposal in the sea are (in order of allocation of funds) the National Oceanic and Atmospheric Administration, the Environmental Protection Agency, and the Department of Energy. The outlays by the Departments of Defense and Interior, the National Science Foundation, and other agencies are smaller. Of the total estimated Federal budget of $186 million for pollution-related work in 1980, some $30 million was allotted for research on marine waste disposal categories. A very rough estimate of $100 million is spent annually by industry for the disposal of 2.2 million tons of industrial waste. Costs for the disposal of dredged material mostly by the U.S. Army Corps of Engineers may be on the order of $75 million.

The operation of large sectors of industry, commerce, and coastal urban communities is still dependent on waste material handling or disposal in the marine environment, in spite of the Marine Protection, Research, and Sanctuaries Act of 1972 (Public Law 92-532). Innovation in disposing of wastes in the coastal area is needed.

2.1. INTRODUCTION

The purpose of this chapter is to review briefly the status of marine waste disposal research and marine dumping, with special reference to the role of U.S. Federal agencies. The distribution of effort, funding, executors of research, sources of information, and literature references are reviewed using some of the major research indices and agency and private industry data sources. Innovation and progress in dumping technology are examined.

2.2. MARINE WASTE DISPOSAL AS A REGULATORY ISSUE

Climaxing the environmental momentum of the 1960s, the U.S. Marine Protection, Research, and Sanctuaries Act (MPRS) of 1972 (Public Law 92-532) (U.S. Congress, 1972a) sent a message that ocean dumping around the Unites States was undesirable and should be ended as soon as possible; the exceptions were limited, but included dumping of dredge spoil (Smith, 1979). Five years later, the U.S. Environmental Protection Agency's (EPA) Final Revision of Regulations and Criteria, 11 January 1977 (Pt. VI), published in the *Federal Register* stated that existing dumpers including municipalities unable to meet the 23 April 1978 deadline could receive interim permits after that date only if they had implementation schedules to allow phasing out of ocean dumping by 31 December 1981, or a special permit. The EPA regulations further stated that "half-hearted attempts at compliance and weak efforts to find alternatives will not be considered to be good faith efforts," and new interim permits would be issued only where other alternatives had been exhausted. Municipalities' inability to raise necessary revenues to meet compliance schedules might be considered at EPA's discretion, but it was not expected that

Table 2.1. U.S. Ocean Disposal of Wastes[a]

Year	Industrial Waste[b] ($\times 10^6$ tons)	Sewage Waste[b] ($\times 10^6$ tons)	Demolition Construction[c] ($\times 10^6$ tons)	Dredged Material[d] ($\times 10^6$ tons)	Incineration at Sea[c] ($\times 10^3$ tons)
1968[e]	4.2	4.5	0.52		
1973	5.0	4.9	0.88	51	9.8
1974	4.1	4.5	0.70	75	13.3
1975	3.1	4.5	0.36	66	5.6
1976	2.4	4.8	0.28	50	7.9
1977	1.6	4.6	0.35	31	13.6
1978	2.2	5.0	0.22	40	16.2
1979				61[f]	

[a] Sources of data for each item are shown in the following footnotes (for 1968, footnote *e* applies). Wasteloadings are approximate metric tons, except as noted.

[b] U.S. Environmental Protection Agency (1978) and earlier equivalent reports.

[c] U.S. Environmental Protection Agency (1978) and earlier equivalent reports. Almost entire total comes from New York City metropolitan area.

[d] U.S. Army Corps of Engineers (1978c) and earlier equivalent reports. Note that U.S. Army Corps of Engineers estimates are usually made in place, and that volumes of actual dredged material will usually increase upon dredging, because of dilution with water. The figures refer to at sea disposal; river-mouth and estuarine disposal may add up to 20% additional dredged material.

[e] National Academy of Science (1976).

[f] Preliminary data for fiscal 1979 (M. Brown, personal communication).

financial reasons would be sufficient justification for noncompliance by industrial dumpers.

Dumping of industrial waste has declined nationwide, and dumping of industrial waste has virtually ceased in the Gulf of Mexico as a result of pressure and permit denials by EPA (Table 2.1). However, we are far from seeing an end to ocean dumping (Palmer and Gross, 1979).

Tonnages of disposed sewage sludge have changed little (Chapter 1). Lacking available land disposal sites, New York City has nowhere to go at present but to the ocean for disposal of sewage sludge, construction and demolition wastes, as well as dredged material. EPA continues active evaluation and certification of some 40 marine disposal sites (Pequegnat et al., 1978), and new research (for example, Section 2.6.2) suggests that for a number of purposes, marine disposal may have advantages over alternative disposal, quite aside from questions of lower cost. Such a conclusion was reached by President Carter's Interagency Task Force set up to formulate U.S. policy for radioactive waste management (Doty and Deese, 1978): "By the end of 1978 the Task Force was considering sub-seabed disposal as the leading option."

EPA's regulations acknowledge complications and ambiguities, an example being that the Ocean Dumping Act (U.S. Congress, 1972a) applies only to hauled or ship-transported waste, not to pipeline effluents that are regulated by the *Federal Water Pollution Control Act* Amendments (U.S. Congress, 1972b), commonly called the Clean Water Act (Public Law 92-500).

Sources of imprecision in studies of marine pollution other than incompleteness in data search include the problem of reporting of funding for a project that grows and wanes, the problem of defining which research should be considered under the classification of pollution studies, and the problem of reporting lumped research funds by agencies that do not split out pollution categories. For example, the National Science Foundation's (NSF) Geochemical Ocean Sections Study (GEOSECS) program, a basic research study of major and trace constituents in the ocean, was initiated partly to identify base line and pollutant levels of chemicals. The Department of Energy (DOE) studies of possible offshore disposal of nuclear wastes are reported by task group or performing laboratory rather than by project or contract, as is the practice with most Federal agencies; thus, calculating the amount of money spent on Federal pollution related research may be difficult. The Bureau of Land Management (BLM) in the Department of Interior has sponsored environmental studies related to offshore oil and gas leasing in amounts as great as $50 million per annum. Some of the areas and parameters studied and methods used in the studies are directly pertinent to ocean dumping and its related problems.

The most complete documentation of both research results and costs seems to have been done by the Dredged Material Research Program (DMRP) of the U.S. Army Corps of Engineers which produced a synthesis plus detailed reports and plans for a searchable data base. A summary of the five-year DMRP indicates that 67% of funding went for research contracts and 33% for facilities, administration, and overhead (U.S. Army Corps of Engineers, 1978b).

2.3. WHO ARE THE SPONSORS OF STUDIES ON MARINE WASTE DISPOSAL?

Recent estimates of total Federal ocean pollution research activity have been made by the Interagency Committee on Ocean Pollution Research, Development, and Monitoring (ICOP, 1979). Summaries of ICOP's estimates of funding by the principal Federal agencies are given in Table 2.2. In this tabulation the Bureau of Land Management's environmental surveys, related to petroleum leasing on the continental shelves (Bureau of Land Management, 1978), and EPA's activities account for the largest monetary outlays.

Estimation of funds spent on the narrower category, "marine waste disposal" is much more subjective. I have chosen to document Federal agencies' research activity by the Smithsonian Science Information Exchange (SSIE) (1979a, b). This bank has grown rapidly in recent years, and though it is still far from exhaustive, now most Federal agencies dealing with earth science and the environment contribute to

Table 2.2. U.S. Federal Ocean Pollution Research, Development, and Monitoring Programs[a] Budget for Fiscal Years 1978–1981

	Fiscal Year (in millions of dollars)		
Agency	1978	1979	1980
Department of Interior	56.4	52.5	76.5
Bureau of Land Management	37.7	30.1	51.2
Geological Survey	16.1	19.7	23.1
Fish and Wildlife	2.6	2.7	2.2
Environmental Protection Agency	35.0	36.4	40.0
National Oceanic and Atmospheric Administration	17.5	19.8	24.7
Department of Energy	21.6	16.8	15.1
Regional programs	6.6	4.9	4.7
Strategic Petroleum Reserve Office	4.5	6.7	4.5
Ocean Thermal Energy Conversion	1.4	2.1	2.8
Other	9.1	3.1	3.1
National Science Foundation	15.1	15.0	13.7
Department of Defense	8.8	7.3	6.6
Navy	6.4	5.2	4.5
U.S. Army Corps of Engineers	2.4	2.1	2.1
Department of Transportation (mainly the U.S. Coast Guard)	8.8	8.4	6.0
Department of Health, and Human Services	3.9	3.7	3.6
Food and Drug Administration	2.6	2.6	2.6
National Institutes of Health	1.3	1.1	1.0
Grand total	167	160	186

[a]Taken from Interagency Committee on Ocean Pollution Research, Development, and Monitoring (1979).

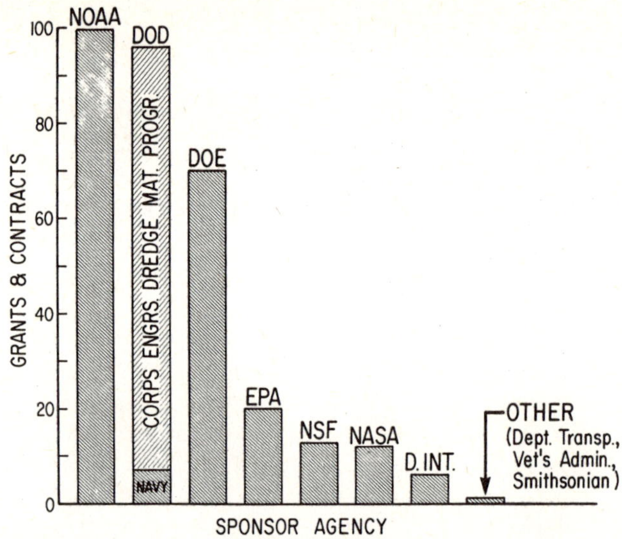

Figure 2.1. Approximate number of contracts and grants in the area of marine and Great Lakes waste disposal for leading Federal agencies, years 1977 and 1978. One item per multiyear contract is included. General sources are cited in Fig. 2.2. An approximate estimate for subcontracts from DOE's Sandia National Laboratories is included [Smithsonian Science Information Exchange (Search No. L30-136), 1979b].

the bank. Data from this source and some supplementary material suggest that in 1978 the National Oceanic and Atmospheric Administration (NOAA) of the U.S. Department of Commerce spent the most on ocean dumping studies (Figs. 2.1 and 2.2). Agencies that awarded grants and contracts for study of marine waste disposal that totaled more than $1 million included DOE, EPA, Department of Interior (U.S. Geological Survey, Office of Water Resources and Technology, Bureau of Fish and Wildlife, and BLM), Department of Defense (principally U.S. Army Corps of Engineers), and the NSF. Other agencies having considerably smaller outlays

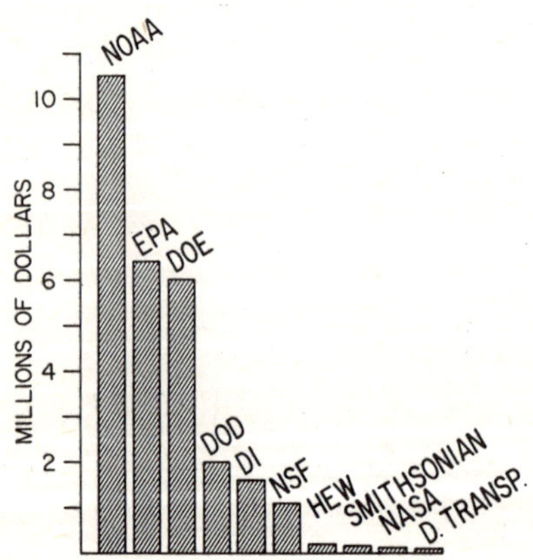

Figure 2.2. Total value of contracts and grants for Federal agencies in the area of ocean and Great Lakes waste disposal and dumping for 1978. Sources: Smithsonian Science Exchange (searches LA 30-136 and LA 15-136), 1979a, 1979b; U.S. Army Corps of Engineers, 1978a; M. Brown, personal communication; W. Musser, personal communication; and T. Landers, personal communication. Prorated estimates are used where exact figures are not available for multiyear contracts and grants.

Table 2.3. Selected U.S. Agencies' Contracts and Grant Research by Performing Organization[a]

Agency	Percentage of Total Expenditure					Total Expenditure
	University	In-House	Government	Industry	Others	($\times 10^6$ \$)
National Oceanic and Atmospheric Administration	18	81	1	–	–	10.5
Environmental Protection Agency	7	25	2	65	–	6.4
Department of Energy, Subseabed nuclear waste disposal, Sandia National Laboratories	56	42	–	2	–	3.1
Department of Energy, Salt dome petroleum reserve	39	–	53	7	1	2.4
Department of Defense (principally U.S. Army Corps of Engineers and U.S. Navy)	33	27	13	26	–	2.0
Department of Interior	8	60	25	–	–	1.6
National Science Foundation	85	–	6	2	2	1.1

[a]Data are for 1978 and are principally from sources cited in Fig. 2.2.

include the Departments of Transportation, Agriculture, Health, Education and Welfare, National Aeronautics and Space Administration and the Smithsonian Institution. Additional information in this area may be obtained from the Office of Management and Budget 1980 Catalog of Federal Domestic Assistance.

The distribution of grants and contracts for the study of marine waste disposal shows that the largest number of individual contract values fell in the range \$32,000–\$64,000 per year. Most of these contracts were awarded to researchers at universities and academic research institutions. The SSIE data do not include many major in-house and other programs, however. Some of these data are incorporated in Fig. 2.2 from supplemental data received from the agencies in question.

NOAA's Sea Grant program distributes many small grants. NOAA's Marine Ecosystems Analysis program efforts are dominated in overall spending by large in-house outlays (Table 2.3). Likewise, whereas EPA's Office of Research and Development lets many contracts to universities, its operational arm (Environmental Protection Division) draws heavily on contracted private firms at a level of about \$4 million per year (W. Musser, personal communication, 1980).

In this survey I have found no comprehensive data on research performed by and for industry, states, and municipalities. The SSIE source contained only one such report. Some estimates on spending for industrial waste disposal are mentioned later, but the largest category for marine waste disposal by far is municipal sewage. Even here in Woods Hole–Falmouth (population 23,000), the town has spent hundreds of thousands of dollars for feasibility studies on sewage disposal, including marine disposal. State regulatory agencies do not necessarily archive such expenditures and studies. The extensiveness and significance of coastal waste discharge in some of the more densely populated coastal zones around the United States is exemplified in Fig. 2.3, showing municipal and industrial waste distribution points that had direct or indirect discharge to the ocean around the State of Florida in 1976. This survey is addressed principally to dumping of wastes, rather than pipeline discharge from local communities. For further background information regarding the latter topic the EPA's Office of Marine Discharge Evaluation (personal communication) advises that some of the most extensive studies have been reported in the Coastal Water Research Project (1979), and reports cited therein.

2.4. COSTS OF DISPOSAL PROBLEMS, SHORT AND LONG TERM

The gross costs of actual seabed waste disposal are usually only a small fraction of the larger values of productive activities that require waste disposal. An important case is that of dredged material disposal in the New York Bight area, where the dredging and cost bookkeeping of the U.S. Army Corps of Engineers is augmented by voluminous gray (unpublished) literature and published economic studies as given by Conner et al. (1979) and references cited by them. The following summary is largely taken from their work.

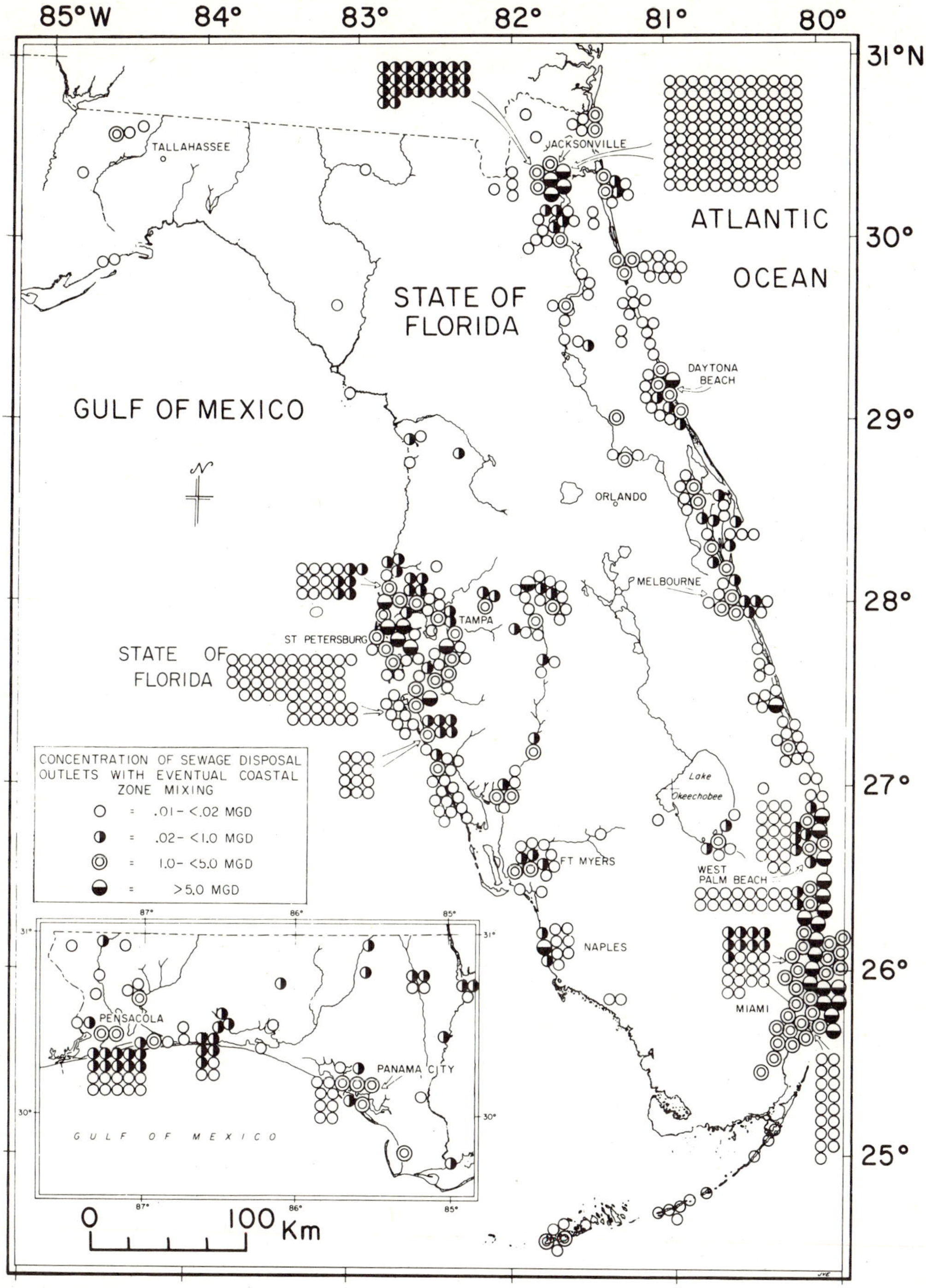

Figure 2.3. Distribution of coastal sewage disposal outlets, State of Florida, 1976 (Haddad et al., 1976) (1 MGD = 3800 $m^3\ d^{-1}$).

Dredging by the U.S. Army Corps of Engineers and contractors work averaged $8 \times 10^6\ m^3 y^{-1}$ between 1970 and 1978 in the New York Bight. Average cost was about $15 million per year. This cost is essential to maintain seaways that support a total value related to foreign trade estimated by the Ports of New York and New Jersey in 1973 to be $130 billion per year. This trade includes 74% of the region's oil supplies, and about 400,000 jobs are related to manufacturing, processing, and shipping of cargo. The cost of transporting dredged material for disposal is estimated to be $\$0.036\ km^{-1}\ m^{-3}$. Thus, it was estimated that increasing disposal distance from present 20-25 km in New York Bight to deep water sites would result in a net disposal cost of $\$3.48\ m^{-3}$, compared with present $\$1.96\ m^{-3}$ (Conner et al., 1979).

In Rhode Island costs for dredged material disposal were $\$13.72\ m^{-3}$ for land disposal and $\$3.92\ m^{-3}$ for sea disposal (Tippie, 1977).

No review of costs of marine disposal of wastes other than dredged material was found, but a crude magnitude (Table 2.4) may be estimated from 1976 waste disposal cost data of the Bureau of Census, cited by T. Rothermel (personal communication).

Comparing total operating cost for water and solids disposal of $2.6 billion (Table 2.4) with minimum industrial wastes of approximately 64 million metric tons (Tables 2.5 and 2.6), one obtains a disposal operating figure of $40 per ton. Applying this to the 1978 industrial total of 2.2 million tons disposal (Table 2.1) at sea, one would obtain about $100 million as disposal costs if marine disposal followed land costs. The use of different year data bases (and possibly differing cost base for sea disposal in this calculation) results in an estimate for orientational purposes only. The actual disposal costs must be measured, on the one hand, against environmental effects of the waste in the sea or risks in transport, and on the other hand, against the productive and indirect benefits of the activity that produces wastes. Finally, one must consider the impact on the disposal problem by inhibition or delay of investment, planning or technological innovation, or commitment of critical staff.

2.5. PUBLICATION AND COMMUNICATION

The four major data bases covering marine waste disposal have been searched to get an idea of the output of information in terms of different types of waste

Table 2.4. Costs, in Billions of Dollars, of U.S. Waste Removal and Disposal in 1976[a]

	Capital Costs	Operating Costs
Air pollutants	1.8	1.9
Water wastes	1.6	1.8
Solid wastes	0.14	0.8

[a]From *Environmental Science and Technology*, April 1978.

Table 2.5. Industrial Waste from Coastal States 1978

EPA Region	State or Commonwealth	Number of Facilities Treating Waste	Metric Tons ($\times 10^3$) of Waste[a]
1	Connecticut	17	4.5
	Maine	4	95
	Massachusetts	18	202
	New Hampshire	1	57
	Rhode Island	2	0.04
2	New Jersey	116	806
	New York	49	736
	Puerto Rico	15	61
3	Delaware	19	242
	Maryland	13	142
	Virginia	24	156
4	Alabama	37	458
	Florida	39	15,850
	Georgia	48	190
	Mississippi	13	66
	North Carolina	49	751
	South Carolina	32	94
5	Illinois	77	3,135
	Indiana	29	1,105
	Michigan	35	408
	Minnesota	4	5.4
	Ohio	86	2,129
	Wisconsin	9	21
6	Louisiana	65	5,153
	Texas	147	3,825
9	California	126	3,379
10	Alaska	1	2
	Oregon	9	3
	Washington	15	19
	Coastal areas, subtotal		39,095
	Total U.S. industrial waste		63,920

[a]From PEDCO (1979).

materials. The results are shown in Fig. 2.4. The National Technical Information Service (NTIS), covering principally unpublished reports on Federal Government-supported research, has 703 items in the combined categories: Ocean/Marine, Sea/Dump, Disposal/Waste/Dredge, Chemical Waste, Sewage, Radionuclide, Nuclear Municipal and Refuse, and Disposal and Waste, compared to 312 for *Pollution Abstracts*, 372 for *Enviroline*, and 81 for *Chemical Abstracts*. Thus, the specialized

Table 2.6. Thousands of Tons of Waste Disposed of by Selected Disposal Methods

	Untreated				Treated						
SIC Division[a]	Landfill	Pit/Pond/Lagoon	Deep Well	(Subtotal)	Incinerated	Recycled	Evaporated	(Subtotal)	Unknown	Others[b]	Total
Organics	2,200	5,237	3,225	10,662	485	539	48	1,072	42	5,067	16,843
Inorganics	2,568	4,702	2,584	9,854	357	77	5,856	6,290	239	3,798	20,181
Plastics	1,112	254	322	1,688	115	385	7	507	11	559	2,765
Agricultural	1,145	351	708	2,204	912	39	2,328	3,279	179	10,686	16,348
Misc. chemicals	132	122	10	264	19	27	10	56	1	8	329
Other	986	4,240	170	5,396	60	60	332	452	0.65	12	5,861
Medicinal and soaps	122	9	0.96	132	85	0.65	0	86	0	135	353
Unknown	33	4	5	42	3	0	179	182	0.90	120	345
Total	8,298	14,919	7,025	30,242	2,036	1,128	8,760	11,924	474	20,385	63,025

[a]SIC refers to Standard Industrial Classification. Data are from PEDCO (1979) summarizing results of a survey of the 53 largest chemical manufacturing companies in the United States, conducted by questionnaire by the Committee on Interstate and Foreign Commerce, U.S. House of Representatives, Honorable Bob Eckhardt, Chairman (known as the Eckhardt Survey). In addition to enumerated results, the survey showed that 822 million tons of waste were disposed of between 1950 and 1979, and that 99% was disposed of in the state of origin; 86.6% went to sites still in use, and 13% to sites now closed.

[b]Includes storage, land application, and ocean disposal as well as other methods, but no further breakdowns are provided.

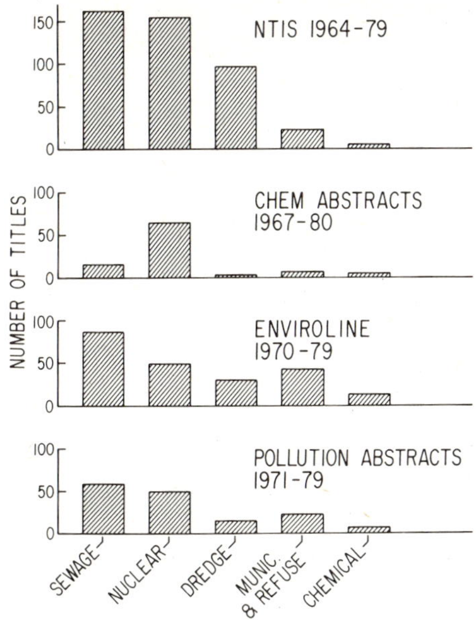

Figure 2.4. Search of four leading information retrieval systems for titles dealing with marine waste disposal. National Technical Information Service (NTIS) deals principally with unpublished reports of government-supported research. Total publications in the category of marine waste disposal were 703, 78, 372, 312, respectively, for NTIS, *Chemical Abstracts, Enviroline,* and *Pollution Abstracts*. The data bases were accessed through Dialog Information Systems, Lockheed Co., Palo Alto, California.

retrieval services are more effective for this kind of search than is the otherwise highly regarded *Chemical Abstracts*. NTIS covers an earlier period of 1964–1970, but relatively few titles are listed. U.S. Government-generated reports exceed documented published or otherwise generally available titles by more than two to one.

A review of items in the titles retrieved from the NTIS data bank, as well as subdivision of marine waste-related items in Edel and Roques (1978), reveals that a new breed of U.S. research organization focused strongly on environmental and applied programs has taken the lead in marine dumping studies. This kind of organization is exemplified by the applied marine efforts of the University of Rhode Island, University of Southern California, University of Delaware, Texas A&M University, and the State University of New York.

The most difficult to document are studies by private industry, municipalities, the activities of the BLM and the DOE. The latter's problems can partly be understood in the light of its buffeting by political remobilization and changes in mandate, which make it hard to establish stable literature dissemination systems. The best paper trails are provided by the U.S. Army Corps of Engineers and NOAA. Of course, it is frequently hard to track work in Federal agencies for which the discipline represents only minor interest. The Smithsonian Science Information Exchange presents a most valuable resource here.

2.6. NATIONAL PRODUCTIVITY, INNOVATION, AND WASTE DISPOSAL

We know that the United States is suffering from economic problems including declining labor productivity, innovativeness, international competitiveness, and industrial investment, as well as increasing inflation and outdated physical plants (Joint Economic Committee, Senate and House of Representatives, 1980). Accordingly, it is appropriate to consider marine disposal not only from the point of view of the health of the recipient environment, but also from the standpoint of economic health of waste-producing industries, services, and communities. Three problems then emerge:

2.6.1. Regulation and Productivity

Until recently lawmakers and economists have assumed tacitly that the U.S. productive system can absorb waste disposal costs without compensatory assistance. The EPA is charged with considering economic feasibility or hardship only in the present. It tries to avoid shutting down plants or otherwise contributing to layoffs or severe economic impacts wherever possible. However, regulatory bodies or enforcers are not required or perhaps even allowed in many cases to consider the effect of regulations on *future* investment, productivity, or innovation. For example, the U.S.'s largest platinum producer, the Goodnews Bay Company, Seattle, Washington, ceased efforts to reestablish platinum production on Goodnews Bay, Alaska, in 1976 after a dispute about handling of waste-water effluent with the EPA. There may be merit on both sides of the dispute. But whereas EPA has the force of law behind it, U.S. industry, unlike Japanese, West German, or South Korean industry, or U.S. agriculture, has no comparable agency to help Goodnews Bay Company overcome problems and provide assistance to meet new regulatory or technical demands.

One environmentally conscious State, Hawaii, has recognized economic realities in consideration of development of a marine manganese nodule processing base in the Islands as a partial replacement for its declining pineapple and sugar industries. It has given attention to the nodule industry's marine waste disposal problem by examining possible hazards and methods of overcoming them at minimum costs. Finally, the issue was opened to public scrutiny at an early stage (Stephen-Hassard et al., 1978). The result apparently has achieved public support.

2.6.2. Failure of Innovation in Waste Disposal and Inhibition of Codevelopment

Five to eight years ago, a bright future was predicted for innovation in sewage disposal, for instance by recycling waste to the soil, using it to stimulate aquaculture, or using its energy potential. However, the National Academy of Science (1978) found problems in achieving "fertilizer-grade sludge." The wastes of even the most innocent-appearing communities contain toxic metals, organic chemical residues, or viruses that bring them afoul of regulation. Even the venerable Milwaukee *Milorganite*, the first U.S. commercial fertilizer product produced from municipal sewage, now features warning labels. Salt-marsh disposal arouses concern for habitats of commercial or other wildlife species. In the Falmouth–Woods Hole, Massachusetts, area spray irrigation experiments with secondary effluent at the nearby Otis Air Force Base have been discontinued because of high costs. These are high partly because high-value plant products from sprayed areas cannot be marketed owing to potential environmental regulatory problems. Likewise, the pioneering and highly promising studies of effluent-fed aquaculture in Woods Hole (Ryther et al., 1972) required 10 permits, even for a study of research scale (J. Huguenin, personal communication). Economic-scale practical application has not been attempted because of uncertainties such as the presence of viruses or fear of plant breakdown. Pyrolysis of waste involves potential air pollution and operational problems. The Monsanto Corporation abandoned its major operation in Baltimore after 7 years of pyrolysis efforts (Council for Environmental Quality, 1977).

In the marine area, progress has been even more limited. The "Oceanlite" patent requests to convert polluted mud and dredged material to usable ceramic and aggregate (Rhoads et al., 1975) were withdrawn in the face of moratoria on offshore mineral recovery and rising fuel costs. The concept of digging excavations in the continental shelf and burying wastes in them, covered with a clean sediment cap, and selling the sand and gravel (Bokuniewicz, 1983) has potential, but likewise cannot now be implemented in the Federal Outer Continental Shelf (OCS) because regulations to permit retrieval of hard minerals from the seabed have not been issued. Ambitious plans to build offshore dredged material islands and terminals have foundered in the face of delays and uncertainties. Ocean disposal of coal waste blocks has shown technical promise, (Duedall et al., 1981; see also Chapters 1, 18, and 19) but appears to be contrary to the intent of the Ocean Dumping Act (U.S. Congress, 1972a) to phase out ocean dumping by 31 December 1981 (Federal Register, 11 January 1977).

The problem is partly that we cannot, apparently, implement new and more efficient waste disposal methods until they are proved "risk-free," or beyond poorly defined "acceptable risk," and we are left, therefore, with expensive, energy-inefficient disposal systems, and known risks, like landfall disposal.

2.6.3. Internal Brain Drain?

The outpouring of Federal research funds in the 1960s caused a burgeoning of scientific establishments and led to major scientific achievements. However, the availability of research funds may have tended also to skim the cream off top technical and administrative talent—as it was indeed designed to do (Hornig, 1965)—at the expense of activities having critical practical importance for the country, in industry and key governmental activities such as regulation.

The scientific research reported in this and similar books and at symposia on ocean dumping or marine pollution must be translated into practical policy by others. Who are these other people? Of the 100,000 persons said to be involved in regulatory activities (Abelson, 1978), do we have comparable talent to that present among our leading scientists? A recent poll of 800 marine science faculty, professional researchers, and administrators shows that, of agencies involved in marine environmental and waste disposal studies, those that have the most critical practical roles, and also the largest budgets, often have the least highly regarded research output (Table 2.7), and vice versa. This poll, which was compiled and not analyzed, does not take into account the specific missions and objectives of

Table 2.7. U.S. Academic Marine Scientists' Assessment of the Quality of Academic Marine Scientific Work Funded by the Federal Agencies

Agency	Percentage of Respondents per Category				
	Excellent or Generally High Quality	Mediocre Quality	Substandard or Not Science at All	Rank	Number of Respondents
National Science Foundation–Deep Sea Drilling Project	95.5	3.7	0.8	1	243
National Science Foundation–Oceanography Section	94.5	4.8	0.8	2	564
Health, Education and Welfare–National Institute, of Health	89.9	6.9	3.2	3	188
National Science Foundation–Environmental Biology	87.9	9.3	2.7	4	182
Office of Naval Research	86.3	12.3	1.5	5	562
National Science Foundation–International Decade of Ocean Exploration	79.6	17.3	3.1	6	515
National Science Foundation–Polar Programs	76.1	17.2	6.7	7	267
U.S. Geological Survey	70.4	24.9	4.6	8	281
Department of Energy	57.2	33.6	9.2	9	339
National Oceanographic and Atmospheric Administration–					
National Marine Fisheries Service	42.1	42.9	15.0	10	273
Sea Grant	34.4	40.6	24.9	11	566
Environmental Protection Agency	32.1	42.7	25.2	12	396
U.S. Army Corps of Engineers	24.0	42.7	33.3	13	342
U.S. Coast Guard	23.0	43.6	33.3	14	204
Bureau of Land Management	13.1	33.4	53.4	15	419

[a]From Ladd et al. (1978).

agencies rated in the poll. It does, however, demonstrate a climate of opinion unconducive to attracting talented younger graduates to important technological decision making roles in society. This unfavorable climate must be rectified. A partial solution may be that competent legal talent be encouraged to have graduate degrees in science or vice versa.

2.7. CONCLUSIONS

Some of our most serious waste disposal problems, such as sewage in the New York Bight, cannot be easily solved without well-designed marine disposal under today's conditions.

Whereas innovation in waste disposal practice has been generally disappointing in comparison with earlier expectations, innovation in marine waste disposal is currently nearly precluded by regulatory limitations. The environmental issues in the marine sphere are inherently complex, and talent on both the scientific and regulatory side will be needed to permit good solutions.

ACKNOWLEDGMENTS

Thanks are due to the many individuals who have provided assistance: R. Anderson, Sandia National Laboratories, Albuquerque, New Mexico; M. Brown, U.S. Army Corps of Engineers, Vicksburg, Mississippi; R. Burroughs, Bureau of Land Management, Washington, D.C.; M. Champ, NOAA, Rockville, Maryland; C. Hurter, Woods Hole Oceanographic Institution, Woods Hole, Massachusetts; J. Landers, Department of Interior, Washington, D.C.; D. Letzkus, U.S. Army Corps of Engineers, Fort Belvoir, Virginia; W. Musser, EPA, Washington, D.C.; and L. Samowitz, EPA Library, Washington, D.C. I am no less grateful to others not cited here.

REFERENCES

Abelson, P. H. 1978. The Federal regulatory machine, *Science*, **200**, 487.

Bokuniewicz, H. J. 1983. Submarine borrow pits as containment sites for dredged sediment. *In*: Dredged Material Disposal in the Ocean, D. R. Kester, B. H. Ketchum, I. W. Duedall, and P. K. Park (Eds.), John Wiley and Sons, New York, in press.

Bureau of Land Management. 1978. Study Design for Resource Management Decisions, OCS Oil and Gas Development and the Environment. U.S. Department of the Interior, Bureau of Land Management, Washington, D.C., approx. 120 pp., (sections paginated separately).

Coastal Water Research Project. 1979. Annual Report, 1978. Southern California Coastal Water Research Project, El Segundo, California, 253 pp.

Conner, W. G., D. Aurand, M. Leslie, J. Slaughter, A. Amer, and F. I. Ravenscroft. 1979. Present practices and candidate alternatives. *In*: Disposal of Dredged

Material within the New York District, Vol. 1. Report MTR-7808. The MITRE, Corp., McLean, Virginia, 215 pp.

Council for Environmental Quality (CEQ). 1977. Annual Report, U.S. Government Printing Office, Washington, D.C., 445 pp.

Doty, P., and D. A. Deese. 1978. Legal, political and institutional implications of the subseabed program for radioactive waste disposal. *In*: Sandia National Laboratories Progress Report on Subsea Waste Disposal, Nuclear Waste, Vol. II, Sandia National Laboratories, Albuquerque, New Mexico, pp. 931–948.

Duedall, I. W., F. J. Roethel, J. D. Seligman, H. B. O'Connors, J. H. Parker, P. M. J. Woodhead, R. Dayal, B. Chezar, B. K. Roberts, and H. Mullen. 1981. Stabilized power plant scrubber sludge and fly ash in the marine environment. *In*: Ocean Dumping of Industrial Wastes, B. H. Ketchum, D. R. Kester, and P. K. Park (Eds.). Plenum Press, New York, pp. 315–346.

Edel, B. M., and C. D. Roques. 1978. Sea Grant Packaged Literature Search on Outer Continental Shelf 1968-MID-1978. National Sea Grant Depository, Pell Library, Bay Campus, University of Rhode Island, Narragansett, Rhode Island, 53 pp. + index.

Haddad, K., F. T. Manheim, and W. Marr. 1976. Waste Disposal in the Florida Coastal Environment. Unpublished Report. Marine Policy Program, Department of Marine Science, University of South Florida, St. Petersburg, Florida, 60 pp.

Hornig, D. F., 1965. Universities and Federal Science Policies. *Science*, **150**, 847–851.

Interagency Committee on Ocean Pollution Research, Development, and Monitoring (ICOP). 1979. Catalog of Federal Ocean Pollution Research, Development, and Monitoring Program, Fiscal Years 1978–1980. Working Paper 1 for the Federal Plan for Ocean Pollution Research, Development, and Monitoring, Fiscal Years 1979–1983. Interagency Committee on Ocean Pollution Research, Development, and Monitoring/Federal Coordinating Council for Science, Engineering, and Technology. Report available from National Oceanic and Atmospheric Administration, Rockville, Maryland, 230 pp + appendices.

Joint Economic Committee, Senate and House of Representatives, U.S. Congress. 1980. U.S. Government Printing Office, Washington, D.C., 114 pp.

Ladd, E. C., Jr., D. O. Palmer, and W. W. Shannon. 1978. Federal Funding of Academic Marine Science; Preliminary Report. *JOIDES Journal* **4**(3), 50 pp.

National Academy of Science. 1976. Disposal in the Marine Environment: An Oceanographic Assessment. National Academy of Science-National Research Council, Washington, D.C., 75 pp.

National Academy of Science. 1978. Multimedium management of municipal sludge. Analytical Studies for the U.S. Environmental Protection Agency, National Academy of Science-National Research Council, Washington, D.C., 187 pp.

Palmer, H. D., and M. G. Gross (Eds.) 1979. Ocean Dumping and Marine Pollution. Dowden, Hutchinson and Ross, Stroudsburg, Pennsylvania, 268 pp.

PEDCO. 1979. PEDCO Analysis of Eckhardt Committee Survey for Chemical Manufacturers' Association. PEDCO Environmental, Inc., Washington, D.C., 3 chapters and 15 appendices plus index, all separately paginated.

Pequegnat, W. E., D. D., Smith, R. M. Darnell, B. J. Presley, and R. O. Reid. 1978. An Assessment of the Potential Impact of Dredged Material Disposal in the Ocean. NTIS No. AD-A053, 183/OST, National Technical Information Service, Springfield, Virginia, 659 pp.

Rhoads, D. C., R. B. Gordon, and J. R. Vaisnys. 1975. Conversion of marine muds to lightweight construction aggregate. *Environmental Science and Technology*, 9, 360–362.

Ryther, J. H., W. M. Dunstan, K. R. Tenore, and J. E. Huguenin. 1972. Controlled eutrophication–Increasing food production from the sea by recycling human waste. *Bioscience*, 22, 144–152.

Smith, D. D. 1979. Dredged material, ocean disposal, and the regulatory maze. *In*: Ocean Dumping and Marine Pollution, H. D. Palmer and M. G. Gross (Eds.). Dowden, Hutchinson and Ross, Stroudsburg, Pennsylvania, pp. 241–262.

Smithsonian Science Information Exchange (SSIE). 1979a, 1979b. Searches: LA 15–136, Effect of Dredging on Freshwater and Marine Environments; LA 30–136, Marine Disposal and Dumping, of Sewage, Ship Waste, Industrial and other Solid Waste. Smithsonian Institution, Washington, D.C.

Stephen-Hassard, Q. D., K. E. Chave, Q. Fernando, K. M. Keith, M. A. Meyland, and W. Miklius. 1978. *In*: The Feasibility and Potential Impact on Manganese Nodule Processing in Hawaii. Center for Science Policty and Technology Assessment, Honolulu, Hawaii, 171 pp.

Tippie, V. K. 1977. Dredging and the Disposal of Dredged Materials in Rhode Island. Coastal Resources Center Report, University of Rhode Island, Kingston, 20 pp.

U.S. Army Corps of Engineers. 1978a. Dredged Material Research Program. Waterways Experiment Station, U.S. Army Corps of Engineers, Vicksburg, Mississippi, 165 pp.

U.S. Army Corps of Engineers. 1978b. Dredged Material Research Program, Executive Overview and Detailed Summary (Final Report). Technical Report DS-78-22, U.S. Army Corps of Engineers, Vicksburg, Mississippi, 227 pp.

U.S. Army Corps of Engineers. 1978c. Report to the International Maritime Commission. Ocean Dumping Reports, Dredged Material. U.S. Army Corps of Engineers, DAEN-CWO, Office of Civil Works, Washington, D.C., 250 pp.

U.S. Congress. 1972a. Marine Protection, Research, and Sanctuaries Act of 1972. Public Law 92-532, 86 STAT. 1052. Washington, D.C.

U.S. Congress. 1972b. Federal Water Pollution Control Act Amendments of 1972. Public Law 92-500, 86 STAT. 1060. Washington, D.C.

U.S. Environmental Protection Agency. 1978. Annual Report to Congress Jan.–Dec. 1978 On Administration of the Marine Protection, Research, and Sanctuaries Act of 1972, as Amended (P.L. 92-532) and Implementing the International Ocean Dumping Convention. U.S. Environmental Protection Agency, Office of Water Programs, Washington, D.C., 33 pp. + appendix.

PART II: PHYSICAL OCEANOGRAPHIC ASPECTS

Modelling and Field Studies

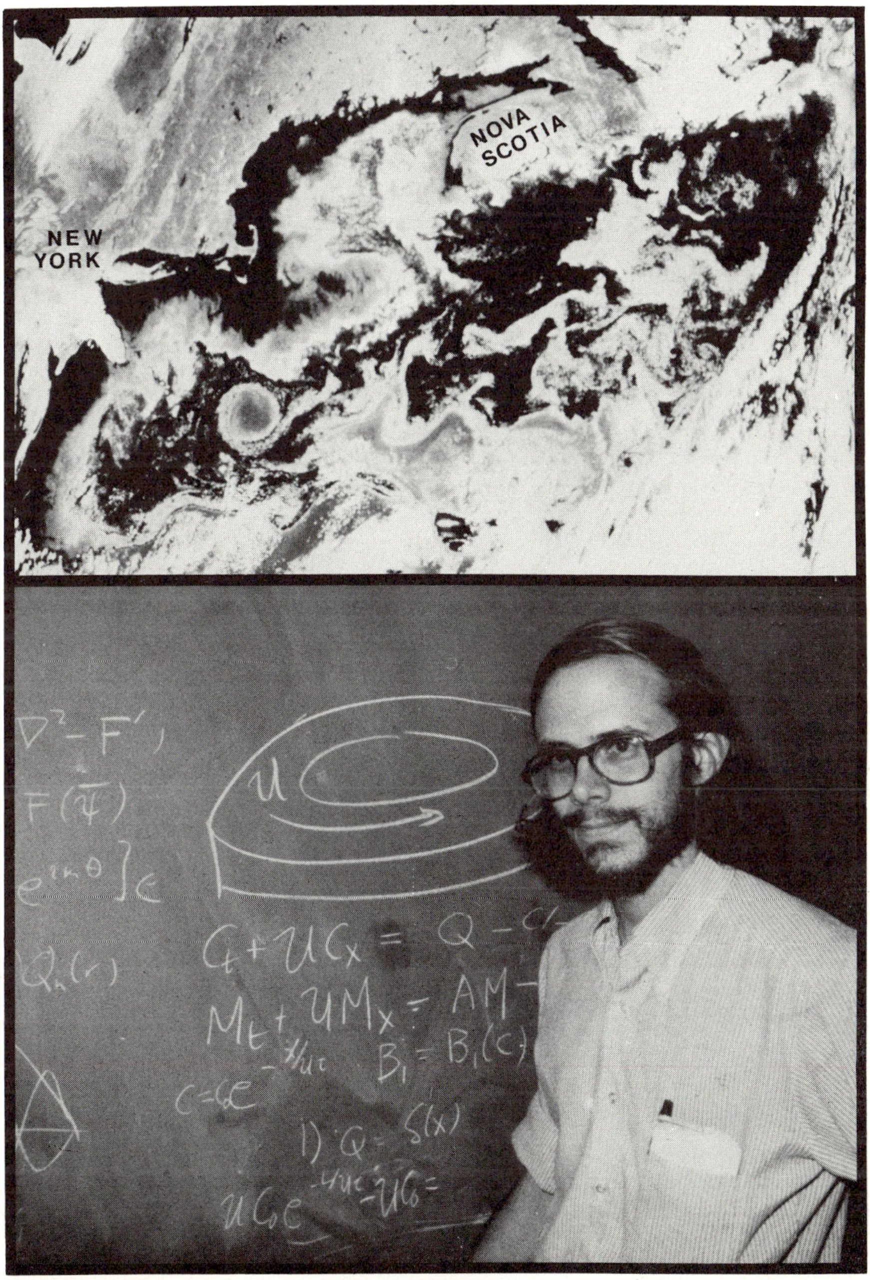

3

SIMPLE MODELS OF WASTE DISPOSAL IN A GYRE CIRCULATION

Glenn R. Flierl

Department of Meteorology and Physical Oceanography
Massachusetts Institute of Technology
Cambridge, Massachusetts

ABSTRACT

We discuss several simple models of the distribution of waste and impact of waste dumping in a region where there is a circulating flow. First, the distribution in the absence of biological feedback effects is considered and the dependence on the flow rate, the scale of the gyre, the decay time of the waste, and the dumping rate are described. Second, a biological model, with a growth rate dependent on the density of organisms and a death rate varying with the waste concentration, is used to examine what the impact on organisms within the gyre might be. Finally, various situations with biological feedback, introduced by making the decay rate of the wastes dependent on the level of biological activity, are considered. When the death rate varies nonlinearly with concentration, "catastrophes" (in the mathematical sense) can occur: the population may suddenly die out with only a small increase in the dumping rate. The models here are intended to illustrate possible behaviors when the physics, chemistry, and biology of a polluted ecosystem are all considered together; no attempts are made to apply these models in detail.

3.1. INTRODUCTION

In order to understand the impact of dumping waste material into the ocean, it is necessary to synthesize the physics, chemistry, and biology of the region. The water motions transport the material from one part of the region to another and disperse the initial plume (Chapters 4 to 6); chemical reactions alter the toxicity of the waste (Chapter 7). Ultimately, the impact of the pollution is measured by alterations in the biology (Chapters 10 to 13). But these topics cannot be treated independently. The concentration of *toxic* material depends not only on physical parameters but also on the rate at which this material is broken down. This rate, in turn, depends on biochemical transformations which may be altered by the presence of the pollutant; indeed, in the extreme case, when the whole population is killed, biochemical breakdown of the waste may stop completely.

This chapter explores a simplified model that illustrates possible impacts of dumping within a part of a large-scale gyre circulation. Such circulations are common in the ocean on many different length scales, from the oceanic wind-driven gyres to tidal flows, which can also cause water to return cyclically to its initial place. For example, it is thought that the Deepwater Dumpsite-106 (DWD-106, Chapter 5) may be embedded in a cyclonic gyre in the slope water (Sverdrup et al., 1942; Wunsch, 1978). Of course, these circulations are rarely completely closed but rather exchange water and organisms with their surroundings. This process could be parameterized in the model.

The model is chosen to be illustrative of possible behaviors and to indicate characteristics that field and laboratory measurements should attempt to determine. No attempt has been made to make the model mimic any particular situation; such exercises must be done in close collaboration with those investigating the chemistry and biology of the wastes in question.

3.2. THE MODEL

Consider a domain of width W, depth H (perhaps the typical mixed-layer depth), and circumference L_g as sketched in Fig. 3.1. The currents flow around this gyre at speed U. Define a downstream coordinate x starting from the upstream edge of the dumpsite ($x = 0$) to the downstream edge ($x = L_d$) and from thence around to just before the upstream edge again ($x = L_g$). Cross-stream variations and depth variations will be ignored. Assume that small-scale processes smooth out the spatial

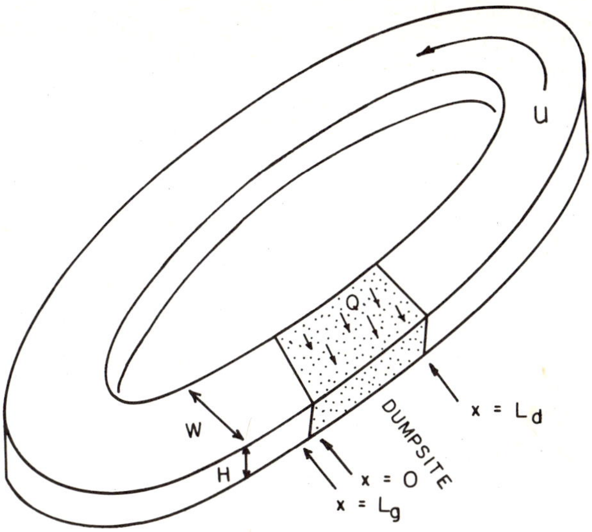

Figure 3.1. Sketch of model geometry. Strippled area represents the dumpsite.

and temporal pattern of dumping and simply specify that the flux into the dumpsite (having volume $L_d WH$) is uniform at rate Q. The concentration (C) of waste therefore satisfies the equation

$$\frac{\partial C}{\partial t} + U \frac{\partial C}{\partial x} = \left\{ \begin{array}{ll} \dfrac{Q}{L_d WH} & 0 < x < L_d \\ 0 & L_d < x < L_g \end{array} \right\} - \frac{C}{\tau} \qquad 1$$

with τ the e-folding time scale for breakdown of the waste. The periodic nature of the circulation enters as a boundary condition

$$C(L_g) = C(0) \qquad 2$$

This model was proposed for DWD-106 in Csanady et al. (1979); the succeeding steps represent new additions to the previous results.

The chemistry and biochemistry of the breakdown of the toxic waste have been represented as a linear decay law with a time scale τ. Loss of waste due to mixing with the deep water could also be considered part of τ and of the natural death rate (D) of organisms described below. Other simple parameterizations of mixing could also be included easily. In general this τ is expected to decrease with increased biological activity so that it can be represented as

$$\tau = \tau(M) \qquad 3$$

where M is the biomass of some indicator species (or larger class of organisms).

For the biology, a logistic growth law (Lotka, 1956) including advection will be used

$$\frac{\partial M}{\partial t} + U \frac{\partial M}{\partial x} = AM - A_1 M^2 - DM \qquad 4$$

Here A is the growth rate at small biomasses, D the death rate, and A_1M represents the density dependent decrease in growth rate at high biomasses due to competition for a limiting resource. It will be assumed that the growth rates are constant but that the death rate is an increasing function of the waste concentration

$$D = D(C) \qquad 5$$

Specification of the functions $\tau(M)$ and $D(C)$ and of the constants U, A, A_1, L_d, W, L_g, and M allows us to predict from equations 1 through 5 the concentration $C(x,t,Q)$ and biomass $M(x,t,Q)$ from initial conditions. Finally, the "pollutant effect" $E(Q,t)$ will be defined as the ratio of the net biomass in the system, given waste input rate Q, to the net biomass that would exist in the absence of the pollutant

$$E(Q, t) = \int_0^{L_g} dx\, M(x, t, Q) / \int_0^{L_g} dx\, M(x, t, 0) \qquad 6$$

The pollutant effect is a nondimensional number ranging from zero (corresponding to 100% extinction of the population) to one (corresponding to no change in the population biomass).

At this point ad hoc functional forms will be chosen for τ and D which, as it turns out, yield fairly dramatic pictures of pollution "catastrophes". These forms are

$$\tau = \tau_0/(1 + M/M_1)$$
$$D = D_0\,[1 + (C/C_1)^{3.5}]$$

The time scale for waste decay in the absence of biochemical action is τ_0; D_0 is the natural death rate; C_1 is the pollutant concentration necessary to double the death rate compared with unpolluted conditions; and M_1 is the biomass necessary to halve the decay time. (When the decay rate τ^{-1} increases linearly with M, it is necessary for the exponent of the C/C_1 term to be >3 to show catastrophic behavior. We illustrate the case in which the catastrophe occurs with little warning; simultaneously, we require that the change in τ from normal M conditions to $M = 0$ conditions be large. These restrictions force us to choose a small exponent.) The equations will be nondimensionalized, scaling t by τ_0, x by L_g, M by $M_0 = (A - D_0)/A_1$ (the population in the absence of pollutant), and C by $C_1\,[(A - D_0)/D_0]^{1/3.5}$ (the concentration of pollutant which gives zero growth rate for a very small population). The resulting equations are

$$vC_x = -C(1 + \alpha M) + q\,\Delta(x) \qquad 7$$

$$vM_x = g(M - M^2 - C^{3.5}M) \qquad 8$$

with

$$\Delta(x) = \begin{cases} \dfrac{1}{\delta} & 0 < x < \delta \\ 0 & \delta < x < 1 \end{cases} \qquad 9$$

$$\begin{aligned} C(0) &= C(1) \\ M(0) &= M(1) \end{aligned} \qquad (10)$$

$$E = \int_0^1 dx\, M(x, q) \qquad (11)$$

and the following nondimensional parameters

$v = U\tau_0 / L_g$	the ratio of τ_0 to the circulation time,
$g = (A_0 - D_0)\tau_0$	the ratio of the nonbiological decay time to the net e-folding time of the population at low biomasses and no pollutant,
$\delta = L_d / L_g$	length of dumpsite divided by the length of gyre,
$\alpha = M_0 / M_1$	normal biomass divided by the biomass necessary to halve the decay time of pollutant,

and the nondimensional dumping rate

$$q = \frac{Q\tau_0}{L_g WHC_1 \left(\dfrac{A_0 - D_0}{D_0}\right)^{1/3.5}}$$

The time-dependent term in equations 7 and 8 have been dropped and only the steady-state solutions will be examined. Some caution must be taken to ensure that the solutions are stable; these calculations are not complete, though some results will be used below.

3.3. SOLUTIONS IN THE ABSENCE OF BIOCHEMICAL FEEDBACK ($\alpha = 0$)

When there is no significant alteration in decay time as the biomass changes (for example, radioactive wastes, physical mixing dominant), the solution to equation 7 can be written explicitly:

$$C = \frac{q}{\delta} \begin{cases} \left[1 - \dfrac{e^{(1-\delta)/v} - 1}{e^{(1-\delta)/v} - e^{-\delta/v}}\, e^{-x/v}\right] & 0 < x < \delta \\[2ex] \left[\dfrac{1 - e^{-\delta/v}}{e^{(1-\delta)/v} - e^{-\delta/v}}\right] e^{(1-x)/v} & \delta < x < 1 \end{cases}$$

However, as described in Csanady et al. (1979), there are three distinct behaviors to this solution. Approximate expressions will be written below for the waste concentration in these cases (see Fig. 3.2 for sketches of the concentration distribution).

First, there is the "local" response, corresponding to very slow motions $v \ll \delta$

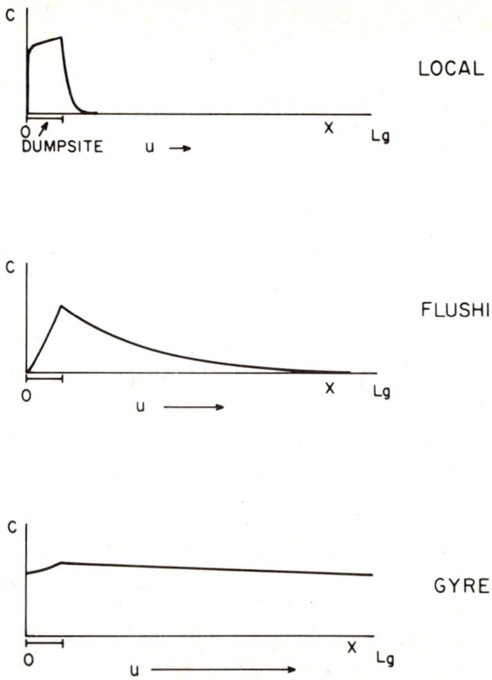

Figure 3.2. Concentration (C) as a function of distance (x) upstream of dumpsite of x for three different conditions: local ($\upsilon << \delta << 1$), flushing ($\delta << \upsilon << 1$), and gyre ($\delta << 1 << \upsilon$). Recall that $x = L_g$ is actually the same point as $x = 0$. The label U and arrows show the direction of flow.

$<< 1$. In this case, the concentration begins at zero (essentially) at $x = 0$ and rises within a boundary layer of thickness υ to the value

$$C = q/\delta$$

corresponding to a local balance of input and decay. This holds until $x = \delta$; beyond this point C decays back to zero in another layer of thickness υ. To a good approximation

$$C = q\Delta(x) \qquad 12$$

C is nonzero only within the dumpsite for this case.

Second, a "flushing" response occurs when $\delta << \upsilon << 1$, so that the waste passes out of the dumpsite in a time short compared to the decay time which, in turn, is short compared to the gyre circulation time. Within the dumpsite, pollutant is being added at a constant rate corresponding to a linear increase in concentration along the flow

$$C = \frac{q}{\delta\upsilon}x \qquad 0 < x < \delta \qquad 13a$$

and beyond it decays exponentially

$$C = \frac{q}{v} e^{-(x-\delta)/v} \qquad 13b$$

Third, there is a full gyre scale response when $\delta \ll 1 \ll v$. In this case $\partial C/\partial x \simeq 0$ and the integral of equation 7 and equation 10 implies

$$\int_0^1 dx\, C = q \qquad 14$$

so that $C = q$ uniformly over the whole gyre.

In addition to these three limits there are two cases useful in the problem with biological feedback. The first is when dumping is uniform over the whole gyre, $\delta = 1$, which gives

$$C = q$$

and the second is the case when the dumpsite is very small so that δ is infinitesimal. In this case $\Delta(x)$ becomes a Dirac delta function, so that

$$C = \frac{q}{v} \frac{e^{(1-x)/v}}{e^{1/v} - 1} \qquad 16$$

which satisfies equation 7 and the requirement

$$C(0^+) = C(1^-) + \frac{q}{v} \qquad 17$$

derived by integrating equation 7 across the dumpsite. (Superscripts + or - refer to just downstream or upstream of the dumpsite, respectively.) Equation 16 can reproduce either equation 13 or 14 in the $v \ll 1$ or $v \gg 1$ limit.

Next, consider the pollutant effect on the organisms within the gyre volume. The measure of this effect, E, can be derived by dividing equation 8 by M and integrating over the domain using $M(0) = M(1)$

$$0 = g\left\{1 - E - \int_0^1 C^{3.5}\, dx\right\}$$

Therefore,

$$E = 1 - \int_0^1 dx\, C^{3.5} \qquad 18$$

which is a general statement independent of the $\alpha = 0$ assumption. For the three limiting cases the response is

$$E = 1 - q^{3.5}\delta^{-2.5} \quad \text{local} \qquad 19$$

$$E = 1 - q^{3.5}v^{-2.5} \quad \text{flushing} \qquad 20$$

$$E = 1 - q^{3.5} \quad \text{gyre} \qquad 21$$

showing the critical dumping rate (q_{crit}), which will kill all of the organisms in the

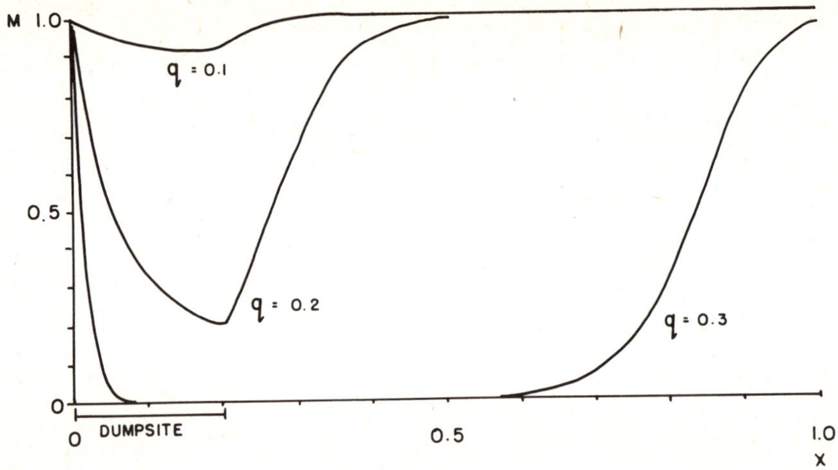

Figure 3.3. Biomass (M) as a function of x for $\delta = 0.2$, $v/g = 0.05$, and various q's. Note the slow recovery when $q > \delta$.

gyre, to be:

$$q_{\text{crit}} = \begin{cases} \delta^{5/7} & \text{local} \\ v^{5/7} & \text{flushing} \\ 1 & \text{gyre} \end{cases} \qquad 22$$

Because of advection, the pollution does not create just a local mortality. This can be seen most clearly in the case in which the conditions are local ($v \ll \delta \ll 1$) so that the waste concentration is limited essentially to the dumpsite. In this case we can solve the biomass equations

$$\frac{v}{g}\frac{\partial M}{\partial x} = [1 - q\,\Delta(x)]M - M^2$$

in terms of the familiar solutions of the logistic equation. In Fig. 3.3, we sketch some solutions for various q values. Note that when q is larger than δ the biomass decays rapidly within the dumpsite and takes a long time to recover downstream. When q is greater than $\delta^{5/7}$, the downstream recovery is insufficient to compensate for the losses within the dumpsite region and the pollution causes 100% mortality. This suggests that dumping is particularly dangerous when the circulation is very sluggish (though not zero), since high local concentrations are achieved and eventually all the population passes through the dumpsite and is destroyed.

3.4. EFFECTS OF UNIFORM DUMPING INCLUDING BIOLOGICAL FEED-BACK

Next, consider the inclusion of the biological feedback term α. Again, an ad hoc choice $\alpha = 8$ will be made so that the decay time is strongly influenced by the biology: it is nine times faster when the population is full strength ($M = 1$) than in the absence of organisms. When the dumping is uniform over the whole region of the gyre $\delta = 1$ so that the solutions are independent of x

$$C(1 + 8M) = q$$

$$M^2 = (1 - C^{3.5})M$$

Since the conditions are uniform over the gyre, the pollutant effect E is just equal to the nondimensional biomass M. (Recall that biomass has been nondimensionalized by its saturation value in the absence of pollutant.) These equations therefore give a relationship between E and q which has two solutions:

$$E = 0 \qquad \text{(no population)}$$

or

$$(1 + 8E)(1 - E)^{1/3.5} = q$$

as shown on Fig. 3.4. The dashed part of the curve indicates unstable conditions in which a small change in q induces a growing or decaying population which in turn causes a decrease or increase in τ. This change in τ then causes the concentration of pollutant to decrease or increase, thus feeding back in such a fashion that the population change is reinforced. Thus, catastrophic changes in which the population crashes can occur. These changes can be reversed only by substantially reducing the input rate–in this case to about 20% of the rate at which the crash occurred. This calculation suggests that it is possible for small changes in pollutant

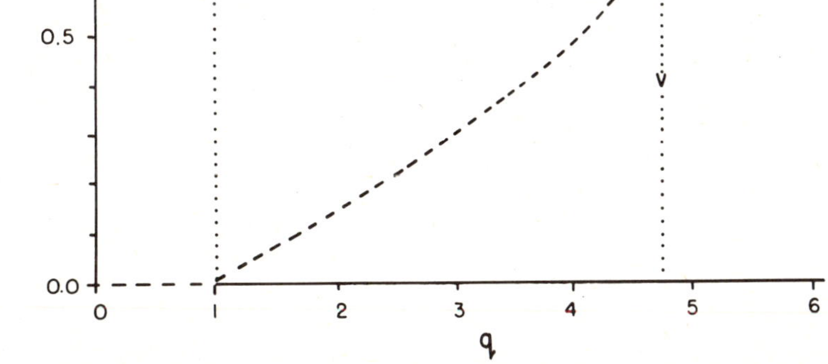

Figure 3.4. The pollutant effect $E(q)$ as a function of the nondimensional dumping rate (q) for the case of dumping uniformly over the whole gyre. The dashed lines indicate solutions which are unstable. The dotted lines represent the jumps that will occur in E for small changes in q near the point where the solid line terminates.

input to induce large changes in population, and furthermore that there may be very little warning that such an event is coming. In the example, very large changes in q induce only a 25% mortality until $q > 4.71$, at which point there is 100% mortality.

3.5. LOCALIZED DUMPING

3.5.1. Effects on Slowly Growing Species

In this section, the case is considered when $\delta < 1$ so that the waste input is confined to a section of the gyre. In addition, suppose that g is very small, corresponding to the case when the population grows at a very slow rate compared to the waste decay time.

In this limit, the population biomass equation 8 shows that M is independent of x to lowest order, so that again $E = M$. Use of the equation for the net dumping effect (equation 11) therefore gives the coupled equations

$$v\frac{\partial C}{\partial x} = -(1 + 8E)C + q\,\Delta(x)$$

$$E = 1 - \int_0^1 dx\, C^{3.5} \text{ or } E = 0 \qquad 23$$

$$C(0) = C(1)$$

The same three limits for equations 23 can be explored. In the local limit

$$C = \frac{q}{1 + 8E}\,\Delta(x)$$

$$1 - E = \int_0^1 \frac{q^{3.5}}{(1 + 8E)^{3.5}}\frac{\Delta(x)}{\delta^{2.5}}$$

so that

$$q = \delta^{5/7}\,(1 - E)^{1/3.5}\,(1 + 8E) \qquad 24$$

In the flushing or gyre limit, for a small dumpsite, Δ can be replaced by a Dirac delta function to find

$$C = \frac{q}{v}\frac{e^{-(1+8E)x/v}}{1 - e^{-(1+8E)/v}}$$

which gives

$$q = v\left[\frac{3.5(1 - E)(1 + 8E)}{v(1 - e^{-3.5\,(1\,+\,8E)/v})}\right]^{1/3.5}\left[1 - e^{-\frac{1+8R}{v}}\right] \qquad 25$$

The flushing limit ($v \ll 1$) is

$$q = v^{5/7}\,[3.5(1 - E)(1 + 8E)]^{1/3.5} \qquad 26$$

and the gyre response ($v >> 1$) is

$$q = (1 - E)^{1/3.5}\,(1 + 8E) \qquad 27$$

When the currents are sufficiently rapid, the pollutant is mixed over the whole gyre, as are the organisms, and the effect is the same as if the same amount of waste is dumped uniformly in the region (equation 27 is identical to the results of Section 3.3.) For smaller v's, the waste becomes concentrated into a smaller portion of the gyre and therefore can reach a higher level. Since all the population circulates through this section of the gyre within the growth time, all the organisms are subjected to high concentrations of waste and all are therefore affected. The extreme form of this is the case when $v << \delta$ in which all waste is confined to $0 < x < \delta$. This case has the lowest critical q value

$$q_{\text{crit}} = 4.71\delta^{5/7} \qquad \text{local}$$

as compared to

$$q_{\text{crit}} = 1.87\,v^{5/7} \qquad \text{flushing} \qquad 28$$

$$q_{\text{crit}} = 4.71 \qquad \text{gyre}$$

Figure 3.5 shows curves of $E(q)$ for various v and δ values. Once again the most dangerous case is the slowly circulating gyre: not only is q_{crit} very small but the

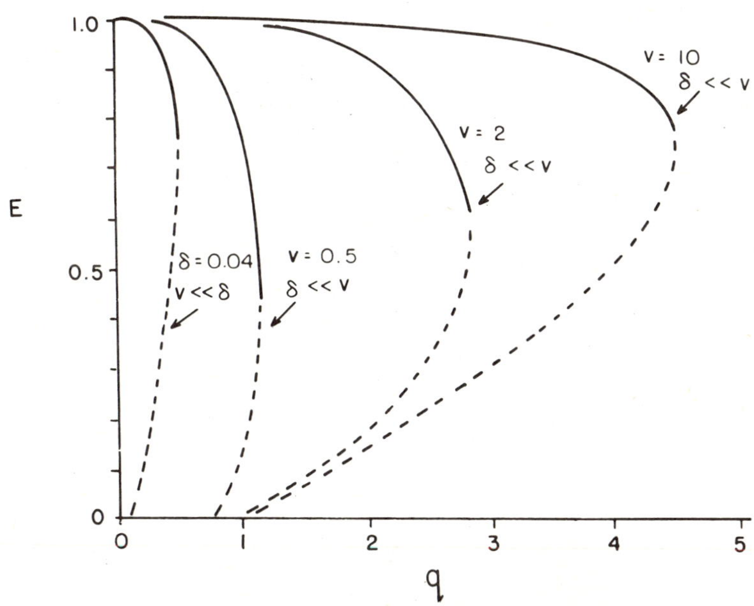

Figure 3.5. The pollutant effect on a slowly growing population as a function of the dumping rate (q) for various dumpsite widths (δ) and flow rates (v). Dashed lines indicate conditions that are probably unstable.

100% mortality occurs with little warning. At higher v's, the tolerance for waste is still smaller than in the uniform dumping case, but at least there is more indication in E of when the catastrophic response will occur.

3.5.2. Effects on Rapidly Growing Species

When the organisms are very rapidly growing so that $g >> 1$, the decay rate τ becomes a variable function of x and the equations become considerably more difficult, especially when C becomes locally greater than the critical concentration 1 (in nondimensional units). In these regions, there is a rapid decrease in population; it may be a long distance downstream after decay of the pollutant before the population recovers.

There are three simple cases that can be addressed here; the full exploration of the effects of pollution on rapidly growing organisms in a gyre will be done elsewhere. Only problems where C is either uniformly <1 or uniformly >1 will be considered. The first such case occurs when the circulation time is small compared to τ_0 so that both C and M are spatially uniform. The response is identical to that described by equation 27 or Fig. 3.4.

The second simple case is when $C > 1$ everywhere. Conditions for this to occur can be derived by noting that M will be zero everywhere so that C satisfies the $\alpha = 0$ form of equation 7. The minimum C occurs at $x = 0$

$$C(0) = \frac{q}{\delta}\left[\frac{1 - e^{-\delta/v}}{e^{(1-\delta)/v} - e^{-\delta/v}}\right]$$

Thus, 100% mortality can occur if

$$q \geq \delta\ [e^{(1-\delta)/v} - e^{-\delta/v}]/[1 - e^{-\delta/v}] \qquad 29$$

It is quite possible that a lower q may give 100% mortality: equation 29 gives only an upper bound on the critical dumping rate. One may also have to exceed this rate in order to actually kill all of the population since this bound is similar in nature to the $q = 1$ and $E = 0$ point in Fig. 3.4. For small δ and small v, this dumping rate is very large; however it decreases to about 1 when $v \geq 1$, as shown in Fig. 3.6. For rapidly circulating gyres, the dumping rate must retreat far below 1 for the population to possibly recover. (Mathematical assurance of its recovery requires solving a difficult stability problem including the spatial variability. This calculation indicates only that there will be local regions where the population could grow.)

The third case which can be solved readily is for a dumping rate sufficiently small so that C is nowhere >1. The population biomass will be locally in equilibrium giving

$$M = 1 - C^{3.5}$$

and the concentration will satisfy the relation

$$v\frac{\partial C}{\partial x} = -C(9 - 8C^{3.5}) + q\Delta$$

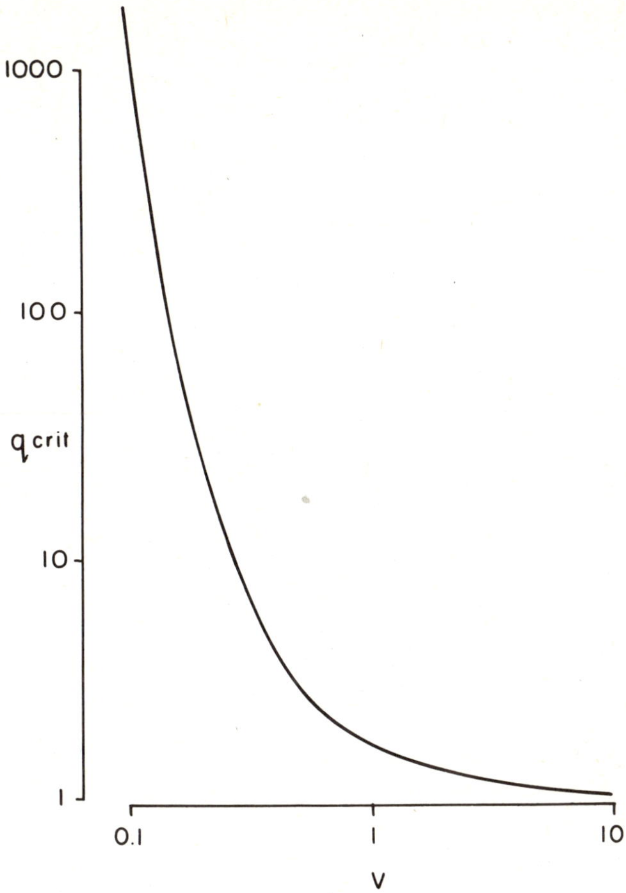

Figure 3.6. Critical dumping rate q_{crit} which will maintain $C > 1$ (lethal) concentrations everywhere in the absence of organisms versus nondimensional flow rate v. Small seed populations will not grow for larger q values. Because of spatial effects, this graph provides an upper bound estimate of the critical q value. $\delta = 0.1$.

which can be solved in the three limits previously used.

In the local limit,

$$E = 1 - \delta(1 - M_0)$$

where

$$(1 - M_0)^{1/3.5}\,(1 + 8M_0) = \frac{q}{\delta}$$

This is again similar to previously solved problems; thus q must be limited to $<4.71\delta$ to maintain a healthy population M_0 within the dumpsite.

When $\delta \ll v$, the spatial distribution of dumping, Δ, can be represented by a Dirac delta function, giving

$$v \frac{\partial C}{\partial x} = -C(9 - 8C^{3.5})$$

$$C(0^+) = C(1^-) + q$$

which can be solved by considering instead the equation for $Z = 1 - M(= C^{3.5})$

$$\frac{v}{3.5} \frac{\partial Z}{\partial x} = -Z(9 - 8Z)$$

$$Z(0^+)^{1/3.5} = Z(1^-)^{1/3.5} + q$$

The first equation in each pair is again the logistic equation; therefore both $Z(1)$ and E can be expressed in terms of $Z(0)$. The second equation then gives a relationship between E and q. In Fig. 3.7 this relationship is shown for various v values. This figure suggests that catastrophes can occur whenever the dumping rate q which can maintain $C > 1$ in the absence of organisms is less than the rate such that C becomes equal to 1 somewhere. For small δ, this happens if

$$v(e^{1/v} - 1) < v\left[1 - \left(\frac{9e^{-31.5/v}}{1 + 8e^{-31.5/v}}\right)^{1/3.5}\right]$$

Therefore, when

$$v > 1.45$$

the dumping rate necessary to first create a local region of lethal concentrations is larger than the rate necessary to maintain lethal concentrations everywhere once the organisms are destroyed. Such conditions may indeed be obtained even at

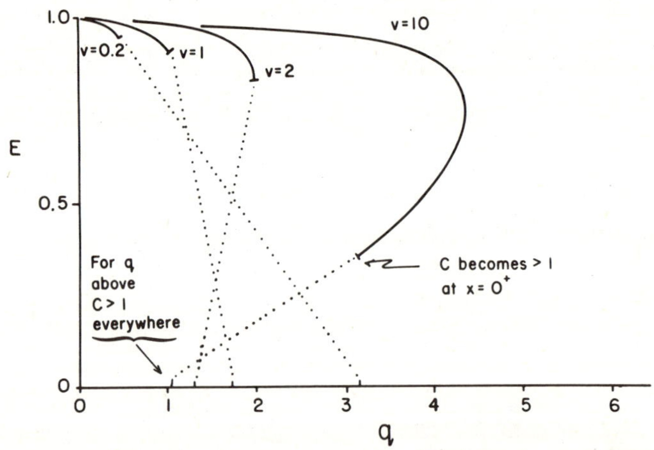

Figure 3.7. Pollutant effect (E) as a function of the dumping rate (q) for various nondimensional flow rates (v) in the case of dumping in a gyre with rapidly growing organisms. $\delta << 1$. The lines terminate at the point where C is greater than 1 as $x = 0^+$. The dotted lines do not indicate solutions but rather serve to connect the labelled curves to the minimum q to maintain lethal concentrations (derived from the $\delta \to 0$ equivalent of Fig. 3.6).

lower v's; analyses of the distributions of C and M in the case not treated here (when $C > 1$ at some points) and of the stability of these solutions is under way.

3.6. SUMMARY

A number of simple models have been described, illustrating the processes by which pollution concentrations and impacts might be determined when physical, chemical, and biological effects all interact. Several general conclusions can be drawn from these models; in addition, they suggest which parameters are important in determining the behavior of the system.

When the biologically induced decay of the pollutant is negligible, the problem of calculating the "pollutant effect" (the net population in the presence of pollutant divided by the net population in natural conditions) can be separated into two parts. First, the concentration of pollutant can be estimated from a physical and chemical model. For the model considered, recirculating gyre flow with a constant decay rate, this calculation is quite simple; however, one can conceive of similar calculations in many other circumstances with many other models. The output from this model, together with a statement of the dependence of biological impact on pollutant concentration and a definition of a metric such as "pollutant effect" E, can be used to calculate E for various dumping strategies.

The "pollutant effect" in this case is a monotonic function of the rate of dumping q. In these circumstances it should be possible to devise an appropriate monitoring strategy for regulating the dumping to maintain levels of effect which are judged acceptable by society, and perceived impacts may be mitigated by reducing the rate of pollution input.

The most severe impacts in this simplest model occur when the circulation time and the time necessary to flush the wastes from the dumpsite are very long compared to the decay time of the waste. In these circumstances, the concentration can build up to high levels within the dumpsite itself causing large impacts upon each group of organisms as they pass through the site. Eventually the whole population has been affected.

When biologically induced decay is a significant contribution to the breakdown process for the waste, very different phenomena can be seen. If the biological breakdown rate is proportional to the number of organisms and the death rate of the organisms in turn increases nonlinearly with waste concentration, then the effect may become a discontinous function of the dumping rate. As the dumping rate is increased slowly, a point is reached at which the death rate increases rapidly enough so that the population decreases quickly and, since the biologically induced decay virtually stops, the concentration of waste increases to high levels. Mathematically this process is represented as a discontinuous jump in the state of the system for an infinitesimal increase in the rate of dumping. Furthermore, because the concentration of waste has increased substantially and the rate of decay has slowed considerably, the situation is not improved when the dumping rate is reduced again. Rather, there must be substantial reductions in dumping before the concentration is decreased to levels at which the population can grow again (a hysteresis effect). Figure 3.4 shows a plot of the effect versus the dumping rate illustrating such behavior. Clearly this behavior poses grave difficulties for any monitoring program since the transition from a basically healthy to a decimated population is very sharp.

These remarks hold when the pollutant input is uniform over the gyre (for example, from an airborne source) or when the circulation time is short compared to the decay time (gyre-type response). When the circulation time is long compared to the decay time, however, more complex behaviors can be observed in the model. Sections of the gyre may be seriously impacted while others are scarcely affected. When the lethal concentration is exceeded (which occurs first at the downstream edge of the dumpsite), local population crashes may occur which then spread out rapidly along the gyre for further increases in dumping rate. Again, the most dangerous situation appears to be the case in which the gyre circulation is very slow.

These model calculations suggest that the most important quantities to measure are the decay time of the waste and its dependence on the population parameters, the rate at which mortality increases given a change in waste concentration, and the circulation time for the gyre. In addition, the rate of loss of material from the gyre and the rate at which fresh populations enter seem also likely to be important information which could be gathered from physical measurements in the field.

The calculations here are clearly incomplete (some cases even of this simple model have not been thoroughly explored and, more seriously, the parameterizations have not been related to real oceanic problems) but do indicate the complex behavior that may occur when dumped waste material is moved by the currents, altered by the oceanic chemistry and biochemistry, and, in turn, inhibits biological processes.

ACKNOWLEDGMENTS

This work was supported by the National Oceanic and Atmospheric Administration, grant number NA 80 AA-D-00057.

REFERENCES

Csanady, G. T., G. Flierl, D. Karl, D. Kester, T. O'Connor, P. Ortner, and W. Philpot. 1979. Assimilative capacity of the oceanic region surrounding Deepwater Dumpsite 106. *In*: Proceedings of a Workshop on Assimilative Capacity of U.S. Coastal Waters for Pollutants, Crystal Mountain, Washington, E. Goldberg (Ed.). Special Report, National Oceanic and Atmospheric Administration–Environmental Research Laboratories Boulder, Colorado, pp. 123–147.

Lotka, A. J. 1956. Elements of Mathematical Biology. Dover, New York, 465 pp.

Sverdrup, H. U., M. W. Johnson, and R. H. Fleming. 1942. The Oceans: Their Physics, Chemistry, and General Biology. Prentice-Hall, Englewood Cliffs, New Jersey, 1087 pp.

Wunsch, C. 1978. The North Atlantic general circulation west of 50° determined from inverse methods. *Reviews of Geophysics and Space Physics*, **16**, 583–620.

4

PHYSICAL OCEANOGRAPHY STUDIES RELATED TO WASTE DISPOSAL IN THE SEA

Gunnar E. B. Kullenberg

Institute of Physical Oceanography
University of Copenhagen
Copenhagen, Denmark

ABSTRACT

Theoretical models of horizontal and vertical mixing of ocean-dumped wastes on small to mesoscales are presented. Observations from ocean dumping and dye tracer experiments are used to test the models and to calculate mixing parameters. These are shown to be highly variable, depending on environmental conditions such as wind, current distribution, and density stratification. The experiments show that the waste material in stably stratified conditions can become distributed in well-defined layers related to the current and density distributions. Such layers can be very persistent.

The mixing in the sea and in the surface layer is often weak. The initial dilution achieved during a disposal operation is therefore an important factor in assessing potential effects of the waste disposal. The physical dispersion characteristics of a potential waste disposal site are often decisive site selection criteria, whereas the biological characteristics often are decisive for not using an area as waste recipient.

For small-scale mixing in the range 1 m to about 10^4 m, parameterization by means of mixing coefficients or diffusion velocities can be useful in describing the combined effects of horizontal and vertical random motion on the dispersion. For the mesoscale range the parameterization of the mixing by means of mixing coefficients does not appear meaningful.

4.1. INTRODUCTION

This chapter is not intended to be a review but rather a discussion of some physical oceanography studies of interest to waste disposal problems. These studies primarily concern the understanding and prediction of the physical dispersion in various parts of the water column. Outstanding features in this context are

1. The generally stable stratification, which usually varies in space and time and which implies relatively weak and intermittent small-scale turbulence and associated mixing. The large-scale motion mainly occurs along isopycnals.
2. The heterogeneity of the physical, chemical, and biological distribution patterns; fronts have been shown to be very common, and mesoscale eddies of sizes in the range 10–100 km contain most of the kinetic energy and influence the general circulation.
3. The energy occurs at a number of frequencies and scales, the fluctuating velocities in the sea are normally larger than the mean, and the motion in the deep sea is driven by wind effects and thermohaline forcing.

The problem we are concerned with in this chapter is the effect of conditions in the sea on the physical dispersion of waste material introduced in different parts of the water column.

4.2. GENERAL CONSIDERATIONS

Factors controlling motion and stratification in the sea are winds (strength and duration), insolation, waves, tides, freshwater runoff, and topographic features. These are the environmental factors considered here. Physical results of waste disposal that affect biological conditions can be related to a change of optical conditions such as transparency and turbidity.

The behavior of the waste in the sea will depend on the properties of the waste material. It is necessary to know if the waste is a liquid or a solid in suspension, the density of the waste as a whole and of any solids it may contain, the shape and size distribution of particulates, and chemical reactions of the waste when being mixed with seawater. Thus, a set of waste material factors is of importance for the dispersion.

The release technique will essentially determine the initial dilution. We may distinguish between release at sea by dumping from a vessel or by discharge directly from land-based sources through pipelines, but the basic physical oceanography problems are the same. For releases at sea, the initial dilution is mainly controlled by the rate of release and the speed of the dumping vessel during release. When waste is released in the wake at speeds of 11–15 km h^{-1} (6–8 knots), a dilution of 1:1000 is normally reached about 500 m astern of the vessel. When the waste is released over a short period of time through large openings (bulk release), the initial dilution varies in the range 1:100 to 1:500 (IMCO/FAO/UNESCO/WMO/WHO/IAEA/UN/UNEP Joint Group of Experts on the Scientific Aspects of Marine Pollution, GESAMP, 1975).

The waste characteristics and the release techniques can be controlled, but the environmental factors can be adjusted only by site selection. Biological characteristics of the environment will often be decisive for the exclusion of a site for waste disposal, and physical oceanography characteristics will often be decisive for the selection of a site as suitable for waste disposal.

4.3. RESULTS OF MIXING IN THE SURFACE LAYER

Liquid waste disposed in the wake of a ship will often become mixed in the surface layer, which is separated from the deep water by a density transition zone. An example of this from a dumping experiment is shown in Fig. 4.1. After the release the waste plume may often be regarded as forming a line source and the subsequent dilution may be parameterized by considering horizontal mixing only. When the vertical current shear is weak, the lateral concentration $C(y, t)$ profiles across the line source can be approximated as Gaussian, or

$$C(y, t) = \frac{M_0}{H_0 L_0 \sigma_y \sqrt{2\pi}} e^{-1/2\,(y/\sigma_y)^2} \qquad 1$$

where y is the lateral coordinate (y = 0 in the center), t is time, σ_y is the standard

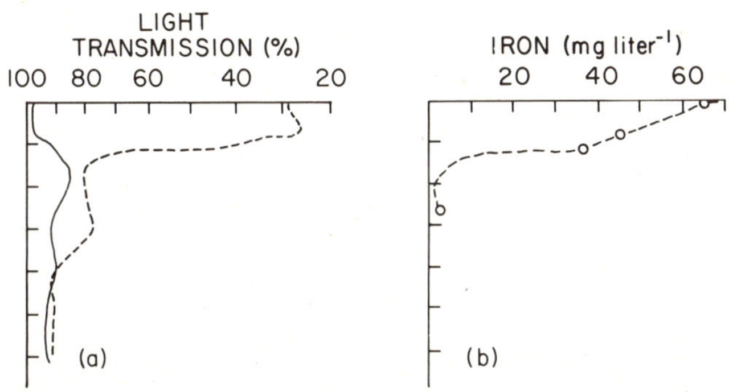

Figure 4.1. (a) Light transmission (%) and (b) iron concentration (mg liter^{-1}) in center of ferrosulfate patch several hours after dumping. The reference profile was made outside the patch at same time (Berge et al., 1971).

deviation of the concentration distribution in the *y*-direction, and M_0, H_0, and L_0 are initial amount, thickness, and length of the line source, respectively.

The lateral variance, σ_y^2, may be related to a horizontal mixing coefficient K_y or diffusion velocity P for reasonably large t but also holding for $t = 0$

$$\sigma_y^2 = 2K_y\, t + \sigma_{y0}^2 \qquad \text{2a}$$

$$\sigma_y = Pt + \sigma_{y0} \qquad \text{2b}$$

$$\sigma_{y0} = \frac{B_0}{\sqrt{2\pi}} \qquad \text{2c}$$

In equation 2c σ_{y_0} is the initial standard deviation and B_0 is the initial width of the line source. The diffusion velocity P (equation 2b) has been determined in a number of areas; it will vary depending on the conditions. In areas of strong tidal currents or in open regions with strong winds, P will be high, whereas in semi-enclosed areas with very weak tides (for example, Baltic Sea) P tends to be low. Generally values of P are of the order of 1% of the advection velocity. Attempts have been made to relate $\sigma_y{}^2$ (equation 2a) to environmental factors such as wind, stratification, and vertical current gradient (Kullenberg, 1978a).

For a dumped waste the subsequent dilution factor is

$$\frac{C_0}{C_{\max}} = \frac{\sqrt{2\pi}\,(Pt + \sigma_{v_0})}{B_0} = \frac{\sqrt{2\pi}\,Pt}{B_0} + 1 \qquad 3$$

where C_0 is the initial concentration of the waste in the vessel and $C_{\max}$ is the maximum concentration in the sea after dumping. With $B_0 = 10$ m and $P = 5 \times 10^{-3}$ m s^{-1} we find a dilution factor of only 46 after 10 h. This illustrates the great significance of the initial dilution that can be achieved during a dumping operation.

However, vertical mixing and vertical current variations (shear) are important features of the surface mixed-layer. Dye studies in the surface mixed-layer and in the seasonal thermocline have shown that the vertical mixing coefficient K_z is related to wind speed W, stratification, and the vertical gradient of the horizontal current vector q (Kullenberg, 1971, 1974)

$$K_z = Rf_c c_d \frac{\rho_a}{\rho_0} \frac{W^2}{N^2} \left| \frac{dq}{dz} \right| \qquad 4$$

In equation 4 Rf_c is the critical flux Richardson number, c_d is the drag coefficient, ρ_a and ρ_0 are the densities of the air and water, respectively, and $N^2 = -g/\rho_0\, d\rho/dz$ is the stratification parameter, with g the acceleration of gravity and z the vertical coordinate axis in positive upward direction.

When waste with a noticeably higher density than the ambient water is dumped at sea it will sink because of its initial negative buoyancy. Often such wastes contain solids in suspension. During the descent of the waste in the sea, ambient seawater is entrained into the waste-contaminated volume. This results in a gradual decrease of the initial momentum and excess density of the waste.

The vertical penetration and subsequent distribution of the waste following a dumping operation is of great interest and has been considered by Koh and Chang

(1973). As an example, consider the dumping of sewage sludge in a linearly stratified body of water, implying that N^2 is constant. Treating the sinking waste as a plume, the following expression for the maximum depth of penetration can be obtained using similarity arguments (Turner, 1973)

$$z_{max} = 1.3\alpha^{-1/2}\left(\frac{F_0}{N^3}\right)^{1/4} \tag{5}$$

where α is the entrainment coefficient. F_0 is the negative buoyancy released per unit time at the source, that is $F_0 = V_0 g\, \Delta\rho_0/(t_0\rho)$ in which V_0 is the volume of waste released over the time period t_0, $\Delta\rho_0$ the initial density difference, and ρ the density of the seawater at the source.

The entrainment coefficient α (equation 5) has been determined by laboratory experiments and by observations on smoke plumes, and lies in the range 0.1–0.6 (Bringfelt, 1969; Turner, 1973). However, in the case of sewage sludge or dredged spoils, lumps of solids disintegrate while falling through the water column and hence the entrainment may vary.

4.4. MIXING IN INTERNAL LAYERS

Experiments in relation to both dumping operations and waste disposal through pipelines show that some waste material in stably stratified ambient conditions will become distributed in layers. Recent studies have shown that the vertical salinity and temperature distributions in the sea are far from smooth, but are characterized by a layered structure on small and microscales (Fedorov, 1978). Suspended matter has also been found distributed in layers, related to the other fields, and bottom nepheloid layers are well known (Gibbs, 1974; Kullenberg 1978b).

Tracers injected internally in a stratified part of the water column have also been observed to become distributed in relatively thin, well-defined layers, which can be very persistent (Fig. 4.2; Kullenberg, 1974, 1977). Waste material would be expected to become distributed similarly in a layered structure. Suspended matter will tend to sink, but the settling velocity for small particles, that is, smaller than 100 μm, can be neutralized by density stratification and vertical turbulent motion. The dilution in such internal layers is the result of a combination of vertical and horizontal mixing processes.

A model on vertical mixing out of such layers can be formulated assuming that diffusion is proportional to the maximum concentration in the layer and inversely proportional to the thickness, with initial and boundary conditions as follows

$$t = 0, C = \overline{C} \text{ for } z < h, C = 0 \text{ for } |z| \geq h \tag{6}$$

$$t > 0, C = 0 \text{ for } |z| \geq h$$

where $2h$ is the thickness of the layer. The latter condition is based on the observation that the layer has very sharp boundaries with virtually zero concentration outside. This may be explained as due to the vertical current shear, implying that material mixed out of the layer is swept away. Considering the conditions in the

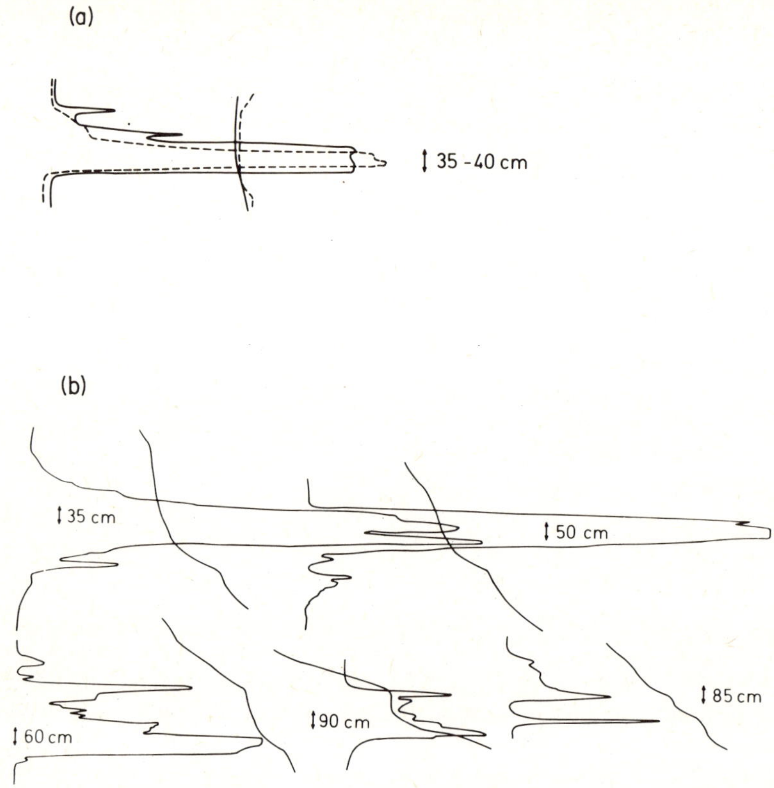

Figure 4.2. Consecutive profiles of dye layers in the same patch several hours after dye release into pycnocline layer. The profiles were obtained by means of an *in situ* fluorometer equipped with a thermistor, towed behind the ship, and passing downward through the dye layer. (a) Patch at 52-m depth in southern Baltic Sea; layer thickness 50–100 cm. (b) Patch at 30-m depth in Lake Ontario; layer thickness 70–200 cm. Marked arrows indicate the vertical scale and the smooth curves through the profiles show the temperature distribution as observed by the thermistor.

center of a diffusing patch in a Langrangian framework, we may use the simplified diffusion equation

$$\frac{dC}{dt} = K_z \frac{d^2 C}{dz^2} \qquad \text{7a}$$

where K_z is assumed to be constant within the layer. The solution of equation 7a under conditions 6 above is

$$C = \frac{4\overline{C}}{\pi} \sum_{0}^{\infty} \frac{(-1)^n}{2n+1} \exp\left[-K_z \, (2n+1)^2 \, \frac{\pi^2 t}{4h^2} \right] \cos \frac{(2n+1)\, \pi z}{2h} \qquad \text{7b}$$

Using only the first term in the series, the concentration of waste in the center of the layer, that is for $z = 0$, is

$$C = \overline{C} \exp\left[-\left(\frac{K_z \pi^2 t}{4h^2}\right)\right] \qquad 8$$

The effect of the horizontal diffusion may be introduced using equations 1 and 2b to obtain an expression for $\overline{C}$ in the center of the waste patch (or line source); thus

$$C = \frac{M_0}{H_0 L_0 (Pt + \sigma_{y_0}) \sqrt{2\pi}} \exp\left[-\left(\frac{K_z \pi^2 t}{4h^2}\right)\right] \qquad 9$$

Alternatively, the initial concentration $C_0 = M_0/(H_0 L_0 B_0)$ in the layer immediately after the dumping event may be introduced into equation 9 giving

$$C = \frac{B_0 C_0}{(Pt + \sigma_{y_0}) \sqrt{2\pi}} \exp\left[-\left(\frac{K_z \pi^2 t}{4h^2}\right)\right] \qquad 10$$

Since we have assumed that K_z and the layer thickness $2h$ are both constant over limited periods of time, we can also write

$$C = \frac{B_0 C_0}{(Pt + \sigma_{y_0}) \sqrt{2\pi}} e^{-kt} \qquad 11$$

where $k = K_z \pi^2/(4h^2)$.

Mixing models that take into account current conditions can also be formulated on the basis of the so-called shear diffusion concept. This is the apparent mixing generated by the combined action of shear and diffusion perpendicular to the shear. Vertical shear, for instance due to tidal currents in combination with a mean current, and vertical diffusion will generate an apparent horizontal mixing (Bowden, 1965; Okubo, 1967; Kullenberg, 1972). In many areas and conditions this shear diffusion can be the dominating mixing mechanism over space scales up to several kilometers and time scales up to weeks.

4.5. OBSERVATIONS

Values of K_z have been determined by means of dye tracer observations on internal layers, such as those illustrated in Fig. 4.2, and by using the above model. Equation 7b converges fairly quickly, and it can be estimated that use of the first term only introduces an error of at most 10%. By determining the concentration in the center of the dye layer, that is for $z = 0$, we find that the ratio of concentrations C_1 and C_2 for the times t_1 and t_2 is (Kullenberg, 1971)

$$\frac{C_2}{C_1} = \exp\left[-K_z \frac{\pi^2}{4h^2} (t_2 - t_1)\right] \qquad 12a$$

$$K_z = \frac{4h^2}{\pi^2 (t_2 - t_1)} \ln \frac{C_1}{C_2} \qquad 12b$$

Equation 12b has been used to determine K_z in the surface mixed-layer, and the thermocline; values of K_z are given in Table 4.1. For comparison, a number of values of K_z found by other methods and in other oceanographic settings are also

Table 4.1. Values of the Vertical Mixing Coefficients K_z

Reference	Area and Method of Determination	K_z ($cm^2\ s^{-1}$)
Kullenberg (1971, 1974)	Pycnocline and various areas, dye	0.02–0.4
Rooth and Östlund (1972)	Sargasso sea, tritium fallout	0.2
Kullenberg (1974)	Surface mixed-layer and various areas, dye	1–200
Woods and Wiley (1972)	Thermocline, Mediterranean Sea; visual, dye	0.2
Peterson and Rooth (1976)	Greenland and Norwegian seas, tritium fallout	3
Sarmiento et al. (1976)	Atlantic bottom boundary layer; radon and radium profiles	10–50

given in Table 4.1. The results show that the magnitude of K_z can vary considerably, which is to be expected because the oceanographic conditions also vary over a large range.

The horizontal diffusion velocity (P) can be determined from tracer observations by applying the following rotationally symmetrical model of Joseph and Sendner (1958).

$$C = \frac{M_0}{2\pi H_0 (Pt)^2}\, e^{-r/Pt} \qquad 13$$

where H_0 is the original layer thickness and r is the equivalent radius, which can be determined as the radius of the circle having the same area as the area covered by the isoline in question at a given time t. Values of P determined from various experiments are given in Table 4.2. Here also a fairly large variation is found.

The above theoretical results will be applied to observations on four experimental dumpings of sewage sludge carried out in the Baltic outside Stockholm (Kullenberg, 1964). Three hundred thirty m^3 of sewage sludge with an average of 8.1% solid material were dumped from a vessel each time over a period of about 1 h through two 15-cm diameter tubes. The densities of sludge and seawater were 1031 kg m^{-3} and 1005 kg m^{-3}, respectively. Water samples were taken from both sides of the dumping vessel to determine the concentration of sludge suspended in the water column during the dumping event; the samples were filtered the same day. Individual observations of suspended material showed a large scatter.

After the dumping operation the suspended material was traced using a 1.5-m path length transmissometer measuring in the red (wavelength = 655 nm) defined by a blocking glass filter. The transmissometer was calibrated in the laboratory by measurements of the light transmission of known amounts of sewage sludge in a fixed volume of seawater.

The research ship cruised over the area taking profiles with the transmissometer. Navigation was by means of parachute drogues drifting at different depths connected to surface markers which gave position relative to the ship's position. DECCA navigation was also used.

Table 4.2. Horizontal Diffusion Velocities *P*

Reference	Area and Tracer	Depth of Observation (m)	Length Scale (km)	P (cm s^{-1})
Joseph and Sendner (1958)	Southwest North Sea, (turbidity)	Mixed-layer	10	0.8
Joseph and Sendner (1958)	Mediterranean Sea, (outflow, S ‰)	–	300–1500	1.3
Joseph et al. (1964)	North Sea, dye	Mixed-layer	0.5–5	0.2–0.4
Weidemann (1973)	North Sea, dye	Mixed-layer	20	0.35
			100	0.45
Schuert (1969)	North Pacific Ocean, dye	300	4	0.13
Ewart and Bendiner (1974)	North Pacific Ocean, dye	1000	0.3	0.08
Kullenberg (1974)	Fjords, dye	20	0.1	0.05–0.1
	Coastal, dye	10–30	0.1–1	0.1–0.3
	Baltic Sea, dye	20–50	1–5	0.1–0.2
	Mediterranean Sea, dye	20–30	2	0.4
	Lake Ontario, dye	25–45	0.5–5	0.03–0.5
Kullenberg (1977)	Baltic Sea, dye	40–50	0.2–1	0.1

The observed concentrations of suspended matter after dumping as a function of time, were used to calculate P and k from equation 11. The initial concentration, C_o, was determined from the direct observations, and the initial width of the line source was given by the dumping conditions. An example of the concentration decrease versus time is given in Fig. 4.3. For this particular series of observations at 19-m depth, $P = 0.3$ m h^{-1} and $k = 0.8$ h^{-1}. Another series of observations (Fig. 4.4) was obtained on a layer around 21-m depth; here $P = 7.3$ m h^{-1} and $k = 0.74$ h^{-1}. These ranges in P and k suggest a considerable variation of diffusion velocity with depth, which may be related to the current structure as well as to the density distribution.

Information on the vertical penetration of the sewage sludge during the dumping operation was obtained in observations carried out from the sludge discharge ship during a dumping event. These observations may be used in equation 5 to determine the entrainment coefficient α. The observed maximum depth of penetration was 27 m beneath the point of injection. Thus, using the values $V_0 = 330$ m^3, $t_0 =$ 75 min, $\Delta\rho_0 = 26$ kg m^{-3}, $\rho_0 = 1005$ kg m^{-3}, $\overline{N^2} = 5.5.10^{-4}$ s^{-2}, we find $\alpha = 0.1$, which is in fair agreement with the results referred to above.

The vertical distribution of concentration determined from all the observations at the dumping vessel can be described by an exponential function of depth for limited depths beneath the source. To explain the observed concentration distribution it is assumed that the fraction of solid material going into suspension in a given layer grows exponentially with depth under the source. If it is further assumed that the amount of solids going into suspension in a given layer is proportional to the

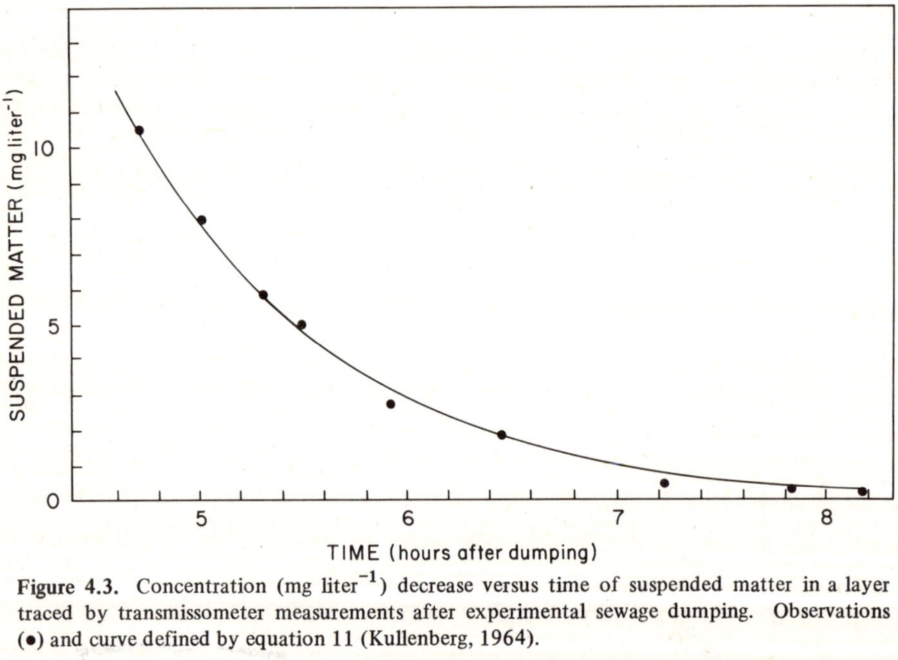

Figure 4.3. Concentration (mg liter^{-1}) decrease versus time of suspended matter in a layer traced by transmissometer measurements after experimental sewage dumping. Observations (•) and curve defined by equation 11 (Kullenberg, 1964).

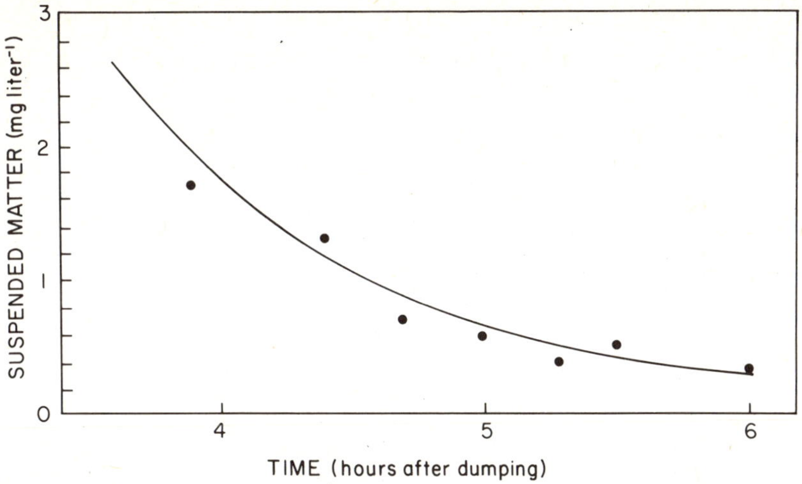

Figure 4.4. Concentration (mg liter^{-1}) decrease versus time of suspended matter in a layer traced by transmissometer measurements after experimental sewage dumping. Observations (•) and curve defined by equation 11 (Kullenberg, 1964).

amount of material reaching the depth in question in nonsuspended form, then

$$C(z) = M\, e^{bz} \qquad \text{14a}$$

$$dQ = -CA\, dz \qquad \text{14b}$$

where Q is the amount reaching the depth z in nonsuspended form, C is the concentration of suspended matter at the depth z, A is the total area covered by the falling waste material, and a and b are constants. The integration of equation 14b yields

$$Q = Q_0 \exp\left[-\frac{aA}{b}(e^{bz} - 1)\right] \qquad \text{15}$$

$$C = aQ_0 \exp\left[-\frac{aA}{b}(e^{bz} - 1)\right] e^{bz} = aQ_0 \gamma\, e^{bz} \qquad \text{16}$$

The observations showed that the factor aA/b is small; thus for small z, the first exponent (γ) in equation 16 is close to 1. This implies that the observed exponential distribution holds for small depths below the injection level. The expression can be compared with observed distributions, using observations at some reference levels to determine the constants a and b, with given values of Q_0 and A. An example from experiments in the Baltic Sea is shown in Fig. 4.5.

4.6. DISCUSSION AND CONCLUSIONS

Mixing in the sea varies greatly with the physical conditions and can be very weak even in the near-surface layer. The dilution of waste material in the environment

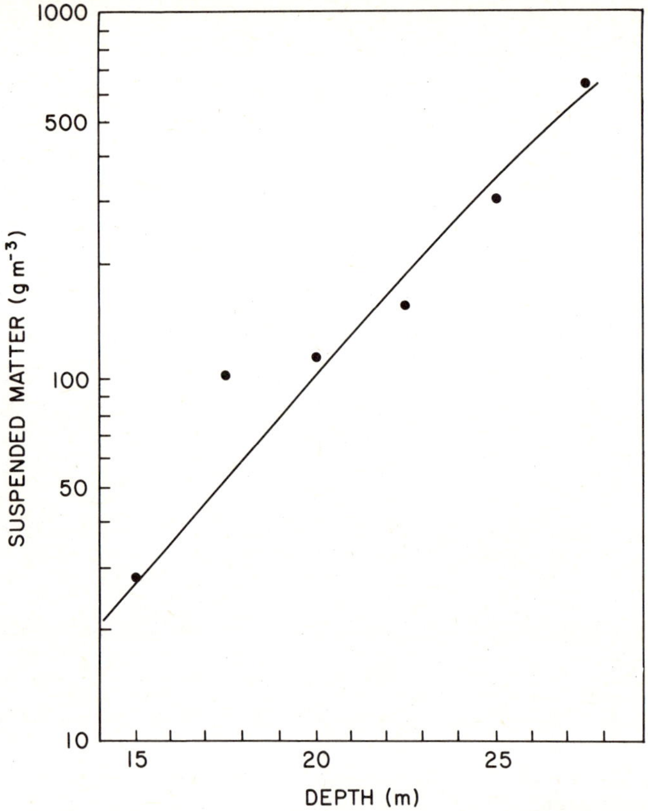

Figure 4.5. Vertical concentration (mg liter^{-1}) distribution (observed means) of suspended matter at dumping vessel immediately after sewage discharge to sea (Kullenberg, 1964).

after the dumping operation is generally much slower than the initial dilution that can be achieved during the dumping.

The dispersive conditions are very important factors in relation to the selection of disposal sites. For many waste materials, chemical and biological processes act to degrade the materials and render them harmless. However, to prevent disturbance of the local environmental balance, it is usually desirable that the waste material be dispersed by the physical processes over as large a volume as possible. It is generally true that appropriate disposal sites are those where rapid mixing and efficient water exchange occur. In semi-enclosed areas, the residence time is normally the crucial time scale, long residence times implying weak water exchange and potential build-up of waste concentrations. The importance of the time scale is also related to the chemical and biological characteristics of the wastes, such as degradability.

Estuaries and coastal zones receive relatively large amounts of wastes. These areas are also generally zones of high biological activity, and as such should be protected. It is difficult or impossible to establish a balance between the assimilative capacity of the coastal zone for wastes of various kinds and a possibly acceptable level of disturbance of the biological system. The lack of observable short-term changes in the biological system does not imply that no subtle changes occur, which may be disastrous after a long time. On the other hand, it does appear that a seriously disturbed system such as an estuary can recover when waste disposal becomes controlled or checked (Holdgate, 1979).

From this point of view one may argue that containment of the waste material in a small volume is desirable, so that control of the material and observation of its harmful effects is facilitated. For some wastes disposed into the sea containment on the bottom is the aim. However, assessments of potential risks cannot assume containment for long periods, but must rather assume that the waste spreads into the ocean. Time scales of decay or degradation, of mixing through the basin volume, and of passage or advection across the basin are then critical.

In stably stratified regions of the sea, such as most coastal zones, some components of the waste material released into the sea will become distributed in internal layers. These layers are generally less than 1 m thick, very well defined with waste concentrations considerably above ambient, and often very persistent. Wastes contaminated with oil can also become distributed in such layers, and certain components of oil released on the surface can sink and form internal layers.

In weakly stratified regions, waste material released either through a pipeline or through a bulk dumping operation can penetrate and contaminate a large part of the water column. The waste then spreads out in layers. The most persistent of these layers will be related to small-scale changes in the density structure of the water column.

The small-scale mixing, in the range 1 m to 10^3 or 10^4 m, may be described as a combination of vertical and horizontal mixing. Parameterization by means of mixing coefficients or diffusion velocities does not appear to be totally unreasonable on these scales, and may be useful in applications. For mesoscales (approximately in the range 10^4-10^5 m) parameterization of the mixing by means of mixing coefficients is not possible; the kinetic energy can vary by several orders of magnitude across a mesoscale eddy. It is still a major research task to achieve a proper understanding and parameterization of the mixing on these and larger scales. The role of advection along isopycnals must be clarified, and research is also needed on the interaction between oceanic boundary layers and the interior of the water column. What are the mechanisms operating on what scales and frequencies to inject waste material from the relatively well-mixed boundary layers into the interior of the water column?

ACKNOWLEDGMENTS

I wish to acknowledge with thanks several suggestions for improvement of the manuscript made by the referees. It is also a pleasure to acknowledge with gratitude the guidance I have received earlier from professor emeritus B. Kullenberg, who introduced me to the ocean mixing problem.

REFERENCES

Berge, G., L. Føyn, R. Ljøen, and K. H. Palmork. 1971. Experimental dumping of waste ferrosulfate (copperas) into the Skagerrak. Paper E: 32, *In*: International Council for the Exploration of the Sea, C. M. 1971. Charlottenlund, Denmark, 13 pp.

Bowden, K. F. 1965. Horizontal mixing in the sea due to a shearing current. *Journal Fluid Mechanics*, **21**, 83–95.

Bringfelt, B. 1969. A study of buoyant chimney plumes in neutral and stable atmospheres. *Atmospheric Environment*, 609–623.

Ewart, T. E., and W. P. Bendiner. 1974. Techniques for estuarine and open ocean dye dispersal measurements. *In*: Rapports et Proces-Verbaux des Réunions, Vol. 167. International Council for the Exploration of the Sea, Charlottenlund, Denmark, pp. 201–213.

Fedorov, K. N. 1978. The Thermohaline Finestructure of the Ocean. Pergamon Press, Oxford, 170 pp.

Gibbs, R. J. [Ed.] 1974. Suspended Solids in Water. Plenum Press New York, 320 pp.

Holdgate, M. W. 1979. A Perspective of Environmental Pollution. Cambridge University Press, Cambridge, 278 pp.

IMCO/FAO/UNESCO/WMO/WHO/IAEA/UN/UNEP Joint group of Experts on the Scientific Aspects of Marine Pollution (GESAMP). 1975. Scientific Criteria for the Selection of Sites for Dumping of Wastes into the Sea. Report and Studies of GESAMP No. 3, United Nations Food and Agricultural Organization, Rome, Italy, 21 pp.

Joseph, J., and H. Sendner. 1958. Über die horizontale diffusion im Meere. *Deutsches Hydrographisches Zeitschrift*, **11**, 49–77.

Joseph, J., H. Sendner, and H. Weidemann. 1964. Untersuchungen über die horizontale diffusion in der Nordsee. *Deutsches Hydrographisches Zeitschrift*, **17**, 57–75.

Koh, R. C. Y., and Y. C. Chang. 1973. Mathematical Model for Barged Ocean Disposal of Wastes. EPA-660/2-73-029, Office of Research and Development, U.S. Environmental Protection Agency, Washington, D.C., 581 pp.

Kullenberg, B. 1964. Experimental Dumping of Sewage Sludge in the Baltic (in Swedish). Technical Report, Oceanographic Institute, Göteborg, Sweden, 76 pp.

Kullenberg, G. 1971. Vertical diffusion in shallow waters. *Tellus*, **23**, 129–135.

Kullenberg, G. 1972. Apparent horizontal diffusion in stratified vertical shear flow. *Tellus*, **24**, 17–28.

Kullenberg, G. 1974. An Experimental and Theoretical Investigation of the Turbulent Diffusion in the Upper Layer of the Sea. Report 25, Institute of Physical Oceanography, University of Copenhagen, 212 pp + 60 pp appendix.

Kullenberg, G. 1977. Observations of the mixing in the Baltic thermo- and halocline layer. *Tellus*, **29**, 572–587.

Kullenberg, G. 1978a. Vertical processes and the vertical–horizontal coupling. *In*: Spatial Pattern in Plankton Communities, J. H. Steele (Ed.). Plenum Press, New York, 470 pp.

Kullenberg, G. 1978b. Light-scattering observations in frontal zones. *Journal of Geophysical Research*, **83**, 4683–4690.

Okubo, A. 1967. The effect of shear in an oscillatory current on horizontal diffusion from an instantaneous source. *International Journal of Oceanology and Limnology*, **1**, 194–204.

Peterson, W. H., and C. G. Rooth. 1976. Formation and exhange of deep water in the Greenland and Norwegian Seas. *Deep-Sea Research*, **23**, 273–283.

Rooth, C. G., and H. G. Östlund. 1972. Penetration of tritium into the Atlantic thermocline. *Deep-Sea Research*, **19**, 481–492.

Sarmiento, J. L., H. W. Feely, W. S. Moore, A. E. Bainbridge, and W. S. Broecker. 1976. The relationship between vertical eddy diffusion and buoyancy gradient in the deep sea. *Earth and Planetary Science Letters*, **32**, 357–370.

Schuert, E. A. 1969. Turbulent diffusion in the intermediate waters of the North Pacific Ocean. Report AD 688443, Naval Radiological Defense Laboratory, U.S. Department of Commerce, Washington, D.C., 20 pp.

Turner, J. S. 1973. Buoyancy Effects in Fluids. Cambridge University Press, Cambridge, 367 pp.

Weidemann, H. (Ed.). 1973. The ICES diffusion experiment RHENO 1965. *In*: Reports et Process-Verbaux des Réunions, Vol. 163. International Council for the Exploration of the Sea, Charlottenlund, Denmark, 111 pp.

Woods, J. D., and R. L. Wiley. 1972. Billow turbulence and ocean microstructure. *Deep-Sea Research*, **19**, 87–121.

5

LONG-TERM MIXING PROCESSES IN SLOPEWATER

Gabriel T. Csanady

Woods Hole Oceanographic Institution
Woods Hole, Massachusetts

ABSTRACT

The fate of industrial waste barged to the Deepwater Dumpsite-106 (DWD-106) in New York Bight can be followed in diffusion experiments for a maximum period of about three days only, after which the waste becomes too dilute and its advection by currents too erratic for successful detection. However, a natural tracer is available to demonstrate the principal mechanisms of long-term mixing in this area: the freshwater originating from land runoff.

Potentially important transport processes for the freshwater runoff are advection by a large cyclonic gyre between the Gulf Stream and the North American continent, an upwelling-like circulation in the vertical plane, a random process of parcel separation at the shelf-edge front, and a similar random process of warm core ring shedding by the Gulf Stream. The random parcel and ring shedding processes are akin to mixing due to mechanical turbulence and are amenable to similar statistical treatment. The landward salt transport by the latter processes constitutes a large fraction of the total salt transport, which compensates for the freshwater flow.

An estimate of the rate at which parcels of anomalous water break down in the slope water region can be reached by considering the balance of mean square salinity fluctuation. The estimated time scale of decay may be supposed to parameterize the long-term diffusion process of barged waste, after the latter has formed a cloud comparable in size to the typical parcels of anomalous water. This decay time scale is estimated to be of the order of five months, on the basis of historical data of mean square salinity fluctuations and the known rate of freshwater inflow.

5.1. INTRODUCTION

North of Cape Hatteras a narrow (~ 100 km) band of deep ocean separates the Gulf Stream from the waters of the east coast continental shelf of North America. The depth gradient in this region is large to moderate, the oceanic waters are distinguishable by their temperature-salinity characteristics, and are described as "slope water" (this name is due to Iselin; see Wright, 1976, whose terminology is adopted here). A region of sharp property gradients separates slope water from shelf water near the 100 m isobath, and a similar region forms the boundary or "wall" of the Gulf Stream (Fig. 5.1). The transition regions are legitimately idealized as "fronts," that is, density interfaces intersecting the free surface.

Advection and mixing processes in slope water are of considerable practical importance because they play a key role in determining coastal concentration levels of any persistent pollutants that may be discharged into the ocean along the populous east coast. Some ocean dumping is practiced directly into slope water off New York, at Deepwater Dumpsite-106 (DWD-106) located 196 km from New York Harbor, the effects of which have been the subject of intensive investigation focusing on the time scale of hours to days after a dump (Park and O'Connor, 1981). Any extrapolation of diffusion parameters determined in such short-term studies to a much longer time scale is not likely to be realistic, because different physical processes are almost certain to come into play (Csanady, 1980a). Direct tracing of pollutants released into a deep, open oceanic region is not a practical possibility after a day or two, because a given tracer cloud is subject to unforeseeable vertical as well as horizontal displacements.

Such problems are especially acute in slopewater, the surface layers of which are subject to a rather extreme variability of temperature and salinity, in comparison with other deep oceanic regions. Various investigators have reported the presence of parcels or "boluses" of shelf water at the surface or at some shallow depth, for example $\simeq$ 50 m (Wright, 1976). At the same time, warm core rings spun off the Gulf Stream, containing water warmer and saltier than slopewater, cross the slope region at random times and even impinge on shelf waters (Morgan and Bishop, 1977). Any larger parcels of shelf water, and certainly warm core rings, induce their own circulation field, consisting presumably of quasi-geostrophic flow along inclined isopycnal surfaces, together with ageostrophic motions, also mostly along isopycnals, associated with internal friction and adjustment to geostrophic equilibrium. Smaller parcels of anomalous water, as well as the fringes of larger parcels,

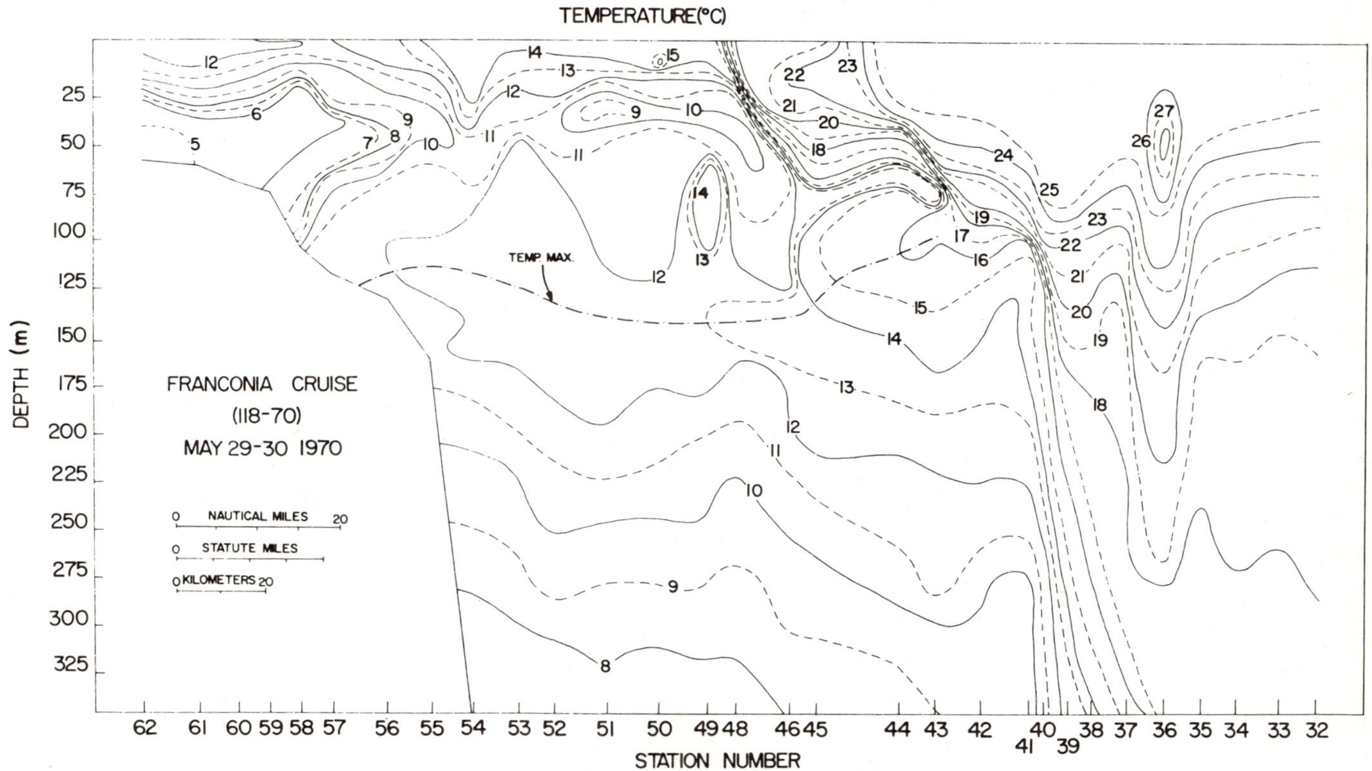

Figure 5.1. Temperature distribution (°C) in a transect of the slope water region southwest of New York. Frontal zone at shelf edge is clear near bottom, around the 100 m isobath. Steep isotherms adjacent to the 18° water of the Sargasso Sea represent the Gulf Stream "wall" (from Bowman and Weyl, 1972).

are presumably subject to considerable mixing due to wind-induced turbulence, or to instabilities of various kinds. Tracer or barged waste caught in a descending branch of circulation along some inclined isopycnal surface is very unlikely to be found in a survey of the near-surface waters following a dump.

5.2. FRESHWATER TRANSPORT AND DIFFUSION IN THE SLOPE REGION

Given the complexity of the environment and the potentially very large spatial scale of the dispersion problem in question, a plausible idea is to look for a natural tracer of large-scale diffusion processes, from the behavior of which something useful could be learned about the likely long-term fate of pollutants in slope water. Such a tracer is freshwater, which enters the ocean in large quantities along the coast and forms shelf water as a result of intense mixing within a narrow coastal region. The most probable sink for this volume of freshwater is the oceanic desert of the Sargasso Sea, where evaporation exceeds precipitation. Figure 5.2 illustrates schematically a two-dimensional conceptual model of the freshwater mixing process in the oceanographic setting of the two frontal zones. Along-isobath current velocities are also indicated in this figure because they may be important in advecting pollutants away from the region. The along-isobath currents are a part of a larger scale oceanic gyre thought to occupy the slope region (Fig. 5.3). Such a gyre was hypothesized by Sverdrup et al. (1942) and was more recently inferred from the

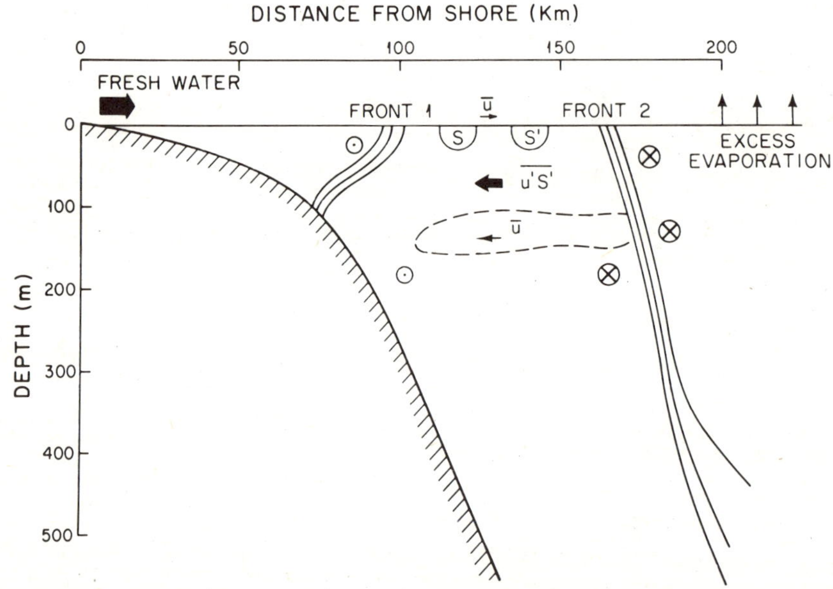

Figure 5.2. Schematic picture of freshwater transport and diffusion in slope water region. Along-isobath velocities to the southwest (forward of transect) are indicated by circled dots, opposite velocities by circled crosses. Mean cross-isobath velocities are offshore in the surface layers, onshore in a layer centered at about 150 m depth.

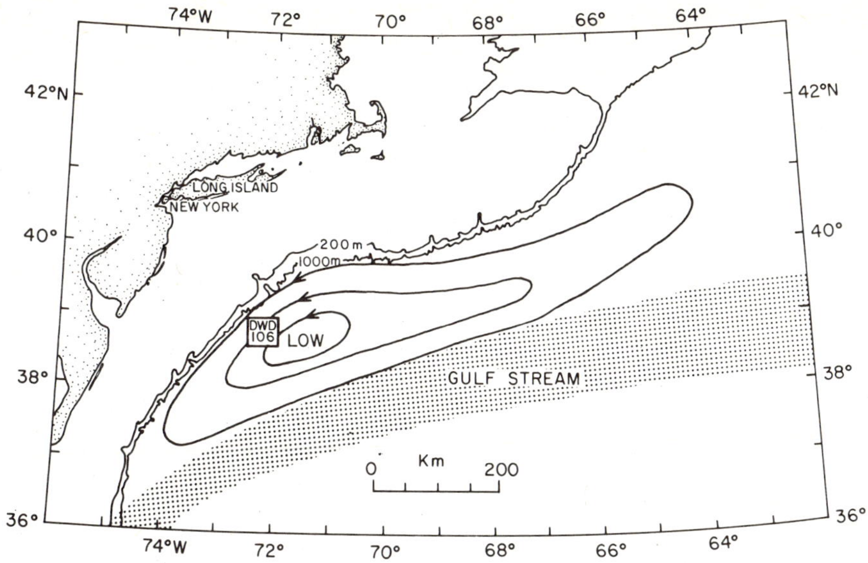

Figure 5.3. Counterclockwise gyre throught to occupy the slope water region between the Gulf Stream and the North American continental plate (from Csanady, 1979).

dynamics of the shelf waters (Csanady, 1979). Flow over the shelf proper is thought to be a boundary layer component of the cyclonic slope water gyre. Shelf waters are certainly entrained into the Gulf Stream north of Cape Hatteras and flow eastward along the northern edge of the Gulf Stream (Ford et al., 1952; Kupferman and Garfield, 1977).

The dynamics of the slope water gyre has not been seriously discussed in the literature so far, although its generation is reasonably attributed to the strong cyclonic wind-stress curl known to characterize the slope region (Leetmaa and Bunker, 1978). A numerical model study due to Semtner and Mintz (1977), which incorporated a cyclonic wind-stress curl region, reproduced something like a slope water gyre (Beardsley and Winant, 1979). In terms of classical ocean circulation theory (Sverdrup, 1957) one may tentatively describe the dynamics of this slope water gyre as follows. The wind-stress curl is balanced above the main thermocline by planetary vorticity advection, that is, northward drift of the upper 200 m or so of the water column. Where this drift runs into the continental slope, a southwestward current develops, the vorticity tendency balance of which is governed by an interplay of vortex stretching over sloping bottom and the curl of bottom stress, in the manner of frictional equilibrium flow over a sloping shelf (Csanady, 1978, 1980b).

This dynamical picture implies a pattern of cross-isobath circulation in the slope region, particularly important for any mass exchange between the shelf and the deep ocean. The cyclonic wind-stress curl implies a divergent surface Ekman layer, which acts as distributed suction on the layers just below. Yentsch (1974) has shown in some illuminating illustrations what the consequences of this are for the distribution of nutrients north of the Gulf Stream wall: massive upwelling in the entire slope region. The northward drift of the slope waters above the main thermocline is also demonstrated by the tongue of warm and saline anomaly reaching to the bottom of the shelf-edge front. This may be seen in surveys by Bowman and Weyl (1972; see Fig. 5.1) and Gordon et al. (1976), recent studies of the shelf-edge front by Mooers et al. (1979), and many other reports. The property distributions in the slope region amply support the notions that (1) the cold and fresh surface layers drift seaward; (2) the remaining layers above the main thermocline, above 200 m or so in the slope region, drift shoreward; and (3) these latter layers are supplied along isopycnals from deeper layers below the Gulf Stream.

Advection processes are thus clearly seen to be important in the horizontal plane, where the large cyclonic gyre is active, as well as in the vertical transect across the isobaths. However, the "diffusion" process consisting of the exchange of large parcels described above also transports freshwater seaward in considerable quantities. Wright (1976) has pointed out that the offshore freshwater transport by various bubbles and boluses is of the same order of magnitude as runoff from land, that is, as the total offshore transport. In a long-term study of fluxes at the edge of the Scotian Shelf, Smith (1978) has demonstrated that a few massive flow events (associated with impinging Gulf Stream rings and similar phenomena) account for most of the total offshore transport of freshwater. Estimates of this kind necessarily remain somewhat coarse, but they have shown beyond doubt that the "diffusion" of freshwater by separated parcels is an important process in the overall mass balance. It is the physics of this peculiar large-scale diffusion process associated with the formation of separated parcels that is of interest in connection with disposal of barged waste in the slope region. Whatever the precise distribution of the total offshore transport of freshwater between advection in a horizontal plane, advection in a vertical transect, and intermittent parcel separation, the last process may be supposed to carry a large fraction of the total, perhaps most of it.

5.3. DIFFUSION OF SALT

The separation of parcels of shelf water or of Gulf Stream water is legitimately regarded as a manifestation of the hydrodynamic instability of the parent current, geostrophic flow along inclined isopycnal surfaces. The implication of this is that the process is random, and that techniques of turbulence theory are conveniently applied to its description. At a given location in the slope region a stochastic mean salinity $\overline{S}$ may be defined, and a given observation of salinity interpreted as the sum of a mean and a fluctuation:

$$S = \overline{S} + S' \qquad 1$$

In a two-dimensional idealization of the process, according to the illustration in Fig. 5.2, cross-isobath velocities are $u = \overline{u} + u'$, and a "turbulent" salt flux arises, $\overline{u'S'}$, in the cross-isobath direction. The total depth-integrated salt flux across any isobath vanishes, but this may be regarded as a balance between a slow offshore advection of freshwater and an onshore diffusive flux of salt. According to the discussion above, a fraction γ, of order one, of the total salt flux neutralizing the freshwater inflow F ($m^2 s^{-1}$, or volume rate of flow per unit length of coastline) is affected by separating parcels, that is

$$\overline{u'S'} = -\frac{1}{h}\gamma F\overline{S} \qquad 2$$

where $\overline{S}$ is local mean salinity. The positive u' direction is offshore, and h is the depth over which the flux is effectively distributed. In the present application h is known to be of the order of 100 m.

The mass balance of salt in an arbitrary fluid element is expressed by

$$\frac{\partial S}{\partial t} + u_i \frac{\partial S}{\partial x_i} = D\nabla^2 S \qquad 3$$

where D is molecular diffusivity, u_i are total instantaneous velocity components (i = 1,2,3), and the summation convention applies. Splitting u_i and S into means and fluctuations and averaging, one finds the equation for the mean salinity distribution:

$$\frac{\partial \overline{S}}{\partial t} + \overline{u}\frac{\partial \overline{S}}{\partial x} = D\nabla^2\overline{S} - \frac{\partial}{\partial x}(\overline{u'S'}) \qquad 4$$

where the mean longshore gradient, the vertical velocity, and the vertical flux divergence have been supposed negligible. In a steady-state problem, and neglecting the very small molecular diffusivity contribution, equation 2 now follows after substituting $\gamma F\overline{S}/h$ for the advective flux.

Subtracting equation 4 from equation 3, an equation for the fluctuation S' arises. Multiplying through by $2S'$ and averaging, one arrives at a "conservation law" for the mean square salinity fluctuation or "segregation" (of fresh and saline waters):

$$\frac{\partial \overline{S'^2}}{\partial t} + \overline{u}\frac{\partial \overline{S'^2}}{\partial x} = -2\overline{u'S'}\frac{\partial \overline{S}}{\partial x} - \frac{\partial}{\partial x}\left(\overline{u'S'^2}\right) + D\nabla^2\overline{S'^2} - 2D\overline{|\nabla S'|^2} \qquad 5$$

where the same simplifications have been made as in writing down equation 4 (fluxes and gradients along x only important). The left-hand side of this equation contains the local and advective rate of change of $\overline{S'^2}$, while the terms on the right-hand side are successively production of segregation, divergence of turbulent flux, divergence of molecular flux, and dissipation (of segregation).

In a steady-state problem, and where horizontal fluxes of $\overline{S'^2}$ are small, the main balance is between production and dissipation. The production term is mean salt flux times mean salinity gradient, that is, it is determined by terms of the first-order equation 4. The dissipation term may be parameterized by a dissipation time scale T such that

$$2D\overline{|\nabla S'|^2} = \frac{\overline{S'^2}}{T} \qquad 6$$

The dissipation term as written on the left of this equation involves the molecular diffusivity and the mean square salinity gradient $\overline{|\nabla S'|^2}$, that is, it depends on the very small-scale details of the mixing process. However, the generation of these small-scale gradients is a consequence of the straining of fluid elements by all the turbulent eddies present, small and large. Just as it may be argued that the turbulent energy cascades down from large- to small-scale eddies (Batchelor, 1953), one may consider the dissipation of $\overline{S'^2}$ as the final step in a process beginning with the handing down of a salinity anomaly from a parcel to a smaller one, then to a smaller one again, and so on. The rate at which $\overline{S'^2}$ is handed down from the largest eddies is necessarily the same as the dissipation rate, and the large-scale eddy breakdown might well be regarded as the process which controls that rate. Adopting this viewpoint one may construct a plausible heuristic argument on the rate of $\overline{S'^2}$ dissipation.

Let the salinity anomaly of a large parcel of characteristic dimension L be S', so that the total excess salt (negative S' for a deficiency) in the parcel is $Q = S'L^3$. Suppose that the parcel breaks down by ordinary turbulent mixing, in the course of which its total salt content remains conserved, but its size grows. It follows that the rate of change of salinity anomaly is

$$\frac{dS'}{dt} = \frac{3S'}{L}\frac{dL}{dt} \qquad 7$$

or, multiplying through by $2S'$ and averaging,

$$\frac{d\overline{S'^2}}{dt} = -2\,\overline{\frac{3}{L}\frac{dL}{dt}}\;\overline{S'^2} \qquad 8$$

where it has been supposed that the correlation of $(S'^2 - \overline{S'^2})$ and $(L^{-1}\,dL/dt - \overline{L^{-1}\,dL/dt})$ vanishes.

Equation 8 is of the form of equation 6 and yields for the decay time scale

$$T^{-1} = 2\left(\overline{\frac{3}{L}\frac{dL}{dt}}\right) \qquad 9$$

Whatever the merits of the argument leading to the last expression, T as defined by equation 6 is a measure of the rate at which individual parcels expand and break down. Suppose that barged waste is discharged into an anomalous parcel in slopewater, containing either shelf water or Gulf Stream water. After a relatively short period (perhaps a day or two) the waste mixes with the volume of the entire anomalous parcel and its subsequent diffusion is governed by the rate of breakdown of the parcel itself. This later stage of mixing may be taken to be parameterized by the time scale T.

To estimate a value for T, the dissipation of $\overline{S'^2}$ may be set equal to the production rate, supposing that in slope water $\overline{S'^2}$ is constant in time, and that its horizontal fluxes are small:

$$\frac{\overline{S'^2}}{T} = -2\overline{u'S'}\frac{d\bar{S}}{dx}$$

$$T = -\frac{\overline{S'^2}}{2\,\overline{u'S'}\;d\bar{S}/dx} \qquad 10$$

Where the flux terms in equation 5 are significant, a horizontally averaged form of equation 5 gives rise to a very similar relationship, with an order one constant multiplying the right-hand side.

Equation 10 contains quantities readily estimated from observation and so yields a method of determining an approximate value of the time scale T. Physically, a known value of segregation leads to a measure of the rate at which isolated parcels break down, an intuitively reasonable proposition.

5.4. ESTIMATION OF DECAY TIME SCALE

Historical data on surface salinity observations have been requested from the National Oceanographic Data Center for the slope water region of interest in the Mid-Atlantic Bight. Figure 5.4 illustrates the density of observations available: not particularly high in the region of the DWD-106, somewhat better south of New England. The boxes shown in Fig. 4 limit two regions chosen, the divisions indicating four and two depth classes respectively, into which the observations could reasonably be grouped. Table 1 lists the mean and mean square salinities (10^{-3} and 10^{-6}, that is, parts per thousand and the same squared) stratified according to season. A lumped fall-winter season is also shown on account of the small number (*N*) of the observations available. The distribution of mean and mean square salinity in the New England slope region, where the density of observations was adequate, is illustrated in Fig. 5.5, assigning the means to the centers of the boxes

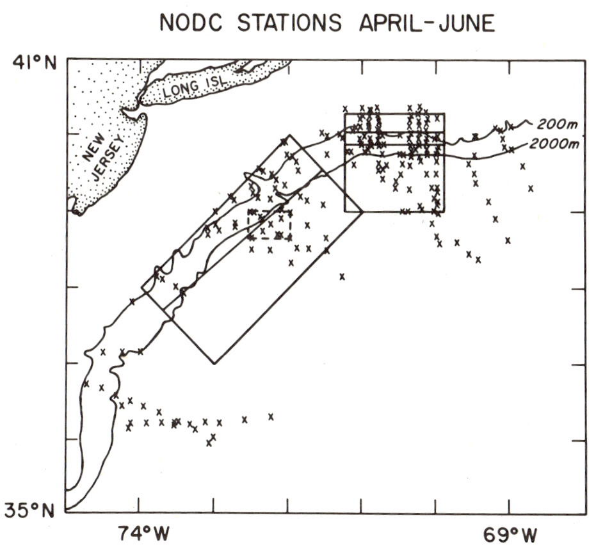

Figure 5.4. Surface salinity observations available in the slope water region, spring period. Each cross represents one reported observation; boxes are the regions for which observations have been averaged.

Table 5.1. Salinity Mean[a] and Variance in Cells Bounded by 72.25°W and 69.9°W in Slope Region

Region	Approximate Depth Region (m)	Winter (January–March)			Spring (April–June)			Summer (July–September)			Fall (October–December)			Fall + Winter		
		N	$\overline{S}$	$\overline{S'^2}$	N	$\overline{S}$	$\overline{S'^2}$	N	$\overline{S}$	$\overline{S'^2}$	N	$\overline{S}$	$\overline{S'^2}$	N	$\overline{S}$	$\overline{S'^2}$
40.05°–40.3°N	100–200	7	33.13	0.16	40	32.70	0.23	44	32.61	1.00	4	34.10	1.28	11	33.48	0.79
39.88°–40.05°N	200–1000	16	33.12	0.42	31	33.00	0.81	40	33.07	1.21	15	33.87	0.62	31	33.48	0.66
39.75°–39.88°N	1000–2000	20	34.35	1.33	19	33.45	1.10	31	33.42	1.14	11	38.43	0.71	31	34.52	1.17
39.00°–39.75°N	2000–3000	62	34.39	1.20	48	34.11	1.26	85	34.30	1.04	53	34.97	0.92	115	34.66	1.14
38.0°N–74.0°W 40.0°N–72.0°W 37.7°N–73.7°W 39.55°N–71.55°W	100–2000	52	34.48	1.06	43	33.59	1.21	46	33.59	1.31	38	34.19	0.79			
37.7°N–73.7°W 39.55°N–71.55°W 37.0°N–73.0°W 39.0°N–71.0°W	2000–3000	23	35.04	1.00	28	34.87	1.13	35	34.81	0.48	23	34.64	1.00			

[a]Units for $\overline{S}$ and $\overline{S'^2}$ are parts per thousand and same squared respectively; N is number of observations obtained from the National Oceanographic Data Center, Washington D.C.

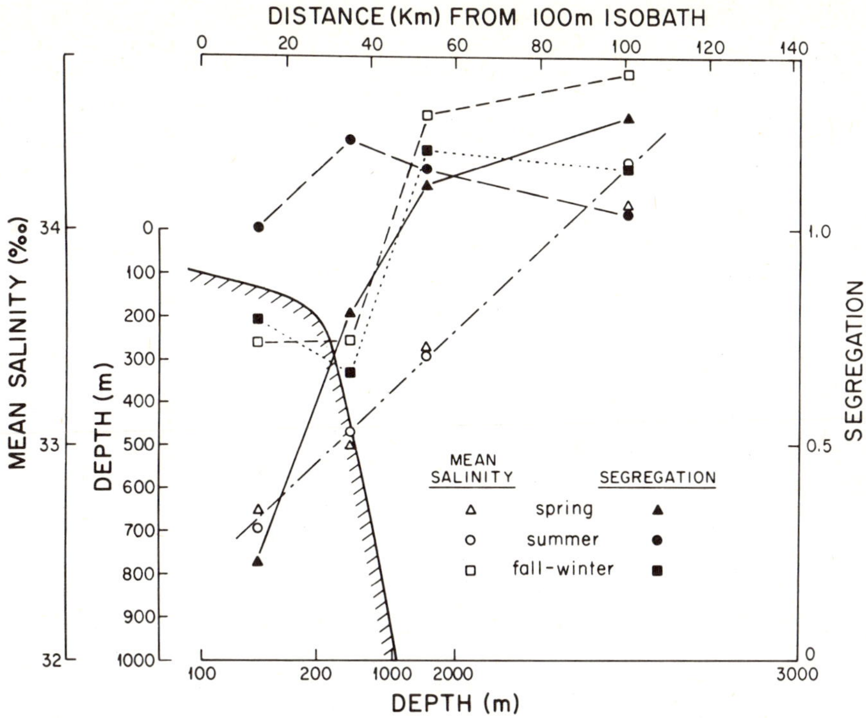

Figure 5.5. Distribution of mean salinity and salinity fluctuation over the New England shelf region (box with 4 divisions in Fig. 5.4) for different seasons. See also Table 5.1.

versus cross-isobath distance. Mean square salinities are consistently around 1×10^{-6}, except at the shelf edge, presumably on the landward side of the shelf-edge front. The mean salinity gradient in spring and summer is close to 2×10^{-10} cm^{-1} throughout the region, while in winter it is about three times higher, but confined to the frontal zone at the shelf edge.

The freshwater inflow into the New York Bight is approximately $q = 70$ cm^2 s^{-1} (volume rate per unit length of coastline; see Ketchum and Keen, 1955). Supposing this distributed over a 100 m deep layer, at mean salinity $\overline{S} = 34 \times 10^{-3}$, and estimating $\gamma = 0.7$ for the fraction compensated for by "diffusive" landward salt flux, one has the estimate

$$\overline{u'S'} = \gamma \frac{q}{H}\overline{S} = 0.17 \times 10^{-3} \text{ cm s}^{-1}$$

Substituting into equation 10, with $\overline{S'^2} = 10^{-6}$ and $\partial\overline{S}/\partial x = 2 \times 10^{-10}$ cm^{-1}, one finds the decay time scale

$$T = 1.5 \times 10^7 \text{ s}$$

or about five months, which is of the order of the observed lifetime of warm core rings (Gotthardt and Potocsky, 1974). The implication is either that most of the salinity fluctuations observed in slope water are due to warm core rings (in the sense that they contribute most to the variance $\overline{S'^2}$), or else that other boluses and parcels, spun off the shelf-edge front, are similarly long-lived. The quantitative estimate of the time scale should be taken seriously only as to its order of magnitude (within a factor of three, say, either way).

5.5. DECAY OF DUMPED WASTE CONCENTRATION

From a practical point of view the results seem to mean that dumped quantities of waste, once dispersed over a "bolus" or a "ring," would be very slow to disperse farther. Like warm core rings, they would presumably travel long distances along the isobaths before being entrained by the Gulf Stream. Their fate could be crudely modelled by an argument similar to that made above to arrive at equation 9. Let a quantity of waste W be dispersed over a bolus of characteristic size L. The waste concentration (χ) is then, by virtue of the definition of L,

$$\chi = \frac{W}{L^3} \qquad 11$$

This reduces in time as L increases, a time differentiation and use of some earlier relationships yielding

$$\frac{d\chi}{dt} = -\frac{\chi}{2T} \qquad 12$$

or

$$\chi = \chi_0 e^{-t/2T}$$

Waste dilution would thus proceed on an e-folding time scale of $2T$.

5.6. CONCLUSION

The results of the above estimates are perhaps somewhat surprising, and they are certainly based on not very detailed data. Specific studies of freshwater dilution in the slope region would presumably improve the reliability of similar calculations. The main intention of this chapter has been to draw attention to the possibility that information on mean square salinity (or the "segregation" of fresh and saline waters) as well as on mean salinity can be used to infer certain properties of large-scale mixing in the slopewater region. These can then be used to determine the long-term fate of waste dumped in the slopewater region.

ACKNOWLEDGMENTS

This work has been supported by the Ocean Dumping Program of the National Atmospheric and Oceanic Administration, and by the U.S. Department of Energy, the latter under a contract entitled Coastal-Shelf Transport and Diffusion. The calculations for Table 5.1 and Figure 5.5 have been performed by Mr. J. Churchill.

REFERENCES

Batchelor, G. K. 1953. The Theory of Homogeneous Turbulence. Cambridge University Press, Cambridge, 197 pp.

Beardsley, R. C., and C. D. Winant. 1979. On the mean circulation in the Mid-Atlantic Bight. *Journal of Physical Oceanography*, **9**, 612–619.

Bowman, M. J., and P. K. Weyl. 1972. Hydrographic Study of the Shelf and Slope Waters of New York Bight. Technical Report No. 16, Marine Sciences Research Center, State University of New York, Stony Brook, 46 pp.

Csanady, G. T. 1978. The arrested topographic wave. *Journal of Physical Oceanography*, **8**, 47–62.

Csanady, G. T. 1979. What drives the waters of the continental shelves? *Oceanus*, **22**(2), 28–35.

Csanady, G. T. 1980a. An analysis of dumpsite diffusion experiments. *In*: Ocean Dumping of Industrial Wastes, B. H. Ketchum, D. R. Kester, and P. K. Park (Eds.). Plenum Press, New York, pp. 109–129.

Csanady, G. T. 1980b. Longshore pressure gradients caused by offshore wind. *Journal of Geophysical Research*, **85**, 1076–1084.

Ford, W. L., J. R. Longard, and R. E. Banks. 1952. On the nature, occurrence and origin of cold low salinity water along the edge of the Gulf Stream. *Journal of Marine Research*, **11**, 281–293.

Gordon, A. L., A. F. Amos, and R. D. Gerard. 1976. New York Bight water stratification–October 1974. *In*: Middle Atlantic Continental Shelf and the New York Bight, M. G. Gross (Ed.). American Society of Limnology and Oceanography, Lawrence, Kansas, pp. 45–57.

Gotthardt, G. A., and G. J. Potocsky. 1974. Life cycle of a Gulf Stream anticyclonic eddy observed from several oceanographic platforms. *Journal of Physical Oceanography*, **4**, 131–134.

Ketchum, B. H., and D. J. Keen. 1955. The accumulation of river water over the continental shelf between Cape Cod and Chesapeake Bay. *Deep-Sea Research*, **3** (Suppl.), 346–357.

Kupferman, S. L., and N. Garfield. 1977. Transport of low-salinity water at the slope water-Gulf Stream boundary. *Journal of Geophysical Research*, **82**, 3481–3486.

Leetmaa, A., and A. Bunker. 1978. Updated charts of the mean annual wind stress, convergences in the Ekman layers, and Sverdrup transports in the North Atlantic. *Journal of Marine Research*, **36**, 311–322.

Mooers, C. N. K., R. W. Garvine, and W. W. Martin. 1979. Summertime synoptic variability of the Middle Atlantic shelf water/slope water front. *Journal of Geophysical Research*, **84**, 4837–4854.

Morgan, C. W., and J. M. Bishop. 1977. An example of Gulf Stream eddy-induced water exchange in the Mid-Atlantic Bight. *Journal of Physical Oceanography*, **7**, 472–479.

Park, P. K., and T. P. O'Connor. 1981. Ocean dumping research: Historical and international development. *In*: Ocean Dumping of Industrial Wastes, B. H. Ketchum, D. R. Kester, and P. K. Park (Eds.). Plenum Press, New York, pp. 3–23.

Semtner, A. J., and Y. Mintz. 1977. Numerical simulation of the Gulf Stream and mid-ocean eddies. *Journal of Physical Oceanography*, **7**, 208–230.

Smith, P. C. 1978. Low-frequency fluxes of momentum, heat, salt, and nutrients at the edge of the Scotian Shelf. *Journal of Geophysical Research*, **83**, 4079–4096.

Sverdrup, H. U. 1957. Oceanography. *In*: Handbuch der Physik, Vol. 48, S. Flügge (Ed.). Springer Verlag, Berlin, pp. 608–670.

Sverdrup, H. U., M. W. Johnson, and R. H. Fleming. 1942. The Oceans: Their Physics, Chemistry, and General Biology. Prentice-Hall, Englewood Cliffs, New Jersey, 1087 pp.

Wright, W. R. 1976. The limits of shelf water south of Cape Cod, 1941 to 1972. *Journal of Marine Research*, **34**, 1–14.

Yentsch, C. S. 1974. The influence of geostrophy on primary production. *Tethys*, **6**, 111–118.

6

DISPERSION OF PARTICLES AFTER DISPOSAL OF INDUSTRIAL AND SEWAGE WASTES

Marshall H. Orr and Lincoln Baxter, II

Woods Hole Oceanographic Institution
Woods Hole, Massachusetts

Dr. Orr's present address: Gulf Research and Development Co., P.O. Drawer 2038, Pittsburgh, Pennsylvania 15230.

ABSTRACT

A high-frequency acoustic backscattering system has been used to study the dispersion of particulates released or formed during the disposal of industrial chemical wastes and sewage sludge at the Deepwater Dumpsite-106 (DWD-106) located 196 km southeast of New York Harbor, New York. The acoustic systems provide real-time data to guide chemical and biological sampling of contaminated and uncontaminated sections of the water column as particles associated with the waste plumes serve as waterborne tracers which can be acoustically tracked. In addition the data sets can be used to determine the temporal variability of the particle vertical and horizontal distribution. The acoustic data have shown that the vertical and horizontal distribution of the particulates released at DWD-106 is seasonally dependent, reflecting both the temporal variability of the magnitude of density gradients associated with the seasonal mixed layer and the water mass variability of the area.

In a qualitative sense, the acoustic backscattering technique is beneficial to an environmental impact monitoring experiment. However, one of the objectives of developing the technique has been to measure the horizontal and vertical dispersion coefficients of particle plumes. The quantitative analysis of the acoustic data has been limited by navigation inadequacies, the inability to measure vertical shear, the lack of knowledge concerning waste particle size distributions and densities (hence settling rates), and the influence of oceanic turbulence on the particle field.

6.1. INTRODUCTION

High-frequency acoustic backscattering systems are being used to detect the position and dispersion of particles released into the ocean during the disposal of industrial chemical waste and sewage sludge (for example, Orr and Hess, 1978b; Proni and Hansen, 1981). There are two reasons for using the acoustic systems. The first is to remotely sense the particulate phase of industrial chemical waste plumes, so that chemical and biological samples can be taken from contaminated and uncontaminated sections of the water column. The second is to evaluate the suitability of using acoustic systems to measure the horizontal and vertical dispersion coefficients of particle plumes. This chapter will be divided into several sections. Section 6.2 outlines the operation of pulsed acoustic systems and the frequency dependence of acoustic backscattering from random distributions of particles. Section 6.3 and its subsections describe the acoustic system used. Section 6.4 describes data obtained from a series of experiments conducted at a deepwater dumpsite over a two-year period.

6.2. THEORY

Acoustic systems project compressional wave pulses from a transducer into the water column (Fig. 6.1). The pulse width (τ) and beamwidth of the transducer

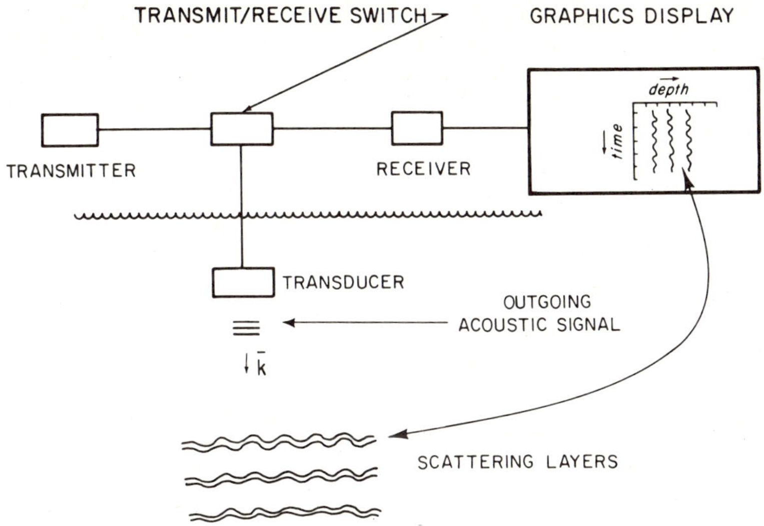

Figure 6.1. Simple block diagram of an acoustic backscattering system. A transmitter applies a voltage to a transducer through a transmit/receive switch. The transducer responds mechanically to the time variability of the applied voltage. The mechanical motion of the transducer produces a compressional wave which propagates through the water column. When the compressional wave interacts with impedance variations at a particular depth, part of its energy is scattered back to the transducer. The transducer responds to the compressional wave with mechanical vibration which produces a voltage across the transducer output leads. This voltage is detected, amplified, and displayed on a graphic recorder. The transmit/receive switch is constructed so that the high voltages produced by the transmitter are applied to the transducer, but not to the receiver.

determine the vertical resolution of an acoustic system. If the compressional wave encounters particles with an impedance (that is, the product of density and sound speed) different from that of water, part of the incident energy is scattered in an angular distribution pattern. This pattern is determined by the physical (density, compressibility, diameter and/or shape) properties of the individual scatters, the spatial distribution of the individual scatters, and the acoustic wave number ($k = 2\pi/\lambda$ where λ is the wavelength $\lambda = c/f$, f is the frequency, and c is the speed of the sound). If the acoustic system is used in a monostatic pulsed mode, that is, with the transducer serving as both the transmitter and the receiver, the backscattered energy will be predominantly from particles of specific size ranges and concentrations. The calculated backscattering intensity levels for incoherent and coherent scattering will be contrasted so that the reader can develop a sufficient background to interpret critically the acoustic data being presented both here and elsewhere in the literature.

The backscattered acoustic energy from spherical particles in a random distribution (incoherent scattering) has been calculated as (Morse and Ingard, 1968)

$$\frac{I_s}{I} \simeq NV \frac{\sqrt{2\pi}}{72} \frac{k^6 a^8}{r^2} (\gamma_\kappa + \gamma_\rho)^2 e^{-2k^2a^2} \quad 1$$

where

I_s is the scattered acoustic energy,
I is the incident acoustic energy,
N is the number of particles per m^3,
V is the volume of water sonified by the acoustic pulse (m^3),
k equals $\frac{2\pi}{\lambda}$,
a is the particle radius (m),
r is the distance from the scattering volume to the transducer,
γ_κ equals $\frac{\kappa_a - \kappa}{\kappa}$; ($\kappa_a$ is the compressibility of the particle, κ is the compressibility of the water), and
γ_ρ equals $\frac{3\rho_a - 3\rho}{2\rho_a - \rho}$, ($\rho_a$ is the density of the particle, ρ is the density of the water).

Backscattered acoustic energy from each particle within a group of particles can be summed assuming there is no interference (coherent scattering). In that case, the backscattered energy is

$$\frac{I_s}{I} \simeq NV \frac{k^4 a^6}{9r^2} (\gamma_\kappa + \gamma_\rho)^2 \qquad 2$$

The backscattered energy shows a stronger functional dependence on the particle radius and acoustic wavelength when incoherent scattering is considered. A comparison of equations 1 and 2 reveals that incoherent scattering from large numbers of randomly distributed particles decreases the backscattered acoustic energy by a factor of $[(2\pi)^{1/2}/8]\ k^2 a^2\ e^{-2k^2 a^2}$. As an example, for a 200 kHz acoustic pulse λ = 7500 μm, the incoherent backscattering term reaches a minimum (that is, backscattered energy is a maximum) for particles of radius about 800 μm (Fig. 6.2). At a particle radius of 100 μm, the backscattered acoustic energy is decreased by 26 dB.

Using equation 1, a family of curves (Orr and Howard, 1980) for 200 kHz and 1 MHz acoustic frequencies has been plotted (Fig. 6.3) to indicate the frequency dependence of backscattering from groups of particles of specified radius and concentration. Note that the curves are referenced to a volume of 1 m^3 at a range of 1 m and are not corrected for the beam characteristics of typical acoustic transducers. Assume, for argument's sake, that an acoustic system operating at either 1 MHz or 200 kHz could detect a particle distribution with a target strength of -100 dB m^{-3} at a stated range. Then, for particles of density 2.6 g cm^{-3} and of concentration 1 mg $liter^{-1}$ a 1 MHz acoustic system could detect particle distributions of diameter about 40 μm and larger, and a 200 kHz system could detect particle distributions of diameter about 300 μm and greater (Fig. 6.3). These statements are for incoherent backscattering from large numbers of particles per unit volume. For individual particles in a volume, equation 2 rather than equation 1 must be applied.

To determine the anticipated sensitivity of acoustic backscattering systems to particles released at an oceanic dumpsite the most appropriate scattering theory

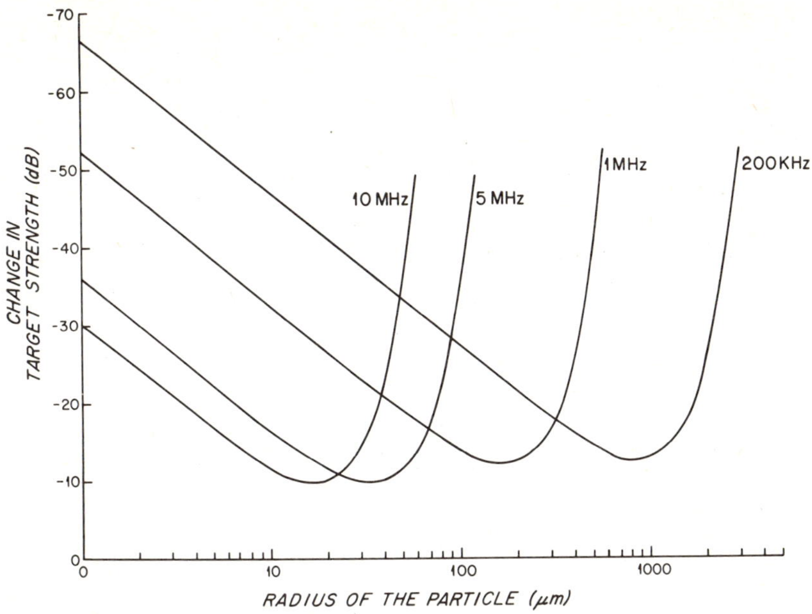

Figure 6.2. Frequency dependence of the target strength adjustment which must be made to the Rayleigh backscattering expression when incoherent backscattering is considered. The Rayleigh backscattering expression is modified by $k^2a^2e^{-k^2a^2}$. The graph plots target strength (TS = 10 log (I_c/I)) decrease versus the radius of the spherical particles that are backscattering acoustic energy.

must be used. To choose the scattering theory the following minimal information concerning the physical characteristics of the disposed particles must be obtained: (1) the density of individual particles and of flocs, if formed, (2) the size distribution of particles, (3) the compressibility of the particles, and (4) the anticipated number of particles per unit volume and their distribution within the volume.

The strong frequency dependence on the particle sizes suggests that high-frequency *multifrequency* acoustic systems should be used to remotely sense particles of random size distributions in the ocean. Acoustic systems consisting of a number of transducers operating at several frequencies are presently being used to distinguish between particle size distributions and acoustic backscattering mechanisms (Proni et al., 1976a, 1976b; Holliday and Pieper, 1980; Orr, 1980) in the ocean.

Acoustic systems are being used to provide qualitative information concerning the temporal and the spatial distribution of particles. Researchers are also attempting to establish the reliability and meaning of quantitative information (dispersion coefficients) extracted from acoustic data. As an example, when wastes are released into the water column, the resulting waste plume contains a particulate fraction and a fluid fraction. Dispersion information extracted from acoustic backscattering should not be used to infer the dispersion rates of the fluid phase unless one establishes that the particles and the fluid fraction do not separate from one another as a result of positive or negative buoyancy effects. How closely the

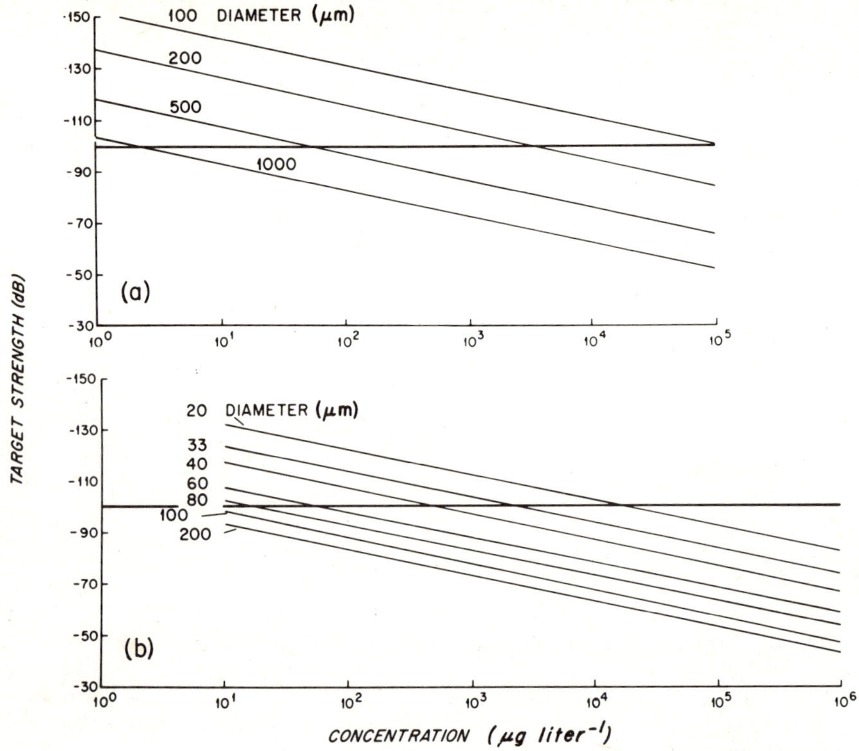

Figure 6.3. Target strength versus particle concentration for incoherent Rayleigh backscattering from spheres of known diameter, density $\rho = 2.6$ g cm^{-3}, and compressibility ($\kappa_e = 1.83 \times 10^{-11}$ cm^2 dyne^{-1}). The calculations were performed for an acoustic frequency of (a) 200 kHz and (b) 1 MHz.

particles of a waste remain associated with the effluent phase of a waste fluid has not been established. It will depend, as may be expected, on the physical characteristics of the individual wastes and the water column at the time of each dump.

The reader should be aware of the fact that acoustic systems provide information concerning the dispersion of that portion or size fraction of the particle phase of a waste plume to which they are sensitive. The sensitivity of a particular acoustic system to a particular plume depends on the operating characteristics of that system's electronics and transducers, the frequency of operation, and the particle physical characteristics (density, compressibility, shape, and spatial distribution). Experimental results obtained from two separate acoustic systems cannot be compared unless both systems are properly intercalibrated.

The results of several experiments which used an acoustic system to study the dispersion of particles will be discussed below.

6.3. EXPERIMENTAL DESIGN AND METHODS

Several experiments to study the dispersion of chemical and sewage sludge waste fields have been performed at Deepwater Dumpsite-106 (DWD-106) (Fig. 6.4), which is located over the continental slope 196 km southeast of New York Harbor.

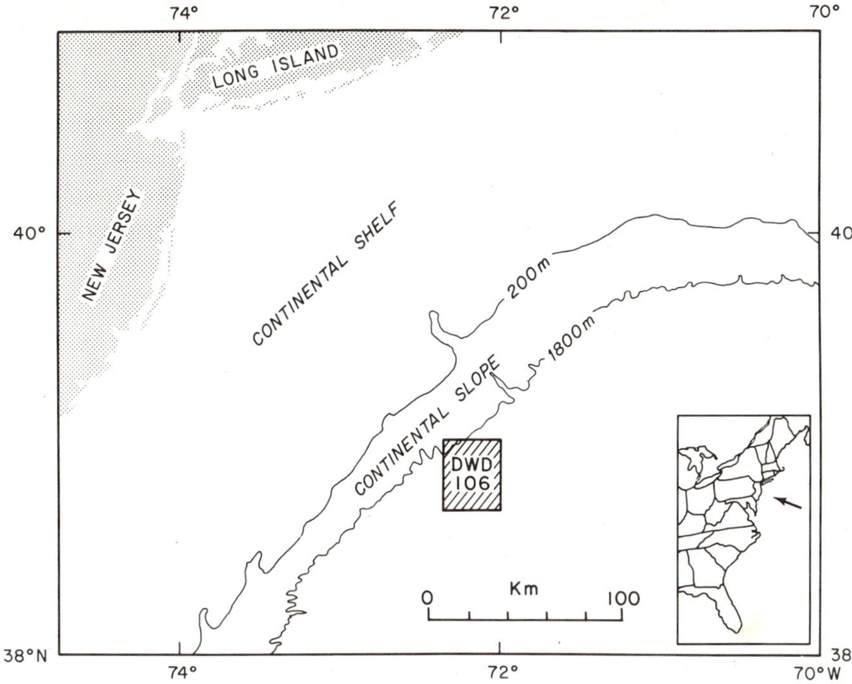

Figure 6.4. Geographic location of Deepwater Dumpsite-106.

The experiments were all performed with the objective of understanding waste plume dispersion. Experiments were conducted in July 1977 when E. I. du Pont de Nemours & Co.'s Edge Moor (Edge Moor, Delaware) (hereafter referred to as DuPont–Edge Moor) acid-iron waste and sewage sludge from Camden, New Jersey were studied independently, in February 1978 when DuPont–Edge Moor acid-iron waste and DuPont–Grasselli waste (Linden, New Jersey) were studied independently, and in April 1978 when a DuPont–Grasselli waste plume was studied. A variety of experiments were performed aboard a research vessel by a number of investigators (for example, see Mukherji and Kester, Chapter 7) to study the waste plume dispersion.

The scientists monitored the waste fields and water column with conductivity-temperature-depth (CTDs) equipment, water sampling equipment, LORAN-C navigation and associated data logging equipment, drogues, real-time chemical analysis equipment, biological sampling equipment, and acoustic backscattering systems. Several of the experimental instruments provided real-time output. This output was used to navigate the ship with respect to the waste plume and to guide sampling.

A typical ocean waste disposal operation will be outlined, followed by a brief description of the acoustic systems used. The section describing the acoustic system will also discuss the meaning of the information which is displayed on a typical acoustic gray scale or graphic record so that the reader can more adequately interpret the acoustic data presented in subsequent sections.

6.3.1. Waste Disposal Procedures at DWD-106

A typical industrial chemical waste disposal monitoring operation proceeds in the following manner. A tug-towed barge carrying about 4.5×10^3 m^3 of industrial chemical waste releases the waste at a rate and in a pattern within a selected quadrant of DWD-106 which is specified by the U.S. Environmental Protection Agency. The chemical wastes are generally released in a line dump. During monitoring experiments, the tug and tow follow a course specified by the scientific party aboard a research vessel which is used as a platform from which the released material is monitored. The course selected is determined from anticipated atmospheric conditions, the sea state, the presence of warm core rings or shelf water intrusions, and the amount of shear present in the water column. Once the barge passes the immediate vicinity of the research vessel, the vessel proceeds across the waste field perpendicular to the barge track while sampling the water column. The research vessel attempts to maintain contact with the waste throughout the experiment. Drogues are often launched in an attempt to determine the effects of currents and shear on the line dump so that the angle at which the research vessel crosses the plume can be determined (assuming a simple shear field), thus allowing reliable estimation of the plume width as a function of time and depth.

6.3.2. Acoustic Backscattering System

The Woods Hole Oceanographic Institution (WHOI) acoustic system (Hess and Orr, 1979), which has been used at DWD-106, operates in pulsed mode at four frequencies (20, 200, 357, and 520 kHz). The transducers are mounted in a hydrodynamic body which is towed from a boom attached to the side of a ship. The tow body is set at 1–3 m depth depending on the sea state and ship speed. All four frequencies are operated at the same time and can provide data (depending on particulate concentration and frequency) to depths of 100 m or greater. The acoustic data are displayed on graphic records, recorded on seven-track analog tape and, on occasion, digitized and stored on nine-track digital tape for subsequent analysis. The graphic records provide real-time shipboard qualitative data concerning the distribution and position of the particle phase of the waste plume. These data are sometimes used to guide chemical and biological sampling of the water column.

The graphic record on which the acoustic data are displayed is created in the following manner. A short acoustic pulse is repeatedly projected downward perpendicular to the water surface. The acoustic energy scattered by particles distributed throughout the water column is detected, amplified, and then plotted with the graphic recorder. The graphic recorder sweeps a stylus across an electric current-sensitive paper once for each ping. The paper displays a gray level proportional to the reflected signal from the transducer depth to a maximum depth which is determined by the recorder sweep speed. The display plots many signals side by side, allowing the eye to discern variations in gray scale as a function of time. The records can be interpreted in terms of the presence or absence of congregations of particulate or other scattering sources in the water column.

Figure 6.6a (see page 130) is an example of an acoustic gray scale record. The darkness of the record at a given point in time and depth is dependent on the amplitude of the backscattered sound from that depth. Time progresses from left to right and the depth scale is noted from the top to the bottom of the record. If a uniform particle concentration exists as a function of depth, the intensity of the reflected sound (hence, gray level on the graphic record) will decrease with increasing depth as a result of attenuation and transmission loss (see, for example, Clay and Medwin, 1977). Time-varying gain normally used in an acoustic system can compensate for transmission loss, but not usually for attenuation loss, which is a function of depth, temperature, and salinity. The gray level at a given depth on the graphic record is roughly proportional to the square root of the number of particles at that depth. Equal gray scales at different depths do not imply that the particle concentration is the same at those depths, unless proper electronic time-varying gain compensation has been used.

6.4. RESULTS

Several interdisciplinary experiments have been performed at DWD-106 to study the dispersion of three types of waste; DuPont acid-iron and DuPont–Grasselli industrial chemical waste, and (in one instance) sewage sludge. Selected sections of the acoustic data set that was acquired during those experiments will be discussed. As the data set was obtained over three seasons (summer, winter, and spring), it will be presented in a seasonal context rather than as individual experiments. The seasonal variability of the vertical and the horizontal distribution of the particle plumes was distinct and will be emphasized in terms of its relationship to the temporal variability of the density structure of the water column. As the vertical density structure of the water column influences the vertical mixing or dispersion rates, it is necessary to describe the density structure encountered at DWD-106 during each of the seasons as well as to discuss possible mechanisms of mixing which have not been extensively studied but which have been detected with acoustic systems. A more detailed discussion of the acoustic data obtained during individual experiments has been presented (Orr et al., 1980).

6.4.1. Seasonal and Temporal Variability of the Pycnocline at DWD-106 and Factors Affecting Mixing

An understanding of the variability of the seasonal pycnocline and the physical processes associated with it is necessary to the understanding of waste dispersion. The seasonal thermocline and its associated density gradients and vertical shear determine the vertical and horizontal dispersion rates of both the particle and the fluid components of a waste field. Because of buoyancy effects, the dispersion rates of the particle and fluid components of a waste field may be different.

The seasonal pycnocline encountered at DWD-106 (National Oceanic and Atmospheric Administration, 1977) changes as a function of time. The temporal variability is caused by the daily and seasonal variability of the heating of the mixed layer, the passage of atmospheric storms and frontal systems of varying intensity, influxes of shelf water and slopewater, and/or the passage of anticyclonic warm core rings (Bisagni and Kester, 1981; Csanady, Chapter 5). Vertical shear, which may be caused by or associated with several of the above processes and also with internal waves, is often associated with high gradient regions in the pycnocline, and can cause horizontal advection to vary as a function of depth and create shear instabilities that can cause periodic and spatially isolated vertical mixing events (Woods, 1968; Koh and Chang, 1973; Orr and Hess, 1978a, 1978b; Thorpe, 1978; Haury et al., 1979).

The late spring and summer months at DWD-106 are normally associated with increased heating of the water column and a decrease in both the intensity and the frequency of atmospheric disturbances. These conditions cause the formation of a shallow seasonal mixed layer (about 10 m) and an associated pycnocline. Depending on their intensity, storms, hurricanes, and atmospheric fronts can significantly disturb the summertime mixed layer and increase the mixed layer depth as well as dispersion rates.

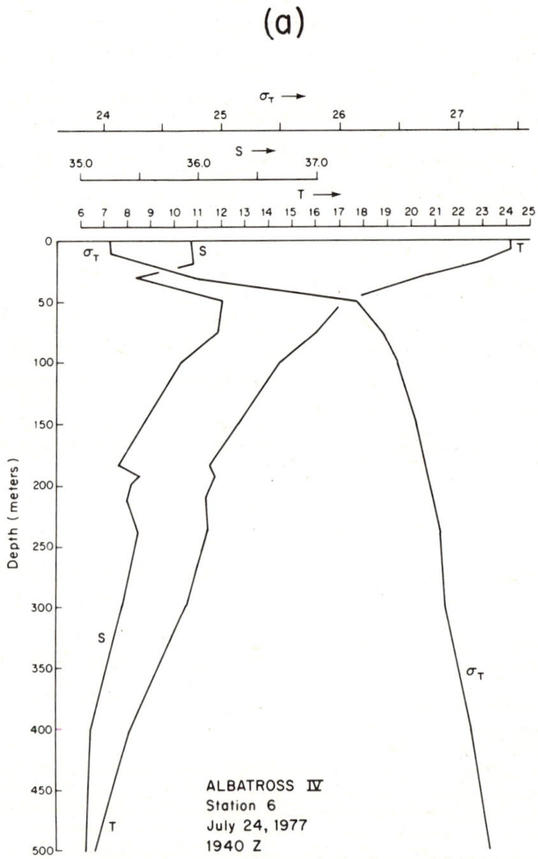

Figure 6.5. (a) Temperature, salinity, and potential density at DWD-106 in July 1977. The data were taken aboard the NOAA ship *Albatross IV* at Station 6, 24 July 1977, 1940Z (Greenwich time).

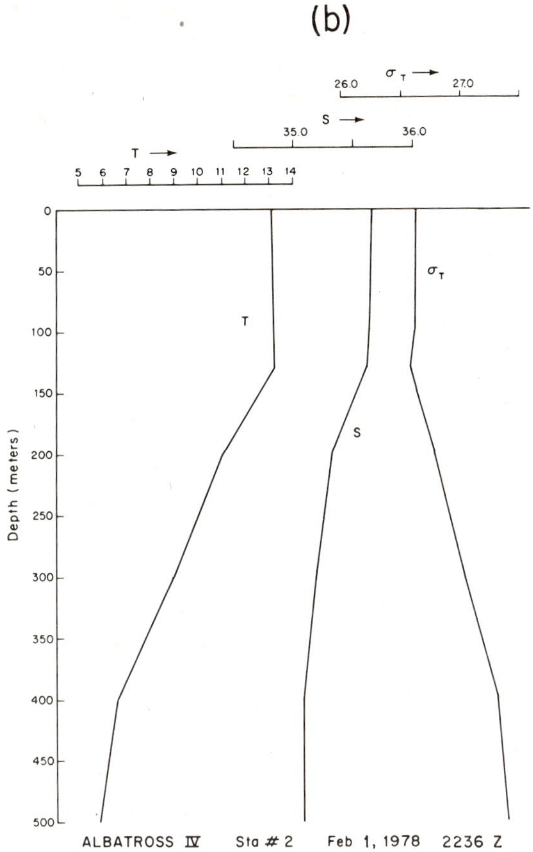

Figure 6.5. (b) Temperature, salinity, and potential density at DWD-106 in February 1978. The data were taken aboard the NOAA ship *Albatross IV* at Station 2, 1 February 1978, 2236Z (Greenwich time).

Wintertime water conditions are characterized by a well-mixed water column which can reach depths in excess of 100 m. The temperature gradient and pycnocline found at the base of the mixed layer are considerably smaller than that found during the late spring and summer months. As a result of the deeper mixed layer, it should be anticipated that the winter water conditions may be associated with a more rapid vertical dispersion of the particulate phase of the chemical waste than found during the late spring, summer, and early fall months.

As an example of the observed seasonal differences in the DWD-106 water column, temperature, salinity (where available), and density (σ_t) plots of the upper 200 m are presented (Fig. 6.5). The shallowest seasonal mixed layers (about 10 m) have been observed in the summertime (Fig. 6.5a) and have a density gradient of 1.4 σ_t units per 10 m. The mixed layers observed during the winter months (about 100 m depth, Fig. 6.5b) have had density gradients of 0.5 σ_t units per 15 m and 0.85 σ_t units per 270 m.

A representative springtime density field is also presented (Fig. 6.5c). These

(c)

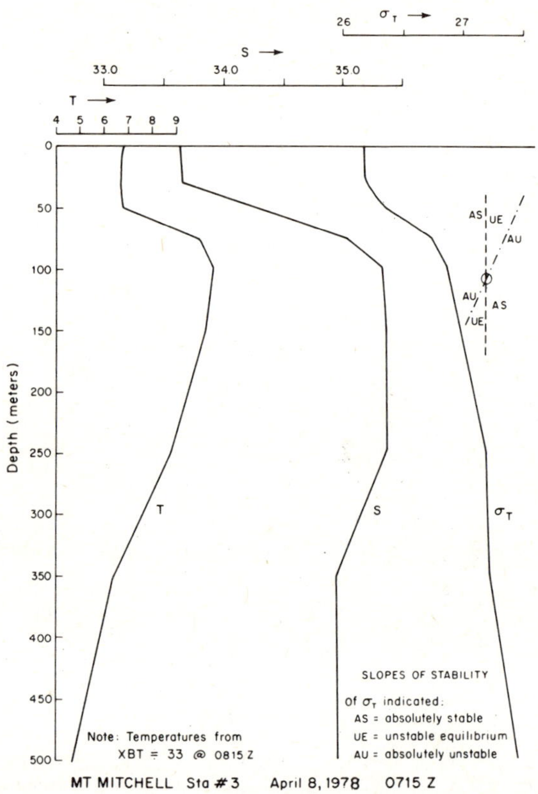

Figure 6.5. (c) Temperature, salinity, and potential density at DWD-106 in April 1978. The data were taken aboard the NOAA ship *Mt. Mitchell* at Station 3 on 8 April 1978, 0715Z (Greenwich time).

data were obtained with a low resolution STD (Bisagni and Kester, 1981) which could not resolve the fine structure often present in the water column. The presence of fine structure with associated high density gradients can cause a reduction in the rate of the vertical transport of particles and must be considered in dispersion studies and calculations. Acoustic data have indicated that particles appear to accumulate on density gradients associated with the seasonal thermocline.

The seasonal thermocline may also be temporally and spatially perturbed by internal wave fields. The magnitude of the vertical displacement of the pycnocline structure can be several meters or more. The minimum value of the internal wave period is determined by the local Brunt–Väisälä frequency, which is dependent on the local water column density gradient. The maximum internal wave frequencies generally encountered are in the range 15–20 cycles h^{-1}. The vertical displacement of the local density structure and associated chemical waste plumes by the internal wave field must be considered in waste disposal impact studies, both during the acquisition of water and biological samples and during data interpretation. As a further example, nonlinear internal waves can develop shear at the base of the mixed layer (Haury et al., 1979) with a resulting vertical mixing of the water column. Nonlinear internal wave packets have been acoustically detected at DWD-106 during the summer (for example, Fig. 6.6a) and springtime. No information on the properties of short-period internal wavefields at DWD-106 is available. Hence, their effects on vertical mixing cannot be estimated.

6.4.2. Seasonal Variability of the Dispersion of Particle Plumes as Detected with High Frequency Acoustic Systems

The seasonal variability of the horizontal and vertical distribution of particles formed or released during the disposal of industrial chemical waste will be illustrated with 200 kHz acoustic data obtained during the independent monitoring of DuPont–Edge Moor acid-iron and DuPont–Grasselli waste particle fields at DWD-106.

The DuPont–Edge Moor acid-iron waste was monitored in July 1977 and in February 1978. The DuPont–Grasselli waste was monitored in February and April 1978. These data sets will be discussed together as the waste fields responded in similar fashion to the variability of the seasonal mixed layer. The effects of vertical shear on a sewage sludge plume that was studied at DWD-106 during July 1977 will also be presented. As only the Lagrangian variability of the waste field is of interest, no Eulerian information concerning the waste field will be provided other than the fact that the wastes were released within the geographic limits of DWD-106.

6.4.3. Seasonal Variability of the Dispersion of Particles Associated with DuPont Wastes

Laboratory studies indicate that DuPont–Edge Moor acid-iron particles are hydrous ferric oxide ($Fe_2O_3 \cdot 3H_2O$, $\rho = 1.2$ g cm^{-3}), and that particles formed during the disposal of DuPont–Grasselli waste particles are magnesium hydroxide [$Mg(OH)_2$, $\rho \simeq 1.1$ g cm^{-3}] (Kester et al., 1981). The density and size distributions of the particles formed in seawater during barge disposal are unknown. It is suspected that they could be flocs. Consequently, the size and the density distributions of the particles in seawater will be difficult to determine because of physical alteration during sampling procedures.

The density differences between the Edge Moor and Grasselli wastes suggest that a difference in the vertical dispersion rates should be anticipated. The vertical distributions of the particles from the two waste fields have, however, been found to be similarly affected by density gradients. Therefore, as stated earlier, the results of independent experiments will be presented together.

The summertime and wintertime distributions of acid waste particles have been detected with the acoustic systems (Figs. 6.6a and b, respectively; note the difference in depth scales). The seasonal difference in the vertical distribution of the

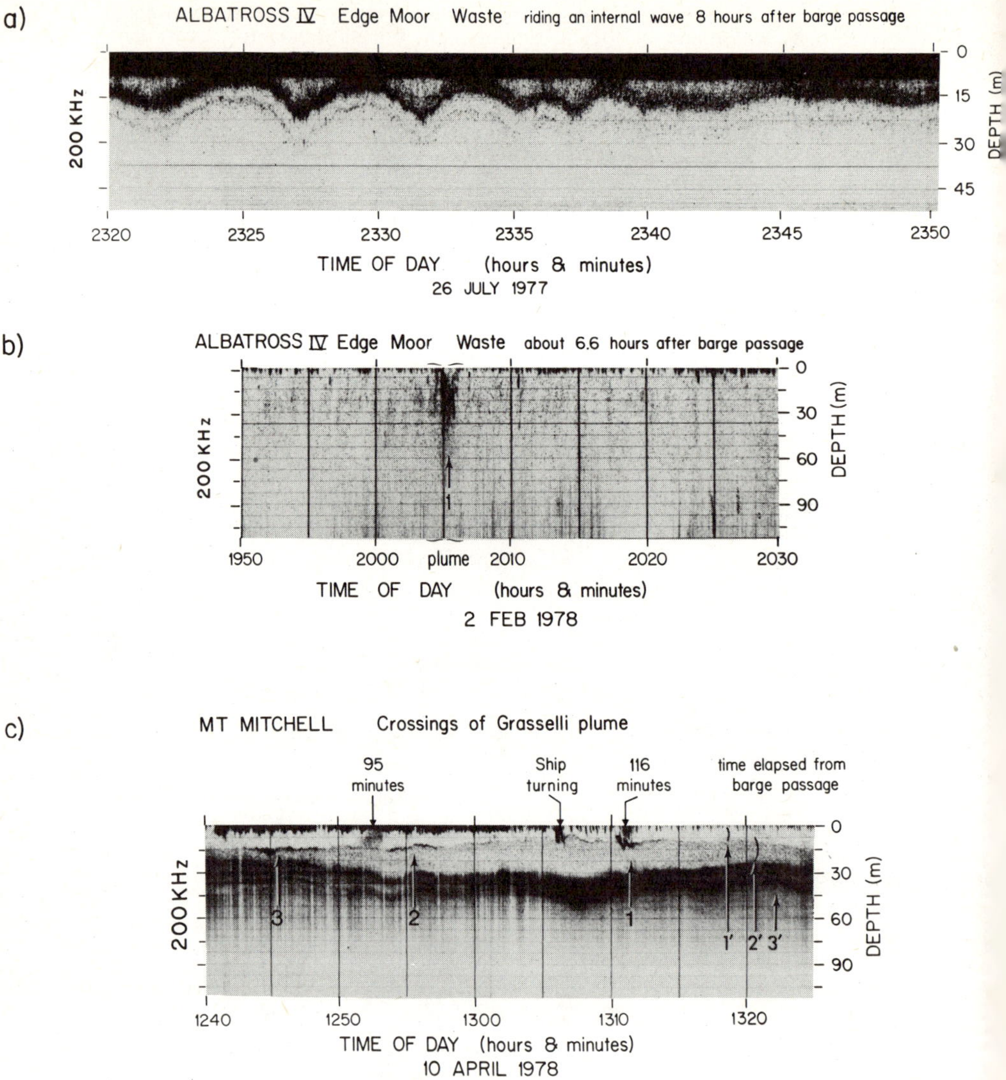

Figure 6.6. (a) Vertical and horizontal distribution of DuPont–Edge Moor acid-iron waste particles acoustically detected during summertime conditions (shallow seasonal thermocline). A 200 kHz acoustic system was used to obtain the data. The particles were being vertically displaced by a conoidal internal wave field. The period of the wave field was Doppler shifted by the motion of the research platform, about 1.9 m s^{-1}. Note the layering of the particles on the density structure in the first 15–30 m of the water column. (b) A 200 kHz record of the vertical and horizontal distribution of DuPont–Edge Moor acid-iron waste as detected during wintertime water conditions (deep mixed layer). The material is distributed throughout the first 82 m of the water column. Note the heavy backscattering areas in the first 30 m of the column. The distributed scatterers suggest differentiation in settling rates among the particles. (c) DuPont–Grasselli alkaline waste detected on 10 April 1978 with a 200 kHz acoustic system. The particles produced when the waste was mixed with seawater are seen to lie on the interface (arrows 1 and 2) between the first (arrow 1′) and second (arrow 2′) water masses and remain suspended in the water column between the 0 to 15 m depth. A time-varying gain was used; thus, the near-surface particle scattering levles were suppressed by the electronics.

particles was caused by the difference in the density gradient between the shallow seasonal thermocline (summer condition) and a deep mixed layer (winter condition). The shallow summer pycnocline traps the particles and appears to limit the vertical distribution to the mixed layer (about 10 to 15 m) and areas of high gradient in the pycnocline. The 200 kHz data were taken about 8 h after the release of the waste into the water. When a deep wintertime mixed layer was present the particles from the acid-iron waste became distributed over the first 60–80 m of the water column 6.5 h after the release.

The influence of an internal wave packet on the vertical displacement of the particle plume during summertime conditions is apparent (Fig. 6.6a). A vertical displacement of the intense acoustic scattering layer by about 10 m and the lighter scattering layer by about 12 m is quite distinct. These data were taken from a ship moving at approximately 1.9 m s^{-1}, and therefore the internal wave period, which is estimated to be 20–40 min, is Doppler shifted on the record. Chemical sampling of the water column in the 15–18 m range with a sampling period of 10–20 min would have yielded distinctly different particle concentrations and interpretations concerning the particle distribution, unless the presence of the internal wave field was accounted for.

A 200 kHz record of the vertical and horizontal distribution of DuPont-Grasselli waste particles in a springtime water mass is shown (Fig. 6.6c). The density structure of the water in the area of the waste disposal operation was more complex than is indicated in Fig. 6.5c. An expendable bathythermograph (XBT) taken in the experiment area detected three overlying water masses (warm water, cold water, warm water) in the first 60 m. The three scattering layers outlining the water masses are distinctly shown (Fig. 6.6c, arrows $1'$, $2'$, and $3'$). Sections of the water column had particles formed with the release of the DuPont-Grasselli chemical waste distributed in the vertical and horizontal. The first density surface limited the vertical distribution of the particles. Detections of the plume up to two hours following the chemical waste release are shown. The particles appeared to be spreading on the density surface at the warm and cold water interface. This spreading is thought to be caused by shear at the boundary zone.

During the spring and summer months, the density structure at the base of the mixed layer (10–30 m) limited the vertical penetration of particles associated with chemical waste disposal (Figs. 6.6a and c). During the winter experiment when a deep mixed layer was present the particles penetrated much deeper (80–100 m, Fig. 6.6b). As the particles are detectable with 200 kHz acoustic systems ($\lambda \simeq 7500$ μm) and remain suspended for considerable time periods, it is anticipated that a portion of the particles or aggregates of particles may be greater than 300 μm in diameter and have a density nearly equal to the surrounding water column.

The temporal variability of the horizontal width of a particle plume associated with the chemical waste fields described above has been calculated for the summer and the winter months using the acoustic records and ship's navigation records. To calculate accurately the horizontal width of a particle plume the following information is needed:

1. Accurate knowledge of the vertical shear in the water column.

2. Accurate navigation of the ship (position known to <10 m).
3. Accurate knowledge of the initial direction of the axis of the line dump and the response of the line dump as a function of depth, to currents, and to vertical and horizontal shear in the water column. It is necessary to know the change of alignment of the line dump as a function of time following the release of industrial chemical waste, because the width of the plume as determined from a ship crossing must be corrected for the ship's angle to the plume axis as well as for the water shear.

To illustrate the impact of vertical shear on the horizontal distribution of particles associated with a waste plume, data obtained in July 1977 showing the dispersion of a sewage sludge plume are presented (Fig. 6.7). These data were obtained when a warm core ring was in the DWD-106 area. The sewage sludge plume was advected nearly 56 km to the east in a 24-h period. A ship crossed a line dump of sewage sludge from opposite sides nearly perpendicular to the dump axis at about 27 and 48 min following the release of the material. Shear in the water column was advecting separate vertical sections of the plume at different speeds. To account properly for the width of the plume as a function of depth and to differentiate between turbulent diffusion processes and shear-related spreading of a plume, the shear in the water column must be measured. The steady sinking of particles into a high-shear zone resulted in a plume of particles in that zone which was of extended width, as compared to the surface seasonal mixed-layer plume width.

The calculated widths of the acoustically detected DuPont-Edge Moor particle

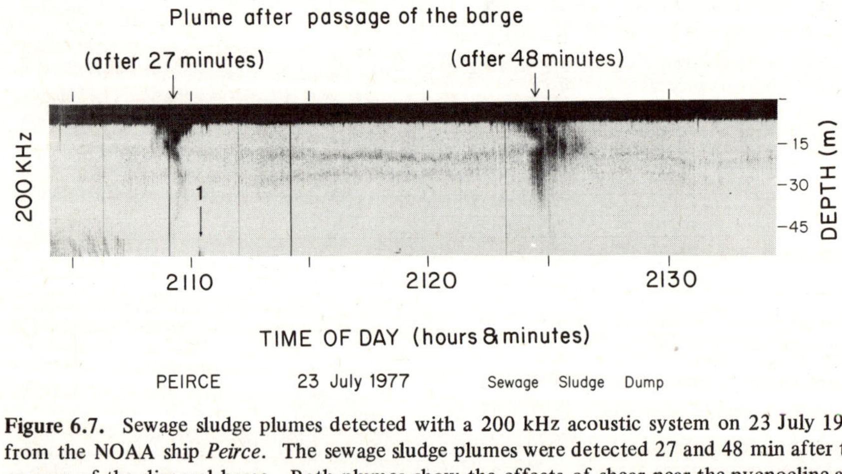

Figure 6.7. Sewage sludge plumes detected with a 200 kHz acoustic system on 23 July 1977 from the NOAA ship *Peirce*. The sewage sludge plumes were detected 27 and 48 min after the passage of the disposal barge. Both plumes show the effects of shear near the pycnocline at a depth of 12 to 17 m. The speed of flow in this region has been calculated to be 2.5 m min^{-1}. Note that a portion of the plume has penetrated the pycnocline to depths of ≈55 m (arrow 1), indicating a separation of the plume by particle size.

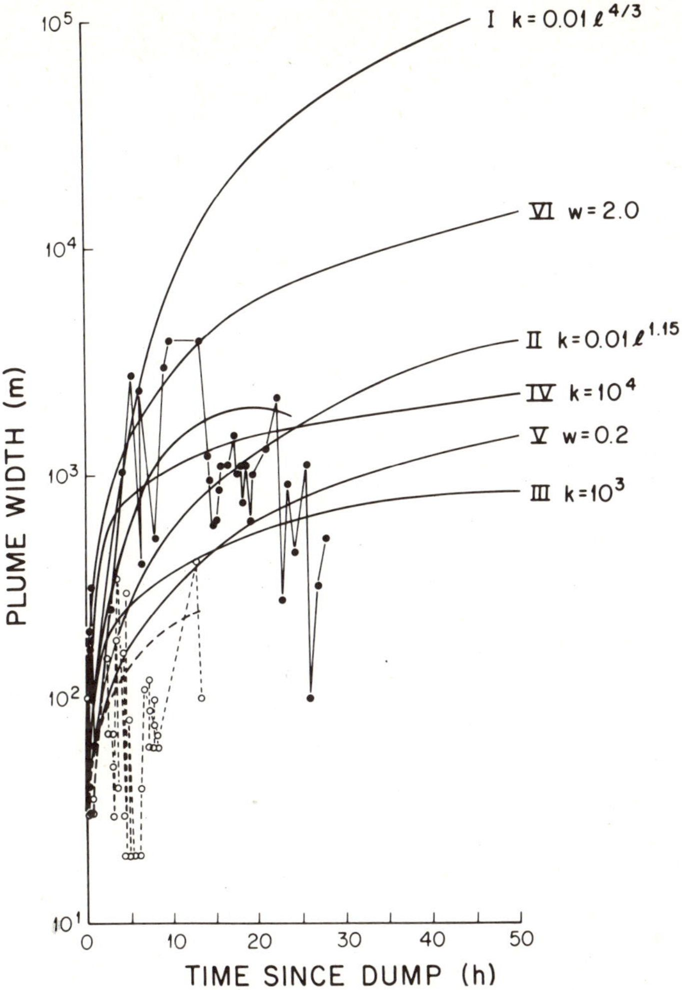

Figure 6.8. Plot of plume width versus time. Empirical relations of plume width versus time using a variety of parameters for the dispersion coefficients have been plotted as in O'Connor and Park (1981). The plume widths as a function of time have been calculated from 200 kHz acoustic backscattering data, LORAN C navigation, and estimated vertical shear. The plume widths are plotted for DuPont–Edge Moor Plant acid-iron waste both for summertime water conditions (•) and wintertime water conditions (○).

plumes for summer and winter water conditions (Fig. 6.8) are plotted against a variety of empirical diffusion models which estimate the temporal dependence of the width of line dumps. The widths taken are for the widest part of the plume as a function of depth. The depth-dependent dispersion has not been calculated to date. The empirical models (O'Connor and Park, 1982) predict the width (l) of a line plume as a function of time as follows:

Case I	$l = l_0 + 4(2Kt)^{1/2}$	$K = 0.01 l^{4/3}$ cm^2 s^{-1}
Case II	same as I with	$K = 0.01 l^{1.15}$ cm^2 s^{-1}
Case III	same as I with	$K = 10^3$ cm^2 s^{-1}
Case IV	same as I with	$K = 10^4$ cm^2 s^{-1}
Case V	$l = l_0 + 4\,\omega t$	$\omega = 0.2$ cm s^{-1}
Case VI	same as IV with	$\omega = 2.0$ cm s^{-1}

where l is the width of the plume, K is the eddy diffusion coefficient, and t is time.

A curve representing the trend in the acoustic plume width as a function of time has been drawn on the same figure. Errors in calculating the plume width from acoustic data are inherent because of the necessity of estimating each of the quantities mentioned above. Approximately 24 h after the waste release the summertime data showed a decreasing plume width. This could be attributed either to a decrease of particle concentration at the plume edges to below acoustic detection threshholds or to the inherent errors in calculations due to navigation and uncertain angle of the ship to the plume axis. Comparison of the horizontal plume widths for summer and winter data indicate that summertime horizontal spreading is an order of magnitude more rapid than winter spreading. Fig. 6.6a and b show that the winter plumes are distributed over a much larger vertical section of the water column. The maximum depth of penetration of the DuPont–Edge Moor waste during the summertime experiment was 15–20 m. The particles reached these depths shortly (<10 min) after the passage of the disposal barge with little apparent deeper penetration over a 24-h period. The depth distribution of the Edge Moor waste in the winter months was surface to 120 m. The maximum depth was reached about 35 min after the particle release. Again, further penetration after the first 35 min was not apparent. Note, the volume of water 10 m deep by 1000 m wide by 1 m thick (summertime particle distribution) happens to be the same as a volume of 100 m deep by 100 m wide by 1 m thick (wintertime particle distribution). Thus, the seasonal variability of plume width may not reflect a seasonal variability in dilution. The vertical distribution of the particles in the winter was, however, patchy, indicating that the above generalization may not be appropriate. The acoustic records have been interpreted only in a qualitative manner, that is, the analog signals recorded during the summer months and the winter months have yet to be digitized to allow comparison of the relative change in signal levels within a specific plume as a function of time so that the rate of change of particle distribution can be estimated.

6.5. CONCLUSIONS

High-frequency towed acoustic backscattering systems have been used in a qualitative manner at DWD-106 to detect and to guide the chemical or biological sampling of the particle plumes created by the release of industrial chemical waste and sewage sludge at DWD-106. Quantitative information concerning the seasonal variability of the plume width and depth has been obtained from these data. Plume widths have been found to broaden quickly during spring and summertime water conditions and remain narrow during wintertime water conditions. In summer and spring the vertical distribution of the particles was limited by the shallow seasonal thermocline and associated pycnocline. The plumes have been found to penetrate much deeper in the winter months, but are not always uniformly distributed in the vertical and show a patchy structure. In particular the wintertime DuPont-Edge Moor acid-iron waste was patchier than the wintertime DuPont-Grasselli alkaline waste field. Further study must be conducted to determine if the particles were diluted more in winter months than in summer months. The plume width versus time calculations have uncertainities caused by navigational limitations. Plots of the plume width versus time for DuPont acid-iron waste for both summer and winter data have been compared with a variety of empirical dispersion models. The data show an order of magnitude difference in the temporal dependence of the plume width as a function of season. This should not, however, be interpreted as a difference in dilution of the waste particles as the vertical distribution of the particles was very different between the two seasons. The acoustic data indicate that the temporal change in plume width measured for the particle phase of the waste field was of the same magnitude as measured during previous experiments (Okubo, 1971).

The suitability of equating the dispersion coefficients of the fluid phase of a chemical waste plume to the dispersion coefficients of the particle phase must be considered. To equate the two coefficients, one needs to measure the particle size distribution produced during the release of industrial chemical waste and the particle density and settling rates in a water column with turbulence levels of the same magnitude as found in the first tens of meters in the oceanic water columns. The turbulence levels or mixing rates are not constant as a function of time or as a function of depth, and there are not many meaningful measurements of turbulence levels in the top tens of meters of the ocean to guide an analysis of the expected settling rates of particles of known buoyancy. If, however, the particle size distributions, buoyancy, water density, and turbulence levels of the water column were known, calculations could be performed (Raudkivi, 1976) that would allow an estimate of the anticipated difference between the horizontal and the vertical dispersion of the fluid and the particle phase of the waste fields.

To perform properly an experiment of the type described here and to interpret properly the acoustic data sets as well as other complementary data sets obtained during the experiment, the following measurements must be made:

1. Measurement of vertical and horizontal shear in the water column.
2. High-resolution Eulerian navigation to be combined with (1) to allow high-resolution Lagrangian navigation of a ship.
3. Measurement of atmospheric forcing, both thermal input and wind speed and direction as a function of height.
4. Measurement of wave height and spectral character.
5. Frequent high-resolution CTD measurements and/or XBTs.

ACKNOWLEDGMENTS

The authors acknowledge the contributions of the following individuals. F. R. Hess designed and supervised the construction, maintenance, and installation of the high-frequency acoustic system used to obtain the data described. Brian Bardsley, Donald Crowell, and William Kucharski participated in the construction, deployment, and maintenance of the equipment. James Doutt assisted in data acquisition aboard the National Oceanic and Atmospheric Administration (NOAA) ship *Pierce*. Stanley Bergstrom put extensive time and energy into preparing the data, as did Elizabeth Howard. Elizabeth Pratt hlped to prepare the figures used in this chapter. James Bisagni of the NOAA National Marine Fisheries Service provided some of the XBT and CTD data which were used to calculate the σ_t profiles presented in this chapter. The manuscript was typed by Angela Sousa and Catherine Herrity. The work presented was supported by NOAA grants 04-7-158-44054, NA 79AA-D-00030, and NOAA's Ocean Dumping Program, grant 04-8-M01-43. The acoustic system development was supported by the Naval Ocean Research Development Activity through contract N0014-77-C-0196.

REFERENCES

Bisagni, J. J., and D. R. Kester. 1981. Physical variability at an East Coast United States offshore dumpsite. *In*: Ocean Dumping of Industrial Wastes, B. H. Ketchum, D. R. Kester, and P. K. Park (Eds.). Plenum Press, New York, pp. 89–107.

Clay, C. S., and H. Medwin. 1977. Acoustical Oceanography: Principles and Applications. John Wiley, New York, 544 pp.

Haury, L. R., M. G. Briscoe, and M. H. Orr. 1979. Tidally generated internal wave packets in Massachusetts Bay. *Nature*, **278**, 312–317.

Hess, F. R., and M. H. Orr. 1979. A Towed, Multifrequency High-Frequency Sonar System for Scattering and Ocean Dynamics Studies. Technical Report WHOI-79-76, Woods Hole Oceanographic Institution, Woods Hole, Massachusetts, 16 pp.

Holliday, D. V., and R. E. Pieper. 1980. Volume scattering strengths and zooplankton distributions at acoustic frequencies between 0.5 and 3 MHz. *Journal of the Acoustical Society of America*, **67**, 135–146.

Kester, D. R., R. C. Hittinger, and P. Mukherji. 1981. Transition and heavy metals associated with acid-iron waste disposal at Deep Water Dumpsite 106. *In*: Ocean Dumping of Industrial Wastes, B. H. Ketchum, D. R. Kester, and P. K. Park (Eds.). Plenum Press, New York, pp. 215–232.

Koh, R. C. Y., and Y. C. Chang. 1973. Mathematical Model for Barged Ocean Disposal of Wastes. EPA-660/2-73-029, Office of Research and Development, U.S. Environmental Protection Agency, Washington, D.C., 581 pp.

Morse, P. M. M., and K. U. Ingard. 1968. Theoretical Acoustics. McGraw-Hill, New York, 927 pp.

National Oceanic and Atmospheric Administration. 1977. Baseline Report of Environmental Conditions in Deep Water Dumpsite 106, Vol. 1, Physical Characteristics. NOAA Dumpsite Evaluation Report 77-1, National Oceanic and Atmospheric Administration, Rockville, Maryland, 218 pp.

O'Connor, T. P., and P. K. Park. 1982. Consequences of industrial waste disposal at the 106-mile ocean waste site. *In*: Ecological Stress and the New York Bight: Science and Management, G. F. Mayer (Ed.). Estuarine Research Federation, Columbia, South Carolina, pp. 675–697.

Okubo, A. 1971. Oceanic diffusion diagrams. *Deep-Sea Research*, **18**, 789–802.

Orr, M. H. 1980. Remote Acoustic Sensing of Oceanic Fluid and Biological Processes. Technical Report WHOI-80-2, Woods Hole Oceanographic Institution, Woods Hole, Massachusetts, 51 pp.

Orr, M. H., and F. R. Hess. 1978a. Remote acoustic monitoring of natural suspensate distributions, active suspensate resuspension, and slope/shelf water intrusions. *Journal of Geophysical Research*, **83**, 4062–4068.

Orr, M. H., and F. R. Hess. 1978b. Acoustic monitoring of industrial chemical waste released at Deep Water Dumpsite 106. *Journal of Geophysical Research*, **83**, 6145–6154.

Orr, M. H., and E. J. Howard. 1980. Acoustic detection of particles in fluids. *EOS, Transactions of the American Geophysical Union*, **17**, 270.

Orr, M. H., L. Baxter, II, and F. R. Hess. 1980. Remote Acoustic Sensing of the Particulate Phase of Industrial Chemical Wastes and Sewage Sludge; Report on the Seasonal Variability of the Dispersion of the Particulate Phase as observed from Three Cruises: July 1977, January-February 1978, and April 1978. Technical Report WHOI-79-38, Woods Hole Oceanographic Institution, Woods Hole, Massachusetts, 153 pp.

Proni, J. R., and D. V. Hansen. 1981. Dispersion of particulates in the ocean studied acoustically: The importance of gradient surfaces in the ocean. *In*: Ocean Dumping of Industrial Wastes, B. H. Ketchum, D. R. Kester, and P. K. Park (Eds.). Plenum Press, New York, pp. 161–173.

Proni, J. R., F. C. Newman, D. C. Rona, D. E. Drake, G. A. Berberian, C. A. Lauter, Jr., and R. L. Sellers. 1976a. On the use of acoustics for studying suspended oceanic sediment and for determining the onset of the shallow thermocline. *Deep-Sea Research*, **23**, 831–837.

Proni, J. R., F. C. Newman, R. L. Sellers, and C. Parker. 1976b. Acoustic tracking of ocean-dumped sewage sludge. *Science*, **193**, 1005–1007.

Raudkivi, A. J. 1976. Loose Boundary Hydraulics, 2nd ed. Pergamon Press, New York, 397 pp.

Thorpe, S. A. 1978. The near-surface ocean mixing layer in stable heating conditions. *Journal of Geophysical Research*, **83**, 2875–2885.

Woods, J. D. 1968. Wave-induced shear instability in the summer thermocline. *Journal of Fluid Mechanics*, **32**, 791–800.

PART III: CHEMICAL EFFECTS

Acid-Iron and Pharmaceutical Wastes

7

ACID-IRON DISPOSAL EXPERIMENTS IN SUMMER AND WINTER AT DEEPWATER DUMPSITE-106

Prithviraj Mukherji and Dana R. Kester

Graduate School of Oceanography
University of Rhode Island
Kingston, Rhode Island

ABSTRACT

The concentrations and distributions of total and particulate iron, copper, cadmium, and lead associated with acid-iron wastes were measured at Deepwater Dumpsite-

Dr. Mukherji's present address: Revere Research, Inc., P.O. Box 1352, Edison, New Jersey 08837.

106 (DWD-106) during July 1977 and February 1978 to study the fate of these wastes in the ocean after dumping. Concentrations were measured by obtaining discrete samples which were analyzed by atomic absorption spectrophotometry. In July the waste material was confined to the upper 20 m of the water column and within the mixed layer for a period of up to 27 h. The waste mixed to greater depths during February when the seasonal thermocline was absent and the permanent pycnocline was below 100 m. Total iron was used as a tracer to follow the waste plume. The results were related to two models for plume dispersion by diffusion processes during the initial 5 h after the dump. A model based on a constant diffusion velocity (as opposed to a constant eddy diffusivity) fit the data best, yielding a diffusion velocity of 1.1 cm s^{-1}.

7.1. INTRODUCTION

Oceanic dispersion of acid-iron waste dumped at Deepwater Dumpsite-106 (DWD-106) (Fig. 7.1) has been studied using a variety of approaches: the tracking of waste

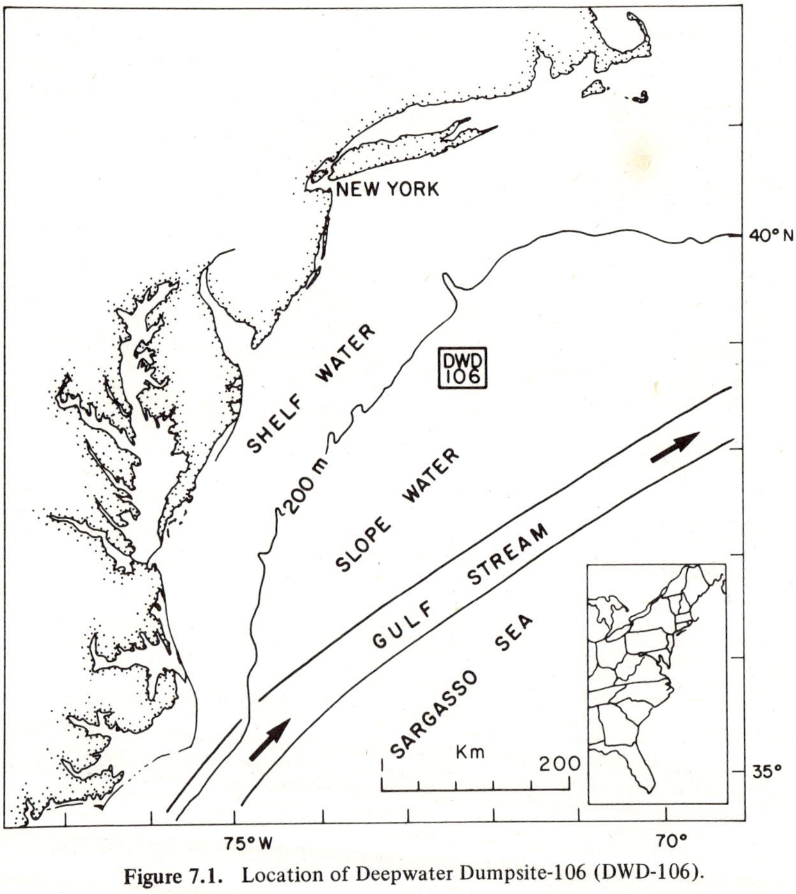

Figure 7.1. Location of Deepwater Dumpsite-106 (DWD-106).

plumes by chemical means (Hausknecht and Kester, 1976; Kester et al., 1981; Chapter 8); by fluorescent dye tracers (Frye and Williams, 1977); by measurement of acoustic scattering intensity by flocs in waste plumes (Orr and Hess, 1978; Orr et al., 1980; Chapter 6); and by multiple frequency satellite imagery (Johnson, 1980).

Several studies of ocean disposal of wastes at shallow water dumpsites (depth <100 m) indicate that waste components become associated with seafloor sediments and benthic organisms (Vaccaro et al., 1972; Gross, 1976; Pesch et al., 1977). However, in the case of deepwater sites such as DWD-106, it is likely that the wastes will be dispersed within the water column before reaching the bottom at depths greater than 2000 m. Consequently, the most likely impact of wastes, if any, is associated with planktonic rather than benthic organisms (see Murphy et al., Chapter 11).

In the present study, total (dissolved plus particulate) iron was used to determine the fate of waste plumes following disposal of acid-iron waste at DWD-106 during July 1977 and February 1978. This study was part of an investigation to establish the extent to which waste disposal has increased the concentration of selected metals (Fe, Cu, Cd, and Pb) at DWD-106. The chemical reactions which occurred when acid-iron waste was mixed with seawater and the chemical forms of these metals at DWD-106 were also addressed in this study. The waste came from E.I. du Pont de Nemours & Co.'s titanium dioxide plant, hereafter referred to as DuPont–Edge Moor, located in Edge Moor, New Jersey. The waste was discharged directly in the wake of a moving barge. On mixing with seawater, a flocculent precipitate of $FeOOH \cdot nH_2O$ forms (Kester et al., 1981).

7.2. EXPERIMENTAL

In this study acid-iron waste dumps from a moving barge were sampled using a rosette sampler with 8-liter Niskin bottles (General Oceanics, Miami, Florida). At the time of sampling the waste was being tracked by acoustic backscattering measurements conducted by M. Orr (Chapter 6) of Woods Hole Oceanographic Institution. The sampling location and depth were selected based on the acoustic measurements which indicated the location of the waste plume (Chapter 6, Fig. 6.6b).

The analytical methods to determine the concentrations of total and particulate metals in seawater were based on a previous study (Kester et al., 1981). Particulate metals were determined by filtering about 6 liters of seawater through acid-cleaned preweighed 0.4 μm Nuclepore filters (Nuclepore Corp., Pleasanton, California) which were then washed at least five times with ultrapure distilled deionized water to remove sea salt. The filters were dried to obtain the weight of suspended material, placed in an acid cleaned polyethylene vial with 3 N HNO_3, and immersed in an ultrasonic cleaner. The Fe, Cu, Cd, and Pb were analyzed using a Perkin-Elmer model 603 (Norwalk, Conneticut) atomic absorption spectrophotometer (AAS) equipped with an HGA 2100 graphite furnace. In some cases it was necessary to dilute the sample with 3 N HNO_3.

Table 7.1. Characteristics of Total Metal Analyses for the February 1978 Dump Experiment at DWD-106

Metal	Detection[a] ($ng\ kg^{-1}$)	Blank[b] ($ng\ kg^{-1}$)	Efficiency[c] (%)	Precision[d] (% RSD[e])
Fe	31	257 ± 25	92 ± 1	9
Cd	1	3 ± 2	90 ± 1	10
Cu	7	10 ± 3	99 ± 3	10
Pb	5	35 ± 20	97 ± 2	42

[a] Two times the instrumental noise.
[b] Obtained from duplicate analyses of blanks based on multiple extractions of seawater.
[c] Determined from the yield of radiotracers.
[d] From analyses of duplicate samples from depth of 97 ± 7 m at station 3.
[e] RSD, relative standard deviation.

Total metals concentrations (dissolved plus particulate) were analyzed using a preconcentration method (Kester et al., 1981). Seawater was drawn into pre-cleaned 250-ml polyethylene centrifuge bottles from 5-liter Niskin samplers. In our shore laboratory the samples were acidified to pH 3.0 with ultrapure HNO_3 and allowed to sit for a period of 48 h. The $CoCl_2$ and ammonium pyrrolidine dithiocarbonate reagents were added to produce the flocculent precipitate which extracts the metals from the seawater (Boyle and Edmond, 1977). After at least 30 min the bottles were centrifuged at an acceleration of 3000 g for 15–20 min and the supernatant seawater aspirated off; 5 ml of 3 *N* HNO_3 were added to dissolve the precipitate and the samples were then analyzed by AAS.

Table 7.1 lists some characteristics for the method used to analyze the samples from the February 1978 experiment. The detection limits and blanks are similar to those found in the analysis of the July 1977 samples (Kester et al., 1981). Except for Pb, the precision of the analytical method was about 10% and was obtained by averaging duplicate concentrations from a depth of 97 ± 7 m at a station occupied 5 h after acid-iron was dumped.

7.3. RESULTS

7.3.1. Total Metal Concentrations prior to Acid-Iron Waste Disposal

Results of analyses of the samples from the July 1977 were published previously by Kester et al. (1981). The data for the February 1978 samples from DWD-106 were presented in greater detail elsewhere by Mukherji et al. (1980). A comparison of total metal levels in "background" seawater at DWD-106 with outer shelf and open ocean concentrations reported in the literature is shown in Table 7.2. Changes in the concentrations of Fe, Cu, Pb, and Cd at DWD-106 resulting from continual disposal of acid-iron waste would be evident. Such a comparison is also useful in determining whether chemical gradients in these metals are present between the continental slope and open ocean waters of the north Atlantic Ocean.

Table 7.2. Comparison of Mean Total Metals Concentrations in the Upper 120 m of Background Seawater at DWD-106 for July 1977, February 1978, and Open Ocean

	July 1977		February 1978		Open Ocean[a]	
Parameter	Mean ± 1σ (ng kg^{-1})	n	Mean ± 1σ (ng kg^{-1})	n	Mean ± 1σ (ng kg^{-1})	n
T(Fe)	670 ± 140	10	589 ± 229	8	900[b]	
					460 ± 289[c]	39
T(Cd)	13 ± 6	5	14 ± 14	7	≤ 10[d]	14
T(Pb)	276 ± 34	5	583 ± 223	3	50[e]	
T(Cu)	170 ± 64	10	246 ± 121	7	120 ± 30[d]	

[a]Upper 100 m only.
[b]Kremling and Petersen (1977).
[c]Danielsson (1980).
[d]Bender and Gagner (1976).
[e]Chow and Patterson (1966).

Except for *T*(Pb), total metal concentrations are comparable in July 1977 and February 1978. The standard deviations are higher in the winter data set relative to those seen in July 1977. The large variability associated with the *T*(Fe) levels for the winter data set can be caused either by contamination from the uncoated steel hydrographic wire used for obtaining samples (Betzer and Pilson, 1975) or by natural variation within the upper 120 m of the water column. Natural variation in *T*(Fe) distribution is more likely in the winter data because we sampled to a depth of 120 m compared to the summer experiment when the samples came only from the upper 25 m.

Results show that except for Pb, total metal concentrations in the upper 120 m of "background" waters at DWD-106 during July 1977 and February 1978 are within the ranges observed for open ocean waters. In the case of Pb, we suspect that the samples were contaminated during collection because metal-free samplers were not used; thus, these results may not be representative of the true concentrations. These values, however, provide upper limits which may be useful for comparison with results from the dump plume.

7.3.2. Particulate Metals prior to Acid-Iron Waste Disposal

Table 7.3 shows a comparison of particulate trace metal data for background waters not influenced by acid-iron waste dumps during the winter dump study. A Student *t*-test of the data presented in Table 7.3 at the $P = 0.05$ confidence level indicates that particulate iron, copper, and lead levels during the winter were significantly higher at station 1 relative to station 5 which was sampled 34 h later. There was considerable change in the hydrography during the winter study. In the upper 60 m, the temperature varied from 12.6 to 14.5°C, and salinities ranged from 35.41 to 35.84‰ between 1 February 1978 and 8 February 1978 (Langone, 1978).

Table 7.3. Variation of Particulate Metal Concentrations (ng kg^{-1}) and TSM (μg kg^{-1}) in Background Waters at DWD-106 in February 1978 at Two Stations Taken 34 h Apart[a]

	Station 1		Station 5	
Parameter	Mean ± 1σ	*n*	Mean ± 1σ	*n*
TSM	123 ± 37	7	98 ± 35	8
P(Fe)	433 ± 118	7	224 ± 84	7[b]
P(Cd)	0.4 ± 0.3	3	0.2 ± 0.2	3
P(Pb)	48 ± 17	6	17 ± 7	7
P(Cu)	63 ± 22	6	16 ± 9	5

[a] Depth, upper 100 m.
[b] Excluded sample from 100 m.

An analysis of temperature-salinity data during this period shows that a hydrographic front in the form of a warm core eddy approached the DWD-106 region (Langone, 1978); the variability in the chemical properties from stations 1 and 5 could be caused by this front in the vicinity of the dumpsite during the winter study. Although the concentrations of trace metals are higher at station 1 when compared to station 5, the suspended material masses remain similar for the two data sets.

The mass ratios of particulate metal to total suspended material (TSM) provide a measure of the particulate matter composition (Table 7.4). At $P = 0.1$ (90% confidence level), particles at station 1 during the winter are enriched in Cu relative to the summer composition and have higher concentrations of Cu associated with particles at DWD-106 relative to the composition observed in other oceanic regions. In the case of Pb, the winter data from station 1 is significantly higher ($P = 0.1$) than the summer and open ocean concentrations of this metal associated with the particulate phase. These compositional differences for Pb and Cu in the winter

Table 7.4. Weight Ratios of Particulate (P) Metal to Total Suspended Matter (TSM) for DWD-106 Summer and Winter Periods and the Open Ocean

	July 1977		Feb 1978			Ocean[a]	
Ratio	Mean ± 1σ	*n*	Station:	Mean ± 1σ	*n*	Mean ± 1σ	*n*
$\frac{P(Fe)}{TSM}$ (x 10^{-2})	0.09 ± 0.03	5	1	0.23 ± 0.12	7	0.13 ± 0.05	7
			5	0.26 ± 0.06	7		
$\frac{P(Cd)}{TSM}$ (x 10^{-5})	0.3 ± 0.2	5	1	5.3 ± 0.2	3	0.3 ± 0.4	5
			5	0.2 ± 0.3	2		
$\frac{P(Pb)}{TSM}$ (x 10^{-5})	8 ± 5	5	1	39 ± 18	6	2.6 ± 1.3	7
			5	17 ± 7	5		
$\frac{P(Cu)}{TSM}$ (x 10^{-5})	15 ± 9	5	1	51 ± 30	6	3.7 ± 1.5	7
			5	15 ± 8	5		

[a] Taken from Wallace et al. (1977).

data could be related to hydrographic changes involving shelf, slope, and Gulf Stream waters during the study. The ratio (by weight) of iron to suspended matter is similar to that observed for open ocean particles and did not vary significantly between the two ocean dumping experiments.

The average TSM concentration for the upper 120 m prior to acid-iron waste dumps was 138 ± 33 (n = 5) and 123 ± 37 (n = 7) $\mu g\ kg^{-1}$ for the summer and winter data sets, respectively. The concentrations of TSM at DWD-106 are within the range of 50–150 $\mu g\ kg^{-1}$ reported by Folger (1970) for surface waters of the Atlantic Ocean. The work of Klenova et al. (1962) and Manheim et al. (1966) for central Atlantic and continental shelf waters of the United States indicate similar values. For continental shelf waters in the Gulf of Maine, average TSM was 230 ± 170 $\mu g\ kg^{-1}$ (n = 90) (Spencer and Sachs, 1970). Brewer et al. (1976), report a range of 50–100 $\mu g\ kg^{-1}$ in the upper 100 m of the midlatitudes of the western Atlantic Ocean. TSM concentrations observed in this study at DWD-106 for the summer and winter months are therefore similar to those observed in open-ocean unpolluted waters.

There are, however, significant differences in the particulate iron, copper, and lead concentrations between background stations for the winter and summer test dumps. In view of the variability that we observe within a data set (Table 7.3), we conclude that the differences between winter and summer are within the natural variability at DWD-106 during one season, and therefore can not be considered to be a seasonally dependent variation.

7.3.3. Total Metal Concentration after Disposal of Acid-Iron Waste

During the winter study a plume was sampled for metal analyses at three different times after a dump of acid-iron waste. The samples collected 5 h after disposal showed total iron concentrations up to 150 $\mu g\ kg^{-1}$ which is more than 100 times greater than control values (0.5 $\mu g\ kg^{-1}$). These high values extended to depths of at least 90 m (Table 7.5). Due to the drift of the ship relative to the acid-iron waste plume, the upcast samples were obtained at the edge or less concentrated portion of the plume than the downcast samples. A second downcast provided a sample at

Table 7.5. Concentrations ($\mu g\ kg^{-1}$) of Total Fe in Samples Collected during Down and Up Casts 5 h after Disposal of Acid-Iron Wastes in February 1978

Depth	First Down Cast	Depth	First Up Cast
0	108	22[a]	60
16	150	22	76
30	152	7	66
75	116	2	22
90	131		56 ± 42[b]
	131 ± 20[b]		

[a]Replicate samples were collected at 22 m.
[b]Mean and standard deviation.

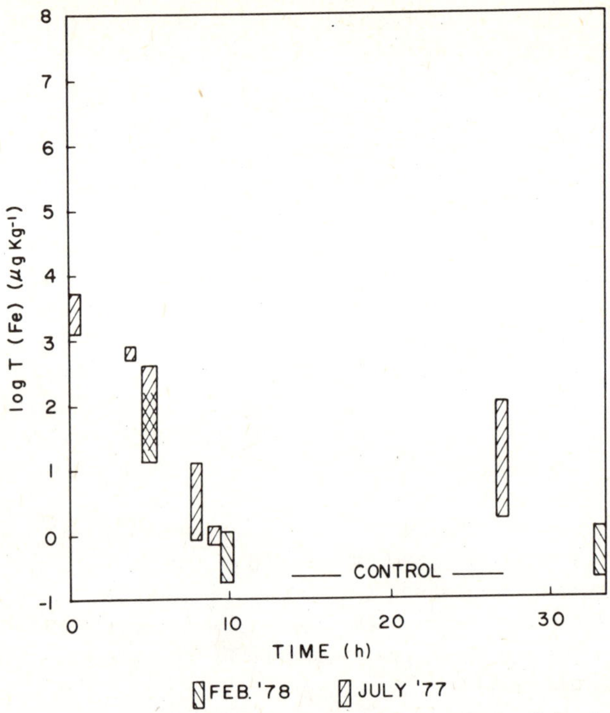

Figure 7.2. Variation with time after the acid–iron waste test dump of the observed *T*(Fe).

12 m with *T*(Fe) = 3 μg kg^{-1} indicating that the ship had drifted nearly out of the plume. Discrete sampling, even when aided by real-time acoustic backscattering measurements (Chapter 6), does not provide adequate information to define the horizontal and vertical gradients in a waste plume (Kester et al., 1981).

A plot of total iron concentrations, *T*(Fe), versus time after a dump provides information on the persistence of waste plumes for the two seasons studied. In both studies, total iron concentrations decreased with time in a systematic manner (Fig. 7.2). *T*(Fe) concentrations for t = 34 h were estimated using the relationship between particulate and total iron concentration from the predump data: *P*(Fe) = 0.6 × *T*(Fe). Figure 7.2 shows that iron levels remain elevated for up to 27 h after dump in the summer. This is in sharp contrast to the winter data when total iron levels were down to near normal levels (0.5 μg kg^{-1}) in the mixed layer 10 h after the dump. Total iron concentrations 100–200 times above normal were observed down to 90 m, the maximum depth sampled during the winter; in summer stratified waters, elevated concentrations were observed primarily in the upper 20 m.

7.3.4. Particulate Metal Concentration after Disposal of Acid-Iron Waste

A comparison of the total metal in particulate form in control and within 5 h after dump waters shows the effect of acid-iron waste on the particulate matter composition at DWD-106 (Table 7.6). The percentage of each metal in particulate form at

Table 7.6. Mean and Standard Deviation for Percentage of Total Metals in the Particulate Phase for Control and Acid-Iron Disposal Stations during Summer and Winter Periods

Metal	Summer	Winter
Control		
Fe	60 ± 50	65 ± 38
Cu	17 ± 8	9 ± 3
Cd	3 ± 4	3 ± 3
Stations within 5 h after disposal		
Fe	46 ± 20	88 ± 25
Cu	78 ± 20	23 ± 8
Cd	9 ± 4	10 ± 20

the control station is indistinguishable for the two seasons. After disposal of the waste, the percentage of Cu in particulate form increased substantially relative to the control station. Particulate Cd remained a small fraction of total Cd; however, the percentage of Cd in the particulate phase is higher after a dump than before it. Copper and iron show differences between the summer and winter data sets which cannot be explained readily, but the amount of metal in particulate phases increases after the dump relative to the control. Particulate lead concentrations were not included in this analysis since the total metal samples probably were contaminated in lead.

A matrix of correlation coefficients for pairs of parameters from the summer data set indicated strong correlations for a number of parameters (Kester et al., 1981). For the winter data set, fewer parameters correlate as strongly. Samples for the summer experiment were obtained using a Teflon coated pump (Kester et al., 1981). Niskin samplers were used for the winter data set. The differences in interparameter correlations may be due to natural variability or to sampling differences. In the winter set TSM has a correlation coefficient $r = 0.87$ with T(Fe) and $r = 0.88$ with P(Fe). P(Fe) and T(Fe) correlate strongly with $r = 0.95$ for the winter data. Partitioning of total metals can be variable from disposal to disposal and is probably dependent on a large number of parameters including the influence of mixing volumes, turbulence at sea surface, temperature, and chemical factors.

7.3.5. Mixing Characteristics of the Waste Plume

Two DuPont-Edge Moor acid-iron waste samples were analyzed (Table 7.7). Fe and Cu were present in comparable amounts in these waste samples; Pb was almost an order of magnitude lower in concentration in the winter sample. Knowing the dumping discharge rate $R = 1.4 \times 10^4$ liters min^{-1} and the barge speed $u = 9.6$ km h^{-1}, and the initial dimensions of the plume formed in the wake, we can calculate the concentration of total iron that must be present within the wake of the barge

Table 7.7. Concentration of Metals in Samples of DuPont–Edge Moore Waste Dumped at DWD-106 in July 1977 and February 1978

	Concentration ($mg\ kg^{-1}$)	
Metal	July 1977	February 1978
Fe	31000	20000
Cd	<0.008	–[a]
Pb	67	0.4
Cu	2.5	3.3

[a]Not determined.

immediately after the waste is dumped. An estimate of the depth to which the plume mixes initially in the wake of the barge is $h = 3.5 \times$ barge draft (Csanady, 1981). The draft of the barge was 3 m; therefore $h = 10$ m is not unreasonable. The value for the plume width, b, soon after a dump was obtained by averaging three cross-plume acoustic measurements (Orr et al., 1980). The volume per unit length of the wake is given by width (b) $\times$ draft (h) $\times$ length (L) and is 40 m $\times$ 10 m $\times$ 1 m = 400 m^3 or 4×10^5 liters m^{-1} of plume length. Using $T(\text{Fe}) = 3 \times 10^7$ $\mu g\ kg^{-1}$ (Table 7.7), we calculate the maximum $T(\text{Fe})$ in the wake to be 5200 $\mu g\ kg^{-1}$. The observed value was 4700 $\mu g\ kg^{-1}$. The good agreement between calculated and observed maximum concentrations of total iron substantiates the above estimates for the initial dilution of the waste.

Concentrations of total Fe in the two wastes (July 1977 and February 1978) and the maximum total iron concentrations from postdump stations for summer and winter (Table 7.8) were used to calculate dilution factors (D) for the waste/seawater mixtures at various intervals of time after dumps. Figure 7.3 shows the plot of log dilution factors calculated from the data in Table 7.8 versus log time and $t^{0.5}$; the former corresponds to a constant diffusion velocity mode, the latter to a constant eddy diffusivity model (Csanady, 1981). The envelopes were calculated using $w = 0.24$–2.0 cm s^{-1} and $K_e = 300$–2000 $cm^2\ s^{-1}$, and represent the range of these two parameters found in earlier studies (Csanady, 1981). In the first 5 h after dumps in winter and summer, the data fit the constant diffusion velocity model better than the constant eddy diffusivity model. Beyond 5 h after disposal of the waste the observed dilutions deviate from that predicted by either model. The concentration of total iron in seawater, which is ~ 0.2 $\mu g\ kg^{-1}$ (Kester et al., 1978), was neglected in calculating D. Dilution factors calculated in this way represent the lower limit for the number of parts of seawater mixing with one part of waste.

In the wake of the barge, mixing occurs rapidly due to the turbulence caused by the motion of the barge. Data from the summer dump taken immediately following a dump, indicate an initial dilution factor of about 7×10^3. This value is fairly typical for dilution factors in the wake of barges (Csanady, 1973).

For time beyond 5 h after dump, the data fall outside the values for the two conventional dispersion models. Application of the constant diffusivity and con-

Table 7.8. Maximum Values for Total (*T*) Iron Concentrations in the Mixed Layer That Were Used to Calculate Dilution Factors[a] at Different Times after Disposal of Acid-Iron Waste

Time after Dump (h)	Maximum T(Fe) (μg kg^{-1})	
	July 1977[b]	February 1978
0.3	4700	
4	630	
5	420	152
8	11.5	
9	1.3	
10		0.95
27	95.0	
34		1.4

[a]Dilution factor (D) = $\dfrac{T(\text{Fe})_{\text{waste}}}{\text{maximum } T(\text{Fe})_{\text{plume}}}$; $T(\text{Fe})_{\text{waste}}$ equals 31,000 and 20,000 mg kg^{-1} for summer and winter, respectively.
[b]Kester et al. (1981).

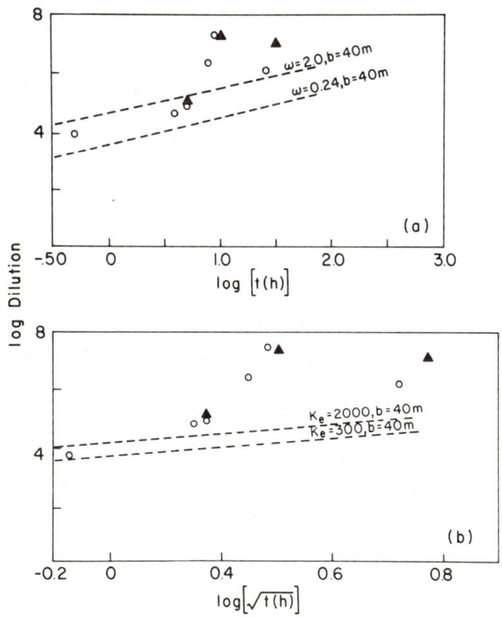

Figure 7.3. Observed dilutions of iron in acid-iron waste plumes: ○, summer; ▲, winter. The dashed-line envelope was obtained using a constant diffusion velocity model in (a) and a constant diffusivity model in (b).

stant diffusion models used by Csanady (1973) to the data for the first 5 h gives reasonable values for K_e and w. The results of this calculation for the summer data are:

$$\text{Maximum } T(\text{Fe}) \text{ in waste plume } (C_w) = 4700\ \mu\text{g kg}^{-1}$$

$$\text{Maximum } T(\text{Fe}) \text{ in plume after 4 h } (C_p) = 630\ \mu\text{g kg}^{-1}$$

$$C_w/C_p = D' = 7.5$$

$$(\text{Variance})^{0.5}\ (\sigma_y) = D'\, b/(2\pi)^{0.5} \qquad 1$$

where b = width of wake = 40 m (Orr et al., 1980). Using $D' = 7.5$ in equation 1 we calculate σ_y = 149 m.

For the constant diffusion velocity model,

$$\sigma_y = wt \qquad 2$$

t is time in seconds and w is diffusion velocity. Using equation 4 at t = 14400 s (4 h) we calculate

$$w = 1.0 \text{ cm s}^{-1}$$

A similar calculation at t = 5 h gives $w = 1.2$ cm s^{-1}.

For constant diffusivity,

$$\sigma_y = (2K_e t)^{0.5} \qquad 3$$

where K_e is the effective diffusivity coefficient. Using σ_y = 149 m as before, and t = 14400 and 18000 s, we calculate

$$K_e = 1471 \text{ and } 2683 \text{ cm}^2 \text{ s}^{-1}.$$

In an earlier study of DuPont-Edge Moor waste dispersion, $w = 0.24$ cm s^{-1} and K_e = 300 cm^2 s^{-1} (Csanady, 1981). The diffusion velocity and K_e calculated from our data, while significantly higher than those reported by Csanady (1981), are similar to those found in earlier studies of waste plume dispersion (Ketchum and Ford, 1952; Csanady, 1969; Csanady, 1973).

The reason for the observed dilution being 1 to 2 orders of magnitude greater than predicted by the diffusion models for $t > 5$ h can be (1) the samples were not taken within the maximum plume concentrations at these times, and (2) the iron concentrations were diminished by processes in addition to oceanic dispersion such as particulate removal from the surface waters. In view of the difficulties associated with sampling in plumes using discrete samplers, the first reason cannot be discounted. However, the anomalously high dilutions between 5 and 34 h could also result from efficient vertical mixing from turbulent entrainment due to changes in sea surface states (Phillips, 1977), or to particulate removal processes.

7.4. CONCLUSION

The observations to date suggest that acid-iron waste is diluted rapidly in the wake of a barge in the initial moments after introduction into the ocean. In the first 5 h, the waste mixes at about the same rate during moderately calm conditions at sea in summer and winter. Waste related material penetrates deeper in the winter water column than when the seasonal thermocline is present in the summer. In the summer, waste concentrations remained above background values by two orders of magnitude for 27 h after dump.

The concentration of particulate copper varied significantly from summer to winter in background waters at DWD-106, the winter concentrations being greater. Significant variations in the Cu and Pb concentrations in background water particulate material were observed for two stations separated by 34 h in time in the winter experiment. These could have been related to hydrographic changes during this time. Particulate copper was present in greater concentrations at DWD-106 relative to open ocean particulate copper for both summer and winter data sets. After acid-iron waste was dumped the concentrations of Fe, Cu, and Pb associated with particulate material increased relative to predump values.

Diffusion velocities can be calculated from the initial 5 h after a dump using *T*(Fe) as a tracer of the waste. The data for times greater than 5 h after a dump indicate that the samples did not come from the main portion of the plume for both the summer and winter experiments; or alternatively there is enhanced dispersion or removal of iron from surface waters over extended time periods.

ACKNOWLEDGMENTS

The authors wish to thank Mr. James Bisagni for the salinity, temperature, and pressure measurements, Dr. Thomas P. O'Connor for assisting with the sampling, and L. R. Lufkin and M. A. Maley for assistance in preparing this manuscript. This work was supported by National Oceanic and Atmospheric Administration grant 04-8-M01-39.

REFERENCES

Bender, M. L., and C. Gagner. 1976. Dissolved copper, nickel, and cadmium in the Sargasso Sea. *Journal of Marine Research*, **34**, 327–339.

Betzer, P. R., and M. E. Q. Pilson. 1975. The effect of corroded hydrographic wire on particulate iron concentrations in seawater. *Deep-Sea Research*, **22**, 117–120.

Boyle E. A., and J. M. Edmond. 1977. Determination of copper, nickel, and cadmium in seawater by APDC chelate coprecipitation and flameless atomic absorption spectrometry. *Analytica Chimica Acta*, **91**, 189–197.

Brewer, P. G., D. W. Spencer, P. E. Biscaye, A. Hanley, P. L. Sachs, C. L. Smith, S. Kadar, and J. Fredricks. 1976. The distribution of particulate matter in the Atlantic Ocean. *Earth and Planetary Science Letters*, **32**, 393–402.

Chow, T. J., and C. C. Patterson. 1966. Concentration profiles of barium and lead in Atlantic waters off Bermuda. *Earth and Planetary Science Letters*, **1**, 397–400.

Csanady, G. T. 1966. Diffusion in the skewed shear flow of lake currents. *Journal of Geophysical Research*, **71**, 411–420.

Csanady, G. T. 1973. Turbulent Diffusion in the Environment. D. Reidel, Boston, Massachusetts, 248 pp.

Csanady, G. T. 1981. An analysis of dumpsite diffusion experiments. *In*: Ocean Dumping of Industrial Wastes, B. H. Ketchum, D. R. Kester, and P. K. Park (Eds.). Plenum Press, New York, pp. 109–129.

Danielsson, L. G. 1980. Cadmium, cobalt, copper, iron, lead, nickel and zinc in Indian Ocean water. *Marine Chemistry*, **8**, 199–215.

Folger, D. W. 1970. Wind transport of land-derived mineral, biogenic, and industrial matter over the North Atlantic. *Deep-Sea Research*, **17**, 337–352.

Frye, D., and G. Williams. 1977. Measurements of the dispersion of barged waste near 38°50′N latitude and 72°15′W longitude at the "106" dumpsite. Unpublished manuscript, EG & G Environmental Consultants, Waltham, Massachusetts.

Gross, M. G. 1976. Waste disposal. MESA New York Bight Atlas Monograph 26, New York Sea Grant Institute, Albany, New York, 32 pp.

Hausknecht, K. A., and D. R. Kester. 1976. Deep Water Dumpsite 106 Chemical Data Report. University of Rhode Island Technical Report (December 1976), University of Rhode Island, Kingston, 16 pp.

Johnson, R. W. 1980. Remove sensing and spectral analysis of plumes from ocean dumping in New York Bight Apex. *Remote Sensing of Environment*, **9**, 197–209.

Kester, D. R., R. C. Hittinger, P. Mukherji, and K. Hausknecht. 1978. Metal studies at DWD-106 (NOAA Grants 04-7-158-44027, 04-7-158-44064 and 04-8-M01-39). Unpublished Report, University of Rhode Island, Kingston, Rhode Island. Submitted to National Oceanic and Atmospheric Administration, Office of Marine Pollution Assessment (Ocean Dumping Program), 49 pp.

Kester, D. R., R. C. Hittinger, and P. Mukherji. 1981. Transition and heavy metals associated with acid-iron waste disposal at Deep Water Dumpsite 106. *In*: Ocean Dumping of Industrial Wastes, B. H. Ketchum, D. R. Kester, and P. K. Park (Eds.). Plenum Press, New York, pp. 215–232.

Ketchum, B. H., and W. L. Ford. 1952. Rate of dispersion in the wake of a barge at sea. *Transactions, American Geophysical Union*, **33**, 680–684.

Klenova, M. V., V. M. Lavrov, and V. K. Nikolayeva. 1962. Distribution of suspended material in the Atlantic Ocean in relation to bottom relief. *Doklady Akademiia Nauk S.S.S.R.*, **144**, 1153–1155.

Kremling, K., and H. Petersen. 1977. The distribution of zinc, cadmium, copper, and iron in seawater of the Iceland Faroe Ridge area. *"Meteor" Torsch.-Ergebnisse*, **19**, 10–17.

Langone, H. K. 1978. Physical Oceanography of Deepwater Dumpsite 106—January–February 1978. Unpublished manuscript, National Marine Fisheries Service, Atlantic Environmental Group, Narragansett, Rhode Island, 25 pp.

Manheim, F. T., R. H. Meade, G. C. Bond, and J. V. A. Trumbull. 1966. Suspended material in Atlantic coastal waters between Cape Cod, Mass. and the Florida Keys (abstract). American Association for the Advancement of Science, Annual Meeting, Washington, 1966, Section E, p. 23.

Mukherji, P., R. C. Hittinger, and D. R. Kester. 1980. Comparison of Summer-Winter Acid-Iron Test Dumps at DWD-106. Technical Report No. 80-3, Graduate School of Oceanography, University of Rhode Island, Kingston, 43 pp.

Orr, M. H., and F. R. Hess. 1978. Acoustic monitoring of industrial chemical waste released at deep water dump site 106. *Journal of Geophysical Research*, **83**, 6145–6154.

Orr, M. H., L. Baxter, II, and F. R. Hess. 1980. Remote Acoustic Sensing of the Particulate Phase of Industrial Chemical Wastes and Sewage Sludge; Report on the Seasonal Variability of the Dispersion of the Particulate Phase as Observed from three Cruises: July 1977, January–February 1978, and April 1978. Technical Report WHOI-79-38, Woods Hole Oceanographic Institution, Woods Hole, Massachusetts, 153 pp.

Pesch, G., B. Reynolds, and P. Rogerson. 1977. Trace metals in scallops from within and around two ocean disposal sites. *Marine Pollution Bulletin*, **8**, 224–228.

Phillips, O. M. 1977. The Dynamics of the Upper Ocean. Cambridge University Press, Cambridge, 261 pp.

Spencer, D. W., and P. L. Sachs. 1970. Some aspects of the distribution, chemistry, and mineralogy of suspended matter in the Gulf of Maine. *Marine Geology*, **9**, 117–136

Vaccaro, R. F., G. D. Grice, G. T. Rowe, and P. H. Wiebe. 1972. Acid-iron waste disposal and the summer distribution of standing crops in the New York Bight. *Water Research*, **6**, 231–256.

Wallace G. T., Jr., G. L. Hoffman, and R. A. Duce. 1977. The influence of organic matter and atmospheric deposition on the particulate trace metal concentration of northeast Atlantic surface seawater. *Marine Chemistry*, **5**, 143–170.

8

AUTOMATED IRON MEASUREMENTS AFTER ACID-IRON WASTE DISPOSAL

Mary F. Brown, Dana R. Kester, and Joyce M. Dowd

Graduate School of Oceanography
University of Rhode Island
Kingston, Rhode Island

ABSTRACT

An automated analytical procedure to measure iron in seawater based on a colorimetric method using the ferrozine reagent was developed. The method, applicable to iron concentrations in the range 0.02 to 10 μmoles kg^{-1}, was used to track an acid-iron waste plume that was formed after an acidic industrial waste highly concentrated in iron was dumped into the ocean at the Deepwater Dumpsite-106 (DWD-106) off the continental shelf of New Jersey. This automated system was used in conjunction with a submersible shipboard pumping system to obtain continuous real-time iron measurements in the waste plume. Real-time iron measurements were used to track an acid-iron waste plume for over 50 h after a dump; a tenfold dilution of iron in the waste plume was observed during this period. Changes in the horizontal distribution of iron in two waste plumes discharged into the ocean within 2 km of each other were monitored for a period of 5 h after the dump. The vertical dispersion of the waste in the ocean observed immediately after the dump and after the waste plume had mixed with seawater for over 50 h suggests that the acid-iron waste does not penetrate the seasonal thermocline.

8.1. INTRODUCTION

The impact on the ocean resulting from the disposal of chemical wastes at deep-water dumpsites is an important environmental problem. In the past decade several industrial wastes have been discharged into the ocean at the deepwater dumpsite off the continental shelf of New Jersey (Fig. 8.1; see also Chapter 14). This dumpsite

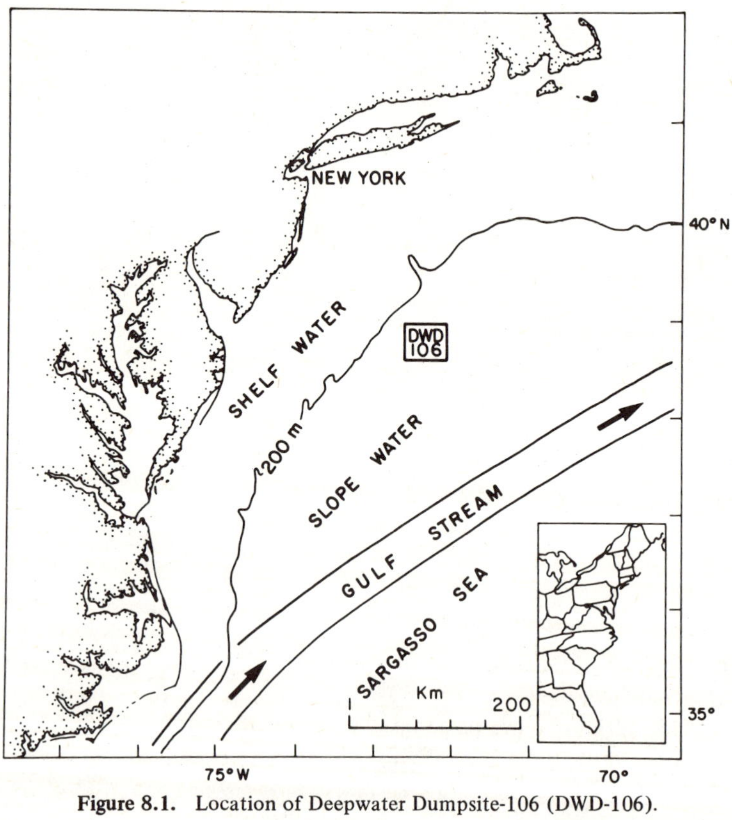

Figure 8.1. Location of Deepwater Dumpsite-106 (DWD-106).

is 106 nautical miles (196 km) southeast of the Ambrose Light Tower in waters 1800–2300 m deep, and it will be referred to as Deepwater Dumpsite-106 (DWD-106). The primary fate of the wastes disposed at this deepwater dumpsite is thought to be dispersion of the constituents in the water column rather than incorporation of the waste constituents in the sediments as was observed at dumpsites located on the continental shelf (Vaccaro et al., 1972; Gross, 1976; Pesch et al., 1977).

In this study the dispersion of an industrial waste from the E. I. du Pont de Nemours & Co.'s Edge Moor facility (hereafter referred to as DuPont–Edge Moor) located in Edge Moor, Delaware, was investigated. The constituents of this waste were mainly dissolved in an acidic (pH < 1.0) and highly concentrated iron ($\simeq 0.8$ M) solution; the nominal composition of the waste is summarized in Table 8.1. As this acid-iron waste mixes with seawater two major chemical reactions occur: the rapid neutralization of the acidic waste by the carbonate and borate systems in

Table 8.1. Comparison of the Chemical Properties of DuPont–Edge Moor Waste and Seawater in Which It Will Be Discharged

Property	Waste[a]	Seawater[b]
Density (g ml^{-1})	1.15	1.02
Total dissolved solutes (g liter^{-1})	240	35
Total suspended solids (g liter^{-1})	1.5	0.001
pH	0.01	8.2
Alkalinity (equivalents liter^{-1})	-4.50	0.002
Chloride (*M*)	5.36	0.55
Sulfate (*M*)	0.0083	0.029
Hydrogen ion (*M*)	0.98	6×10^{-9}
Iron (*M*)	0.90	$<1 \times 10^{-8}$
Titanium (*M*)	0.063	$<1 \times 10^{-8}$
Aluminum (*M*)	0.063	$<1 \times 10^{-7}$
Manganese (*M*)	0.031	$<1 \times 10^{-8}$
Vanadium (*M*)	0.004	$<4 \times 10^{-8}$
Chromium (*M*)	0.004	$<2 \times 10^{-8}$
Lead (*M*)	0.0002	1.4×10^{-10}
Total organic carbon (mg liter^{-1})	960	1

[a]The data in this table for the waste are not internally consistent. Charge equivalents of cations and anions are not equal and the alkalinity is not fully accounted for by the hydrogen ions and iron. The pH may be inaccurate if it was measured by an instrument which does not permit pH < 0. Consequently, the hydrogen ion molarity which appears to be derived from pH may be in error. If the alkalinity is accepted as being correct the hydrogen ion concentration may be calculated by difference. The Fe contributes -2.70 equivalents liter^{-1} and other metals (Al, Mn) provide about -0.2 equivalent liter^{-1}. Thus H^+ could be about 1.6 equivalent liter^{-1} in which case the pH would be -0.2. These data based on a 10 September 1976 U.S. Environmental Protection Agency (Region III), Discharge Permit Application (Kester et al., 1981).
[b]Brewer (1975); Wilson (1975).

seawater, and the formation of hydrous ferric oxide precipitate according to the reaction

$$Fe^{3+} + (n + 2)H_2O \longrightarrow FeOOH \cdot nH_2O + 3H^+$$

where n represents the variable extent of hydration of the solid phase. Within a few hours after the dumping of the acid-iron waste, there is a rapid 10^4 to 10^5 dilution of the $FeOOH \cdot nH_2O$ in the seawater due to the turbulence created by the wake of the barge. Secondary, oceanic mixing processes are much slower and subsequent ten- to hundredfold dilution of the waste plume requires several days (Csanady, 1981).

To evaluate the effects on the ocean environment produced as this acid-iron waste mixes with seawater, one would like to track and map the acid-iron waste plume for an extended period of time after a dump. This objective requires an automated shipboard system for measuring iron in seawater with high spatial resolution and minimum time delay. This system must be able to handle samples highly concentrated in iron (10 μmoles kg^{-1}) that are observed immediately after the dump and also be able to analyze samples with iron concentrations as low as 0.02 μmole kg^{-1}, which are observed after the iron has been dispersed in the ocean for several days. In this chapter we report the methodology developed and its application to an acid-iron waste dump.

8.2. EXPERIMENTAL

8.2.1. Materials

The analytical system used to determine iron in seawater was an automated method based on a colorimetric procedure in which ferrous iron is complexed with ferrozine (Stookey, 1970); the ferrozine method has been used in the determination of iron in natural water samples (Boyle et al., 1977; Murray and Gill, 1978; Gibbs, 1979). The real-time iron measurements were performed at sea using a continuous flow analyzer (CFA) (Scientific Instruments, Pleasantville, New York) that was operated in conjunction with a submersible pump.

The difference between the refractive index of distilled deionized water (DDW) and seawater causes systematic absorbance errors when using a flow cell with curved windows (Froelich and Pilson, 1978). To minimize the absorbance error caused by the differences in the refractive indices of the seawater samples, iron standards were prepared in acidified seawater with the same salinity as the samples. The reagents used in the analytical cartridge were prepared in 0.60 molal NaCl because this medium has the same refractive index as seawater with 37‰ salinity (Horne, 1969; Wolf and Brown, 1969). The reagents used were prepared gravimetricaly (Table 8.2). The concentration of the standards and samples used in this study are expressed in micromoles of Fe per kilogram of seawater (Table 8.3).

As with other continuous flow analyses the analytical blank is incorporated into the calibration curve by using a "zero" iron standard solution to establish the

Table 8.2. Preparation of Reagents Used in the CFA Analysis of Iron

Reagent	Method of Preparation
0.60 *m* NaCl[a]	Dissolve 35.00 g NaCl (Baker Analyzed Reagent, J. T. Baker Chemical Co., Phillipsburg, New Jersey in 1 kg DDW.
Ferrozine	Dissolve 2.00 g ferrozine, 3-(2-pyridyl)-5,6-bis(4-phenyl sulfonic acid)-1,2,4,triazine monosodium monohydrate (Hach Chemical Co., Ames, Iowa) in 1 kg of 0.60 *m* NaCl.
Reducing agent $NH_2OH \cdot HCl$	Dissolve 20.00 g $NH_2OH \cdot HCl$ in 160 ml of 0.60 *m* NaCl. Purify using a bathophenanthroline extraction (Lee and Stumm, 1960). After purification acidify with 40 ml doubly distilled HCl (G. Frederick Smith Co., Columbus, Ohio). Dilute one part of this concentrated solution with 10 parts 0.60 *m* NaCl for use in the analytical cartridge.
Buffer CH_3COONH_4/NH_4OH	Dissolve 3.00 g CH_3COONH_4 (Baker Analyzed Reagent) in 50 g 0.60 *m* NaCl. Add 45.0 ml concentrated NH_4OH (Baker Analyzed Reagent) and dilute with 0.60 *m* NaCl to give a final weight of 1 kg.

[a] *m* = molal.

instrumental base line. In this work we used surface water from the Sargasso Sea or surface water from the dumpsite region, collected prior to dumping, for the "zero" iron standard. Atomic absorption spectrophotometric (AAS) analyses of these waters using a preconcentration method showed that the natural iron concentrations were $< 5 \times 10^{-9}$ mole kg^{-1} (Kester et al., 1981); thus, this seawater was considered zero Fe for the CFA analyses in the 0–10 μmoles kg^{-1} range. The

Table 8.3. Preparation of Standards Used in the CFA Analysis of Iron

Stock Solution	Method of Preparation
Fe^{2+} stock	Dissolve 0.1381 g of $Fe_2(NH_4)_2(SO_4)_2 \cdot 12H_2O$ (Baker Analyzed Reagent) in DDW which was acidified to 0.1 equivalent H^+ per kg H_2O with doubly distilled HCl prior to the addition of the iron salt. Ferrous iron standards are prepared by the appropriate dilutions of this stock solution in acidified seawater ($[H^+] = 0.12$ mole kg^{-1}).
Fe^{3+} stock	Dissolve 0.1399 g of $FeCl_3 \cdot 6H_2O$ (Mallinckrodt Analytical Reagent) in DDW which is acidified to 0.1 equivalent H^+ per kg of H_2O with doubly distilled HCl prior to the addition of the iron salt. Ferric iron standards are prepared by the appropriate dilutions of this stock in acidified seawater ($[H^+] = 0.12$ mole kg^{-1}).

detection limit was 1% of the full scale concentration or 0.2 μmole kg^{-1} in the concentration range of 0.0–2.0 μmoles kg^{-1}.

8.2.2. Techniques

Two submersible pumping systems were used for sampling. One was developed by M. Orr and C. Winget at the Woods Hole Oceanographic Institution (WHOI), Woods Hole, Massachusetts (*Mt. Mitchell* cruise, April 1979) and the other was developed by the National Oceanic and Atmospheric Administration, National Ocean Survey Engineering Development Laboratory, Rockville, Maryland (*George B. Kelez* cruise, September 1979). Both pumps were designed to minimize contamination in trace metal analysis; the interior portions of the pump in contact with seawater and all tubing used to transfer the seawater were made with Teflon. Both pumps could be towed to a depth of about 50 m making it possible to obtain vertical as well as horizontal distributions of iron in the acid-iron waste plume.

The seawater from the pump sampling system entered a predigestion chamber (Fig. 8.2), which was used to dissolve the iron precipitate. The flow of seawater into the predigestion chamber was regulated by a Teflon and glass flowmeter. The seawater was acidified to a pH of 1.0 with concentrated HCl (Baker Analyzed Reagent) which was added continuously with a syringe pump. The pH of the effluent from the chamber was continually monitored with an in-line electrode connected to a digital pH meter. The pH of the solution in the digestion chamber was maintained by varying either the flow rate of seawater or the flow rate of acid from the syringe pump. The residence time of seawater in the predigestion chamber was about 2 min. The predigestion chamber was continuously ultrasonicated in a

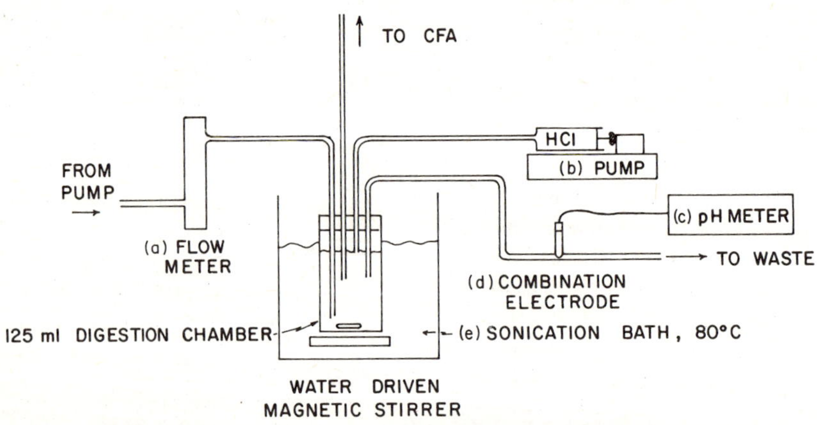

Figure 8.2. Schematic predigestion chamber used to dissolve particulate iron: (a) Flowmeter, Gilmont Instrument Co., Great Neck, New York; (b) Sage syringe pump, Orion Research Corp., Cambridge, Massachusetts; (c) Corning model 112 pH meter, Corning Glass Works, Corning, New York; (d) Combination electrode, Broadly-James Corporation, Santa Anna, California; and (e) Ultrasonic cleaning bath, Branson Ultrasonic Cleaning Equipment Co., Shelton, Connecticut.

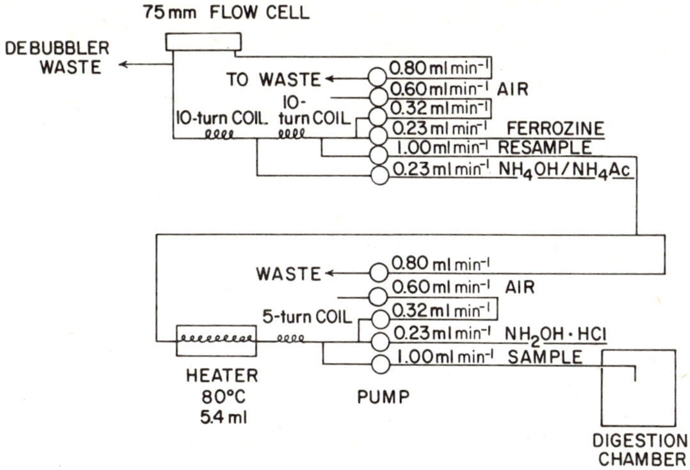

Figure 8.3. Schematic diagram of the analytical cartridge used in the continuous flow analyzer.

ultrasonic cleaning bath heated to 50°C to promote dissolution of the iron waste particles.

Iron colorimetric analyses were performed on the CFA at a wavelength of 560 nm using a 75 mm flow cell. A schematic diagram of the analytical cartridge used in this study is shown in Fig. 8.3. The AAS analyses were made on a Perkin-Elmer Model 5000 spectrophotometer (Perkin-Elmer Corp., Norwalk, Conneticut) using an air–acetylene flame by the method of additions.

8.3. RESULTS AND DISCUSSION

8.3.1. Laboratory Testing of the Analytical Cartridge

The sensitivity range of the CFA varied from a lower sensitivity of 0–2 μmoles kg^{-1} to a higher sensitivity of 0–10 μmoles kg^{-1}. The precision of the analysis was determined from five replicate analyses of a sample on a high and low sensitivity range (Table 8.4). The same reagents were used on both sensitivity ranges; the concentration of the ferrozine reagent was adjusted to provide complete complexation of the most concentrated sample. Because ferrozine will form a colored complex only with the Fe^{2+} oxidation state of iron, it was necessary to establish that there was complete reduction of Fe^{3+} to Fe^{2+} in the analytical cartridge. Standard solutions prepared from a $FeCl_3 \cdot 6H_2O$ stock solution were analyzed using a calibration curve obtained from $Fe_2(NH_4)_2(SO_4)_2 \cdot 6H_2O$ solutions. The reduction of Fe^{3+} to Fe^{2+} in the analytical cartridge was greater than 95% for several different sensitivity ranges (Table 8.4).

Statistical analyses of the calibration curves for the high and low sensitivity scales are given in Table 8.5. Excellent linearity was observed in both ranges.

Table 8.4. Summary of the Precision of the CFA Analysis and the Percentage Reduction of Fe^{3+} to Fe^{2+} for Various Sensitivity Scales[a]

Sensitivity Scale (μmoles kg^{-1})	Reduction (%)	Precision[a] (μmole kg^{-1})
0–2	95.8	0.99 ± 0.01
0–4	97.0	–
0–6	96.4	–
0–8	96.7	–
0–9	–	4.53 ± 0.03

[a]The ± are standard deviations for $n = 5$ replicates.

To determine the ability of the analytical cartridge to completely recover the acid-iron waste from seawater, a comparative analysis of laboratory acid-iron waste mixtures was made using the ferrozine colorimetric technique and an atomic absorption analysis for iron. Four different acid-iron waste samples previously discharged at DWD-106 were analyzed: a waste dumped in July 1977, a waste dumped in April 1979, a waste dumped in September 1979, and a waste dumped in August 1980. Each acid-iron waste was initially diluted to give a final *T*(Fe) concentration of approximately 4×10^{-4} mole kg^{-1} in DDW which was acidified to $[H^+] = 0.1$ mole kg^{-1} with HCl (Baker Instra-Analyzed) prior to the addition of the waste. The waste samples analyzed on the CFA were prepared by adding 500 μliter of the waste-DDW mixture to 100 g acidified seawater, $[H^+] = 0.12$ mole kg^{-1}. The waste samples analyzed by AAS were prepared by adding 5 ml of the waste-DDW mixture to 100 ml 0.12 M HCl. The results of these comparative analyses are summarized in Table 8.6. The agreement between the AAS and the CFA-ferrozine method for the four different waste samples is within the experimental error of the analysis. The iron concentration of these wastes varies considerably among different waste samples and these particular wastes were less concentrated than the composition reported in Table 8.1. Variation in the composition among different batches of waste is an important consideration when determining a mass balance of the dispersion of the acid-iron waste in a particular field observation.

The analytical method used in this work approximates a total iron, *T*(Fe), measurement in that Fe^{3+} and Fe^{2+} are determined and the sample stream is

Table 8.5. Statistical Analysis by a Least Squares Linear Regression of the Calibration Curve for the 0–2 μmoles kg^{-1} and 0–8 μmoles kg^{-1} Range[a]

Sensitivity Scale	0–2 μmoles kg^{-1}	0–8 μmoles kg^{-1}
Slope	45.6 ± 0.2	11.3 ± 0.1
Intercept	0.72 ± 0.31	0.37 ± 0.30
Correlation coefficient	1.00	1.00

[a]Five points were used for each calibration; the ± values are one standard deviation.

Table 8.6. Comparison of Iron Analyses using CFA-Ferrozine Method and Flame AAS for Four DuPont–Edge Moor Acid-Iron Waste Samples[a]

	Fe (moles liter^{-1} in waste)	
Waste Batch	CFA	AAS
July 1977	0.814 ± 0.009	0.844 ± 0.026
April 1979	0.222 ± 0.002	0.237 ± 0.008
September 1979	0.785 ± 0.008	0.796 ± 0.010
August 1980	0.039 ± 0.001	0.043 ± 0.003

[a]The ± values are one standard deviation based on $n = 3$ analyses.

"digested" without filtration, by acidifying to pH = 1, ultrasonicating, and heating to 50°C for 2 min. This procedure provides essentially total recovery of iron added to seawater by the acid-iron waste. It probably does not completely liberate all the Fe bound in detrital mineral phases that may be suspended in ocean waters. The concentrations of Fe which can be determined by this method (0.02 μmole kg^{-1}) are at least an order of magnitude greater than the concentrations of natural particulate iron (Betzer and Pilson, 1970), and therefore it is reasonable to consider the values as "total iron" for the concentrations found in the acid-iron waste plume.

8.3.2. Shipboard Analyses

The CFA-ferrozine method was used at sea for real-time iron measurements of two DuPont–Edge Moor acid-iron waste dumps (April 1979 on the *Mt. Mitchell* and September 1979 on the *George B. Kelez*). The iron measurements were performed continuously on line with the shipboard pumping system in order to provide high resolution profiles of the horizontal and vertical distribution of iron in the water column and its variation with time.

Changes in the horizontal distribution of the waste plume with time are illustrated in the iron profiles obtained from four successive crossings of two acid-iron waste plumes that were initially discharged into the ocean within 2 km of each other (Fig. 8.4). The maximum iron concentration observed in the waste plume at a depth of 6 m was 3 μmole kg^{-1} with a minimal dispersion of the waste plumes observed in the 4 h sampling time. The two legs of the waste plume drifted closer together during each successive plume crossing and 4½ h after the dump they appeared to coalesce into one continuous plume of acid-iron waste. These crossings of the plume show the value of real-time continuous measurements. A complete distribution of the horizontal dispersion of iron in the waste plume was obtained in each crossing of the plume. Furthermore, changes can be observed in the horizontal distribution of the plume with time.

Shipboard analyses of discrete samples were used to construct a vertical profile of iron in the waste plume. These samples were obtained from a conductivity-temperature-depth cast using a General Oceanics rosette sampler (General Oceanics,

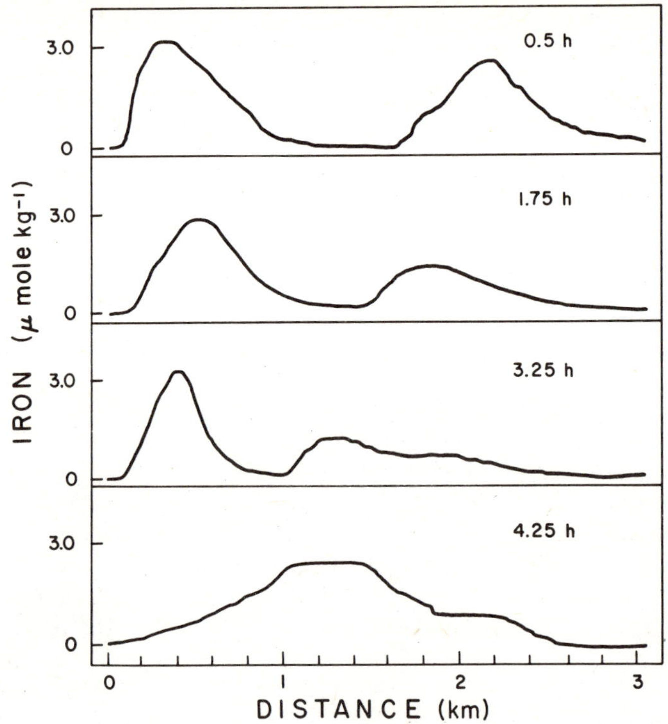

Figure 8.4. Horizontal distribution of iron (continuous, real time measurements) in four successive crossings of two adjacent Dupont–Edge Moor acid-iron waste plumes at 6 m.

Inc., Miami, Florida) with 8-liter Niskin bottles. The samples were obtained 8 h after the dump when the research vessel was steaming in the waste plume. The vertical distribution of iron with depth and the variation of temperature with depth are shown in Fig. 8.5. The iron in the waste plume was not observed below the seasonal thermocline at 25 m. Above the thermocline we observed a gradual decrease in the iron concentration with depth.

The vertical and horizontal profiles illustrate iron gradients that are observed during the first 12 h after the dump. As the waste plume dispersed in the ocean, the concentration of iron in the plume decreased and the width of the plume increased. To detect these low levels of iron a higher sensitivity range of the CFA was used making it possible to track the acid-iron waste plume for over 53 h after a dump. This was the first time a waste plume had been monitored for such an extended period of time. The WHOI pump used in this work was not designed for vertical profiling; however, by stopping the ship and allowing the pump to fall through the water column a vertical profile was obtained from 3 to 30 m. Three successive falls of the pump through the water column are shown in Fig. 8.6. As with the previous data the acid-iron waste was not observed below the seasonal thermocline at 28 m after 53 h dispersion.

The ability to track continuously an acid-iron waste plume in real-time will lead

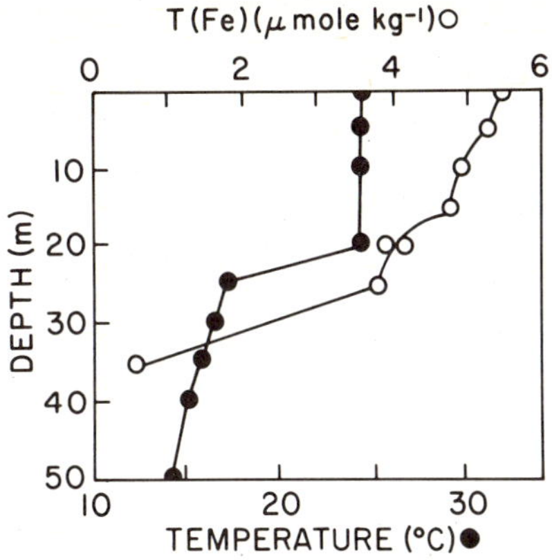

Figure 8.5. Vertical profiles of iron and temperature in the DuPont–Edge Moor acid-iron waste plume 8 h after the dump.

to a better understanding of the oceanic processes that control dispersion of wastes in the ocean and to determine the ultimate fate of the acid-iron waste in the ocean. In previous studies designed to define the horizontal and vertical gradients of trace metals in a waste plume (Kester et al., 1981) and to determine the horizontal dispersion of a waste plume with time after the dump (see Mukherji and Kester, Chapter 7), a limiting factor has been that the data analyses were made on discrete samples obtained using conventional bottle sampling techniques. Discrete samples

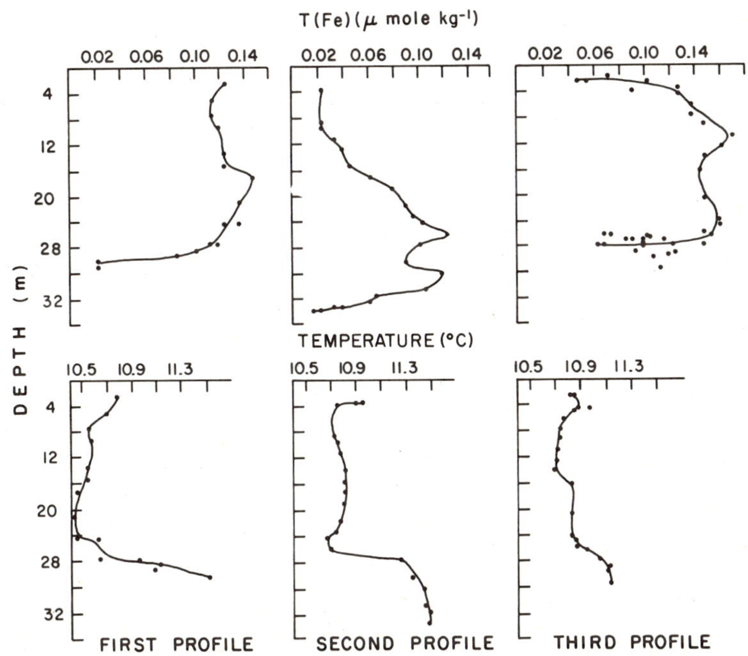

Figure 8.6. Vertical profiles of iron and temperature in the acid-iron waste plume 53 h after the dump obtained during three consecutive vertical falls of the pump through the water column.

were drawn with uncertainty of the position of the research vessel relative to the most concentrated region of the waste plume, a criterion necessary for correct interpretation of the data (Okubo, 1971). Continuous real-time on-line analysis of the iron concentration in the waste plume made in conjunction with a shipboard pumping system allows samples for chemical and biological measurements to be taken when the exact position of the research vessel relative to the waste plume is known. In addition to improved sampling, continuous real-time on-line iron analyses will provide a more complete understanding of the vertical and horizontal distribution of iron in the water column after an acid-iron waste dump.

ACKNOWLEDGMENTS

We wish to acknowledge Gerald Epstein who performed the atomic absorption analyses. We are also grateful to Thomas P. O'Connor for his help with the shipboard sampling. This work was supported by National Oceanic and Atmospheric Administration grant NA79AA-D-0003.

REFERENCES

Betzer, P. R., and M. E. Q. Pilson. 1970. Concentration of particulate iron in Atlantic open-ocean water. *Journal of Marine Research*, **28**, 251–267.

Boyle, E. A., J. M. Edmond, and E. R. Sholkovitz. 1977. The mechanism of iron removal in estuaries. *Geochimica et Cosmochimica Acta*, **41**, 1313–1324.

Brewer, P. G. 1975. Minor elements in sea water. *In*: Chemical Oceanography, Vol. 1, 2nd ed., J. P. Riley and G. Skirrow (Eds.). Academic Press, London, pp. 415–496.

Csanady, G. T. 1981. An analysis of dumpsite diffusion experiments. *In*: Ocean Dumping of Industrial Wastes, B. H. Ketchum, D. R. Kester, and P. K. Park (Eds). Plenum Press, New York, pp. 109–129.

Froelich, P. N., and M. E. Q. Pilson, 1978. Systematic absorbance errors with Technicon Autoanalyzer II colorimeters. *Water Research*, **12**, 599–603.

Gibbs, M. M. 1979. A simple method for the rapid determination of iron in natural waters. *Water Research*, **13**, 295–297.

Gross, M. G. 1976. Waste disposal. MESA New York Bight Atlas Monograph 26, New York Sea Grant Institute, Albany New York, 32 pp.

Horne, R. A. 1969. Marine Chemistry. Wiley-Interscience, New York, 568 pp.

Kester, D. R., R. C. Hittinger, and P. Mukherji. 1981. Transition and heavy metals associated with acid-iron waste disposal at Deep Water Dumpsite 106. *In*: Ocean Dumping of Industrial Wastes, B. H. Ketchum, D. R. Kester, and P. K. Park (Eds.). Plenum Press, New York, pp. 215–232.

Lee, G. F., and W. Stumm. 1960. Determination of ferrous iron in the presence of ferric iron with bathophenanthroline. *Journal of the American Water Works Association*, **52**, 1567.

Murray, J. W., and G. Gill. 1978. The geochemistry of iron in Puget Sound. *Geochimica Cosmochimica Acta*, **42**, 9–19.

Okubo, A. 1971. Oceanic diffusion diagrams. *Deep-Sea Research*, **18**, 789–802.

Pesch, G., B. Reynolds, and P. Rogerson. 1977. Trace metals in scallops from within and around two ocean disposal sites. *Marine Pollution Bulletin*, **8**, 224–228.

Stookey, L. L. 1970. Ferrozine: A new spectrophotometric reagent for iron. *Analytical Chemistry*, **42**, 779–781.

Vaccaro, R. F., G. D. Grice, G. T. Rowe, and P. H. Wiebe. 1972. Acid-iron waste disposal and the summer distribution of standing crops in the New York Bight. *Water Research*, **6**, 231–256.

Wilson, T. R. S. 1975. Salinity and the major elements of sea water. *In*: Chemical Oceanography, Vol. 1, 2nd ed., J. P. Riley and G. Skirrow (Eds.). Academic Press, London, pp. 365–413.

Wolf, A. V., and M. G. Brown. 1969. Concentrative properties of aqueous solutions: Conversion tables. *In*: Handbook of Chemistry and Physics, 50th ed. R. C. Weast (Eds.). Chemical Rubber Co., Cleveland, Ohio, pp. D171–D210.

9

VOLATILE ORGANIC WASTES AT THE PUERTO RICO DUMPSITE

James M. Brooks, Denis A. Wiesenburg, Guy Bodennec, and Theodor C. Sauer, Jr.

Department of Oceanography
Texas A&M University
College Station, Texas

Dr. Wiesenburg's present address: Naval Ocean Research and Development Activity, NSTL Station, Mississippi 39529. Dr. Bodennec's present address: Centre National Pour L'Exploitation Des Oceans, Boite Postale 337, 29243 Brest Cedex, France. Dr. Sauer's present address: Exxon Production Research Company, P.O. Box 2189, Houston, Texas 77001.

ABSTRACT

Waste materials from pharmaceutical production are discharged into the ocean at a dumpsite which is located approximately 74 km north of Arecibo, Puerto Rico, overlying 6000–8000 m of water. Volatile organic material constitutes the major nonaqueous fraction of the approximately 3.0×10^8 kg y^{-1} (wet wt) discharged into the surface waters at this site. Three samples of composite waste discharged at different times indicate significant changes in the waste composition over a two-year period (1978–1980). Waste samples obtained in February 1978 consisted mainly of dichloromethane, butanol, toluene, and dimethylaniline. These compounds were measured in the discharged waste plume and were also found in the dumpsite area three days after dumping. Dimethylaniline was observed in background measurements at the dumpsite at a concentration of approximately 100 ng $liter^{-1}$. Background concentrations of dichloroethane, benzene, chloroform, toluene, ethylbenzene, and xylene at the dumpsite area were between 5 and 30 ng $liter^{-1}$. These concentrations are six orders of magnitude less than the 1–10 mg $liter^{-1}$ observed in the plume sampled 30 min after the waste was dumped.

Composite waste samples taken in October 1979 indicated significant compositional differences from the February 1978 samples. The majority of the volatiles in the 1979 waste consisted of ketones and alkyl substituted one-ring aromatic compounds (toluene, xylenes, C_3-benzenes, and C_4-benzenes) with no significant amounts of alcohols or substituted anilines being observed. The volatile hydrocarbon "fingerprint" in the waste plume, in the surface waters north of Puerto Rico, and in the 1979 waste consisted mainly of these alkyl benzenes. A volatile organic "fingerprint" from the discharged waste was observed over a 23,000 km^2 area north of Puerto Rico at concentrations ranging from 0.1 to 8 μg $liter^{-1}$. The observed "fingerprint" suggested that volatile organics from previous dumps still remained in the area. Samples taken from depths down to 200 m indicated that the dumpsite wastes had penetrated the thermocline.

9.1. INTRODUCTION

Volatile organic compounds are those compounds that exhibit a significant vapor pressure at 25°C and one atmosphere total pressure. These compounds generally range in boiling point up to 254°C (*n*-tetradecane). With high vapor pressures and low boiling points, volatile organic compounds are capable of being purged from a heated seawater sample by a flowing gas stream. When discharged at sea, volatile organics are continually lost to the atmosphere, as turbulent mixing brings them into contact with the atmosphere at the air–sea interface. The rate of loss to the atmosphere depends on the degree of turbulent mixing, sea surface condition, and the specific compound solubility, which in turn is dependent on the seawater temperature and salinity as well as on the hydrophilic or hydrophobic nature of the compound.

Concentrations of most volatile organics are well below the detection limit in the open ocean (10^{-9} g liter^{-1} of seawater; Sauer et al., 1978; Schwarzenbach et al., 1978; Sauer, 1980). For this reason, volatile organics are very useful in tracking oceanic, organic, industrial waste discharges. These industrial discharges often contain many volatile organic compounds used as industrial solvents or chemical intermediates. At the Puerto Rico deepwater industrial waste dumpsite (hereafter called the Puerto Rico dumpsite, Fig. 9.1), industrial wastes from several chemical plants, which primarily produce pharmaceuticals, are dumped into the ocean. The nonaqueous portion of this waste is composed mainly of volatile organics.

Determinations of volatile organics in the surface waters were made at the Puerto Rico dumpsite at two different times. The first sampling was during February 1978 aboard the *R/V Knorr*. At this time, volatile organics were sampled at a few stations in the dumpsite area unrelated to any barge dumping, and an effort was made to obtain a series of samples in the barge discharge plume at various times after the dump. Samples for volatile organic determinations were transported to the laboratory for analysis. Shipboard analyses of hydrographic, chemical, and biological samples were also made for characterization of the dumpsite. High concentrations of volatile organics found in the waste, in the barge discharge plume, and in background samples were the most significant finding from this initial investigation (Schwab et al., 1981). Comparisons of waste concentration with

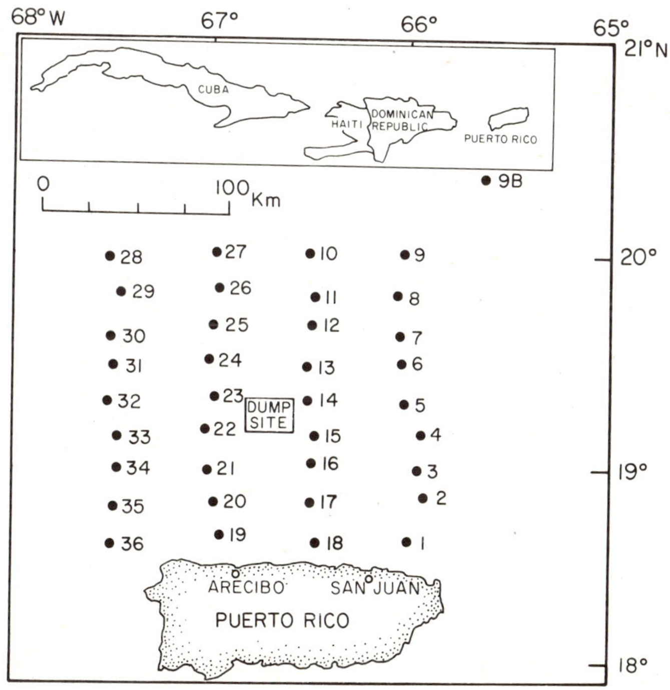

Figure 9.1. Location of the Puerto Rico industrial waste dumpsite. Numbers near dots represent stations sampled during the October 1979 cruise.

waste marker dye measurements allowed calculations of partitioning of volatile organics to the atmosphere.

Our second sampling of the Puerto Rico dumpsite was in October 1979, aboard the National Oceanic and Atmospheric Administration (NOAA) ship *Mt. Mitchell.* This cruise was undertaken to confirm our earlier findings (February 1978), and those of Dr. G. Harvey (personal communications) at NOAA, from two sets of samples taken in October 1978 and May 1979. The earlier findings indicated that detectable volatile organic waste extended over a wide geographic area away from the dumpsite. High concentrations of several volatile organic compounds were also found below the surface mixed layer. One goal of our October 1979 cruise was to use the waste volatile organic components as a means of tracking waste dispersion and to verify possible penetration of waste through the thermocline. A secondary objective was the determination of possible temporal changes in the waste composition between our two samplings. To accomplish these goals, a 36-station grid covering approximately 23,000 km^2 north of Puerto Rico was sampled for volatile organics (Fig. 9.1). Samples were taken from a depth of 10 m. A vertical profile of samples for volatile organic analysis was also taken in association with one barge dumping at the site. Volatile organic measurements were made both aboard ship and in the laboratory, where gas chromatography–mass spectrometry (GC–MS) was used to identify the major compounds in the barge wastes. This chapter reports the findings from our grid study of the dumpsite and compares the different characteristics of the dumped material from the two sampling periods.

9.2. THE DUMPSITE

The Puerto Rico dumpsite was established in 1972 as an interim disposal location for wastes resulting primarily from pharmaceutical production. The dumpsite overlies the Puerto Rico Trench, which has a water depth of 6000–8000 m (Fig. 9.1). The site is 74 km north of Arecibo, Puerto Rico (between 19°10′ and 19°20′N, and 66°35′ and 66°50′W) and occupies an area of approximately 500 km^2. The material dumped there is a variable combination of batch wastes from several pharmaceutical production facilities. The combined wastes are collected for six or seven days in a large on-shore storage tank. On two or three consecutive days each week, the combined wastes (referred to hereafter as the composite waste) are transferred to a special dumping barge for ocean discharge. The composite wastes are dumped into the surface waters in 2.5×10^3 m^3 lots over a distance of about 30 km. Approximately 3.0×10^8 kg y^{-1} (wet wt) are dumped into the surface waters at this site.

Figure 9.2 shows a vertical profile of temperature, salinity, and density in the dumpsite area. The surface mixed layer in October 1979 extended down to about 30 m, where increasing salinity and decreasing temperature result in a seasonal pycnocline. This pycnocline could inhibit transport of materials from the surface mixed layer to depth. Thus, it has been suggested that dumped material may not penetrate the thermocline, but rather is transported from the site and mixed by

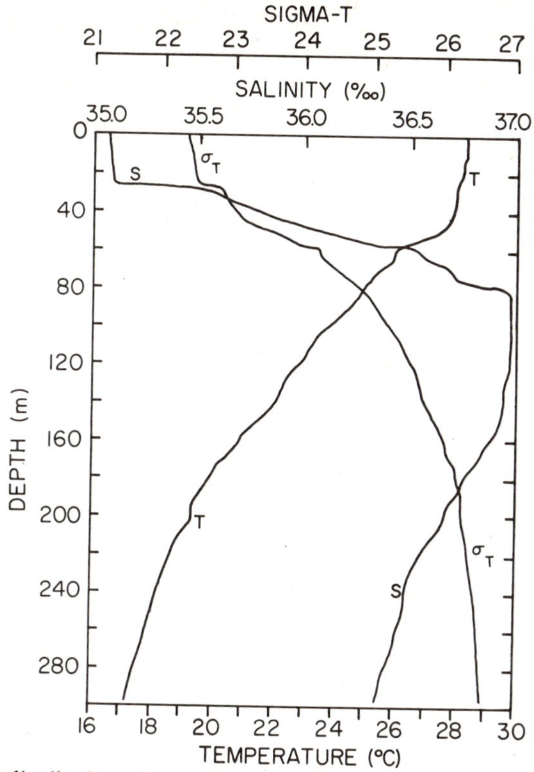

Figure 9.2. Depth distribution of temperature, salinity, and σ_t at the Puerto Rico dumpsite during October 1979. Data were compiled from a STD profile obtained at grid station 14 (see Fig. 9.1). Total water depth below this station was 7050 m.

rapidly moving surface currents. The Antilles Current moves through the site in a westerly direction (Wust, 1964). However, current drogues and dye dispersion studies indicate that the surface current pattern is not a consistent westward flow, but is variable in the dumpsite area (T. Ichiye, personal communication). A variable, rather than directional, current regime would cause reduced dispersion of discharge waste. The dilution process would be less effective in this situation, and enhanced concentrations of dumped material would be expected. Volatile organics are a major component of the waste, but they are found in very low concentrations (order of ng liter^{-1}) in open ocean seawater. Thus, analysis of volatile organic compounds in the surface waters of the dumpsite area provides an effective technique for examining waste dispersion.

9.3. METHODS

9.3.1. Sampling

Seawater samples for measurement of salinity, temperature, and various chemical and biological parameters were collected on the 1978 cruise with a 12-bottle

Niskin [polyvinyl chloride (PVC)] rosette sampler (General Oceanics, Inc., Miami, Florida), fitted with a Plessey 9040 STD (Grundy Environmental Systems, Inc., San Diego) for measurements of salinity, temperature, and depth. Volatile organic water samples were collected on the cruise with glass-lined Niskin bottles attached to the rosette sampler. Volatile organic samples were drawn under pressure (purified N_2) from the bottles through copper tubing into volatile-free (combusted) 2-liter glass bottles. Sodium azide was added to prevent biodegradation of the organic material, and the stoppered bottles were stored at ~4°C for later analysis. Water samples for analysis of the rhodamine-B marker dye were taken from both glass-lined and PVC Niskin bottles. Water samples were taken at depths of 1,5, and 10 m in the waste plume during the 1978 sampling cruise.

During the 1979, 36-station grid study (Fig. 9.1), samples for volatile organic analysis were taken with a specially designed sampling device. This device collects the seawater sample in the same bottle that is used for sample stripping and analysis. The bottles used were glass, hydrocarbon-free (combusted) 2-liter bottles. The closed sample bottles were lowered into the water in a stainless steel cage suspended by a nylon line. The Teflon bottle top was connected via a second nylon line to the deck of the ship. When the sample bottle was in place at depth (down to 10 m), the top was opened from the deck, the bottle filled and returned to the deck rapidly. Sodium azide was added as a preservative and the sample bottle was sealed and refrigerated in the dark until analysis, either aboard ship or in our shore-based laboratory at Texas A&M University. Sauer et al. (1978) conducted several comparisons between standard sample bottles and this glass bottle sampling technique. The glass bottle sampling technique is a proven, contamination-free method.

Individual waste samples from the different companies dumping at the site were collected in May 1979 and February 1980 and analyzed. These samples were shipped in glass containers at room temperature. Composite samples were taken from the fill pipe leading to the barge, corresponding to the dumps on 5 and 8 February 1978, and on 7 October 1979. Two additional composite samples were taken 28 February 1980. These composite samples were stored in dark glass, air-tight bottles for transport to the laboratory. Concentration measurements were made on pure waste samples using gas chromatography, and compound identifications were made by mass spectroscopy.

9.3.2. Volatile Organic Analysis

Volatile organics were extracted from the dumpsite seawater samples using the dynamic headspace sampling technique developed by Sauer (1978). This method is a modification of the technique presented by Bertsch et al. (1975) and May et al. (1975). The volatile organics were stripped from a heated (70°C), stirred, 2-liter seawater sample by purging with ultrapure helium using a coarse frit. Volatile organics were trapped on a 0.5 cm i.d. × 15 cm column of solid polyphenyl ether absorbent (Tenax-GC, Enka Research Institute, Arnhem, Netherlands), which has excellent retention characteristics for liquid hydrocarbons, retains very little water and exhibits negligible column bleed at temperatures under 300°C (Butler and

Burke, 1976). The all-glass headspace sampling apparatus is designed so that the stripping of the volatiles occurs in the sample bottle itself. Contamination from the atmosphere and loss of volatiles due to adsorption on sample bottle walls are thus minimized. Samples were stripped with helium at 70°C for 90 min at 120 ml min^{-1}, a time sufficient for complete stripping of most volatile organics.

After stripping of the volatile organics from the water sample, the volatile components trapped on the Tenax-GC were desorbed by heating (250°C), using an on-line heating unit. They were transferred by helium flow to a liquid nitrogen-cooled trap for sample consolidation (Fig. 9.3). This trap is attached to a six-port sample valve that can be isolated and then heated in order to transfer the trapped hydrocarbons (using the GC carrier stream) as a plug onto a gas chromatographic column for sample separation and analysis. The separating column for the 1978 cruise samples was a 0.3 cm X 460 cm copper chromatographic column packed with 10% SP-2100 on 80/100 Supelcoport (Supelco, Inc., Bellefonte, Pennsylvania). The column was temperature programmed at 0°C for 2 min, 0° to 180°C at 4°C min^{-1}, and 180°C for 16 min. For the 1979 samples, the separating column was a 0.3 cm X 305 cm stainless steel chromatographic column packed with 10% SP-2100 on 80/100 Supelcoport, temperature programmed from 50° to 180°C.

The eluted organics from the dumpsite samples were analyzed with a Hewlett-Packard (HP) 5830A gas chromatograph (Hewlett-Packard Co., Avondale, Pennsylvania) using a flame ionization detector. Initial sample identification and quantitative measurements were accomplished by comparing retention times and peak areas with standard compounds. Standards were injected in prepurged water samples which were then processed in an identical manner to the seawater samples.

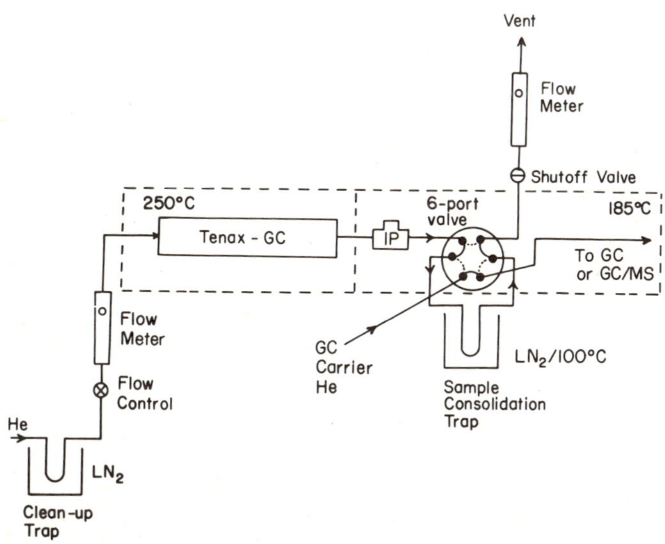

Figure 9.3. Desorption system designed to remove adsorbed volatile organics from the Tenax-GC, consolidate them on a liquid nitrogen trap, and inject them as a plug onto a GC column for separation and analysis.

Barge waste samples were analyzed using an automated purge-and-trap system (HP 7675A) similar to the dynamic headspace analysis technique described above. A 0.2-ml sample of barge waste (individual or composite) was added to 5 ml of prepurged seawater for analysis. The sample was stripped for 12 min, and the volatiles were trapped on a 60/80 mesh Tenax-GC column (0.64 cm o.d. × 10 cm) at room temperature (~25°C). The trapped volatile organics on the Tenax-GC column were desorbed by heating to 200°C and injected directly onto the chromatographic column for separation and analysis. A 12-min stripping time will not remove all the volatile organics from the 5.2-ml samples, so two purges were run on each waste sample in order to determine the actual component concentrations.

9.3.3. Concentration Calculations

The solubility of many volatile organics in seawater depends to a large extent on the hydrophobic or hydrophilic nature of the various compounds. Because of large solubility differences, the partition coefficients between seawater and headspace vary for each volatile organic compound. The partition coefficients for the dynamic headspace stripping technique need to be known in order to calculate concentrations for specific compounds.

Partition coefficients for various volatile organics were determined by repeated stripping of samples. Some organics, such as toluene and xylene, are completely removed in one purging (90 min for seawater, 12 min for wastes). Other compounds (for example, acetone) remain even after several purgings. Individual partition coefficients were calculated for each volatile organic compound using the relation

$$K_i = 1 - (X_2/X_1)_i \qquad 1$$

where K_i is the partition coefficient for a given compound and $(X_2/X_1)_i$ is the ratio of integrated peak areas generated from the first (X_1) and second (X_2) purgings. Total integrated peak area (T_i) of a given compound in a sample can then be calculated from

$$T_i = (X_1/K_i) \qquad 2$$

K_i should be constant for any compound purged under identical conditions. Thus, second purges need not be run on every sample; they were run on only enough seawater samples to establish partition coefficients for the compounds of interest. All wastes samples were purged at least twice. Standard compounds were either injected directly into the GC for analysis or injected into prepurged seawater and analyzed like samples. Concentrations of each compound can be calculated from the direct injection standards from the equation

$$C_i = (C_{STD}/X_{STD}) \times T_i \qquad 3$$

where C_i is the concentration of component i in ng liter^{-1}, C_{STD} is the concentration of standard, and X_{STD} is the integrated peak area of the standard component. Compound concentrations for which we had no standards were calculated from equation 3 using carbon compounds with the same chain length to estimate the X_{STD} and C_{STD} values.

The reliability of our calculation procedure was verified by stripping and trapping a waste sample ten times, until most major volatiles were completely removed. The sum of all peak areas from the ten purges for each compound was within 10% of the response calculated from equation 2. The effects of particulates and other organics in the aqueous matrix on K_i were not considered.

9.3.4. Identification of Compounds by GC–MS

The purge-and-trap techniques used for compound analysis by gas chromatography were also used for component identification using GC-MS. The February 1978 composite waste and one seawater sample were analyzed by GC-MS using an HP 5980A dodecapole mass spectrometer. This instrument was interfaced to an HP 5710A GC with a single stage jet separator and supported by a HP 5933A Data System. The techniques and procedures for the 1978 GC–MS analysis were described by Schwab et al. (1981).

Individual and composite waste samples from October 1979 and February 1979 cruises were subjected to component identification using a HP 5992 GC–MS with jet separator. For the waste samples, the HP 7675A automated purge-and-trap system was used to deliver the volatile organics to the GC–MS, using the same operating conditions as described above. Mass spectra were recorded at a rate of one per 2.25 s from 33 to 250 atomic mass units (amu) with an electron ionization source of 1600 volts. Each spectrum was recorded using a HP 9885 M/S flexible disk data system. This system allows playback of the complete total ion chromatogram and plotting of individual ion abundances for the run. Mass spectra identifications were assisted by Heller and Milne (1978) and McLafferty (1973). Many of the component identifications were confirmed by GC retention times.

9.4. RESULTS AND DISCUSSION

9.4.1. Waste Characterization

The composite waste discharged at the Puerto Rico dumpsite is a variable mixture of the seven individual chemical wastes delivered to the onshore storage tank in Arecibo, Puerto Rico (Fig. 9.1). Composition of the waste depends, therefore, not only on the relative proportion of the contribution of each company to the total waste, but also on periodic changes in waste composition from individual plants. Our data show great variability, not only in the compounds in the composite waste, but also in individual chemical companies' contributions.

Samples of composite waste were obtained from the barge-loading facility for both the 5 February and 8 February 1978 barge dumps, which corresponded to our sampling cruise aboard the *R/V Knorr*. Waste samples for the seven contributing chemical companies were also obtained at this time. Another composite waste sample was obtained from the barge before an October 1979 dumping, correspond-

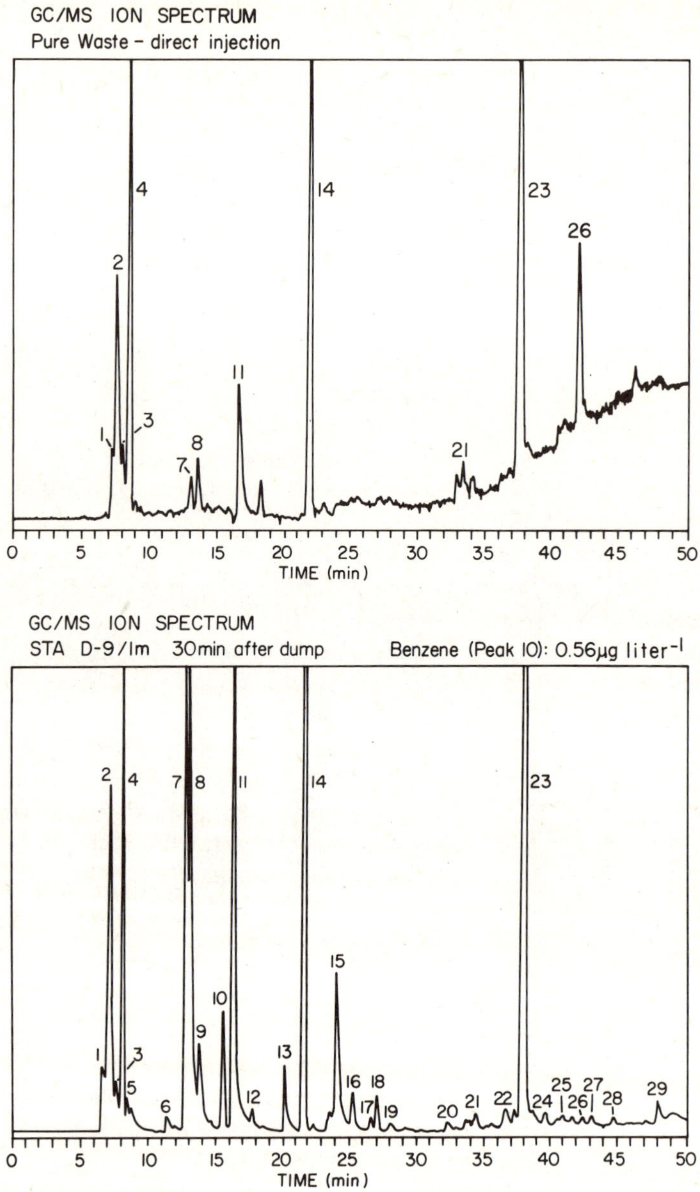

Figure 9.4. GC-MS spectrum of barge waste material taken 8 February 1978 and a surface sample at station DS/9 (9 February 1978) taken 30 min after dump (peak numbers correspond to components in Table 9.1).

ing to our cruise aboard the *Mt. Mitchell*. A composite waste and seven individual company waste samples were again obtained for analyses in February 1980.

Figure 9.4 shows the GC-MS total ion spectrum of a composite waste sample obtained from the barge which dumped on 9 February 1978 and the GC-MS ion spectrum of a surface sample 30 min after dumping. Fewer than 15 compounds (Table 9.1) comprised over 98% of the volatile organics in the pure waste. Dichloromethane, *n*-butanol, toluene, dimethylaniline, naphthalene, and an unknown alcohol were the major components of the waste. A sample (not shown) taken three days earlier on 5 February 1978 showed a similar compound distribution, but the concentration of dimethylaniline was less the 25% of that in the 8 February sample.

Figure 9.5 shows a gas chromatogram of volatile organics in the composite waste obtained on 7 October 1979. This chromatogram indicates that the volatile

Table 9.1. Volatile Organic Peak Identifications from February 1978 Puerto Rico Dumpsite Cruise

Peak Number	Compound
1	Acetone
2	Many possible alcohols and ketones
3	Ethylmercaptan
4	Dichloromethane
5	*n*-Propanol
6	Unknown
7	Chloroform
8	Ethylacetate
9	Tetrahydrofuran
10	Benzene
11	*n*-Butanol
12	Trichloroethylene
13	Methylisobutylketone
14	Toluene
15	*n*-Butyl or isobutylacetate
16	Column bleed
17	Ethylbenzene
18	*m,p*-Xylene
19	*o*-Xylene
20	C_3-Benzene
21	Column bleed
22	Dimethylpyridine
23	Dimethylaniline
24	C_4-Benzene
25	Alkenyl (C_4) benzene
26	Naphthalene
27	C_5-Benzene
28	Ethylcumene
29	Unknown

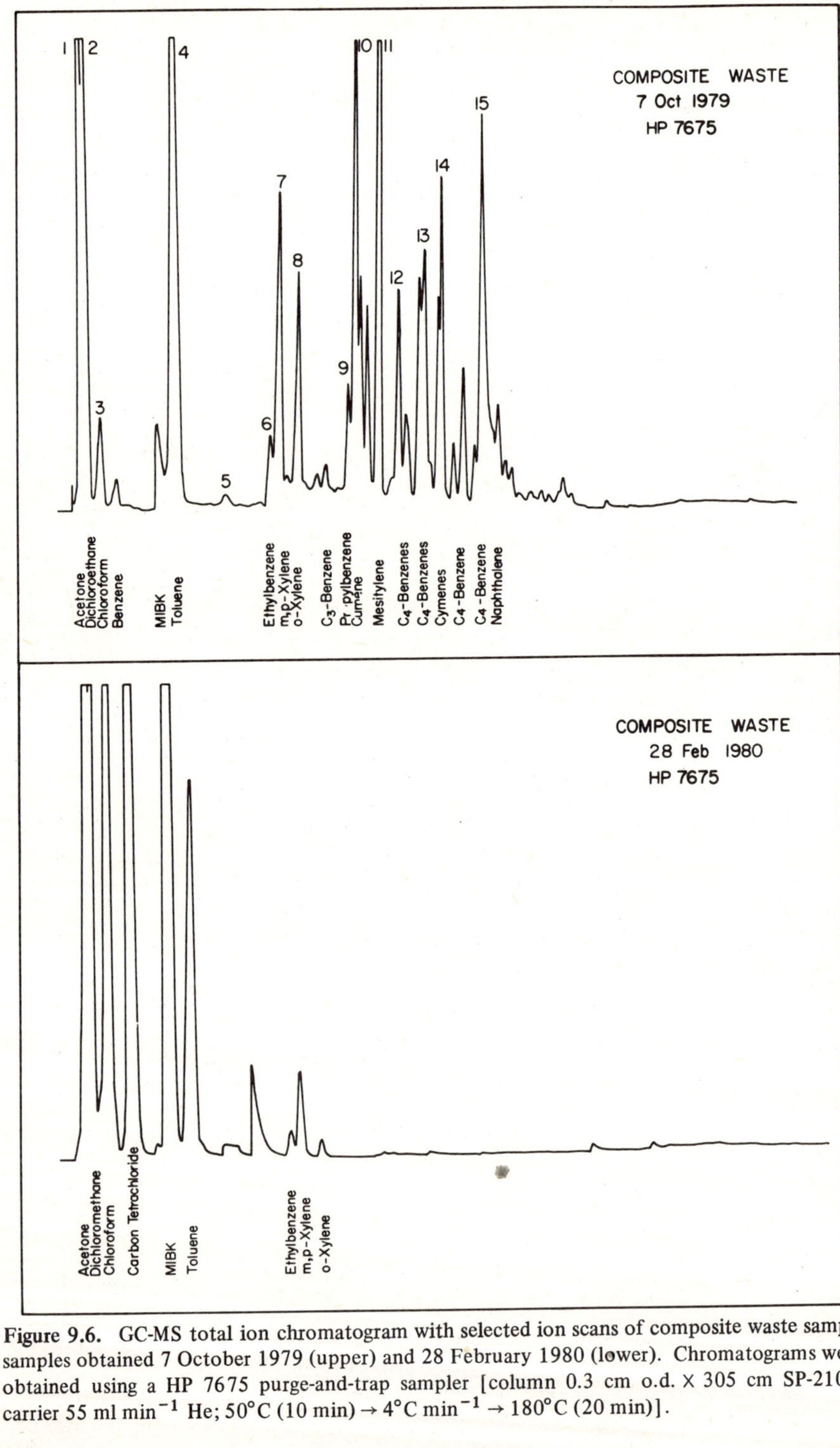

Figure 9.6. GC-MS total ion chromatogram with selected ion scans of composite waste sample samples obtained 7 October 1979 (upper) and 28 February 1980 (lower). Chromatograms were obtained using a HP 7675 purge-and-trap sampler [column 0.3 cm o.d. × 305 cm SP-2100; carrier 55 ml min^{-1} He; 50°C (10 min) → 4°C min^{-1} → 180°C (20 min)].

organics in this sample are significantly different from the composite waste of 8 February 1978. Neither butanol nor dimethylaniline is present in this waste. Table 9.2 presents concentrations of compounds identified in this composite waste. The waste is characterized by a series of methyl-, ethyl-, propyl-, and butyl-substituted benzenes. A GC–MS total ion chromatogram and selective ion scans from this same waste are shown in Fig. 9.6. The presence of ions with masses of 50, 53, 57, 78, 91, and 106 amu in most of the total ion scan peaks indicates the strong aromatic nature of this waste. Nonaromatic compounds found in the waste, as in the February 1978 waste, include acetone, dichloromethane, chloroform, and methylisobutylketone (MIBK).

Figure 9.5 also presents a gas chromatogram of a composite waste sample obtained during February 1980. The chromatogram indicates that this waste includes several chlorinated compounds and does not include any of the C_3 and C_4 substituted benzenes observed in the October 1979 sample. The major volatile components in the individual contributing companies' waste samples taken at this time are given in Table 9.3.

9.4.2. Plume Studies

Samples were taken in the waste plume during both the February 1978 and the October 1979 cruises. On both cruises, the research vessel moved into a portion of

Table 9.2. Concentrations of Compounds Identified in the 7 October 1979 Composite Waste

Compound	Concentration (mg liter^{-1})	Peak[a]
Acetone	~6,000	1
Dichloromethane	110	2
Chloroform	5	3
Benzene	2	
Methylisobutylketone	20	
Toluene	50	4
Ethylbenzene	0.4	6
m, *p*-Xylenes	2	7
o-Xylene	1.5	8
Propylbenzene	0.4	9
Cumene	2	10
Mesitylene	3	11
C_3-Benzene	1	12
C_4-Benzene	0.9	13
C_4-Benzene	1	13
C_4-Benzene	0.8	14
Cymene	1	14
Butylbenzene	1	15
Naphthalene	0.3	

[a]Peak number refers to Figs. 9.5 and 9.10 to 9.12.

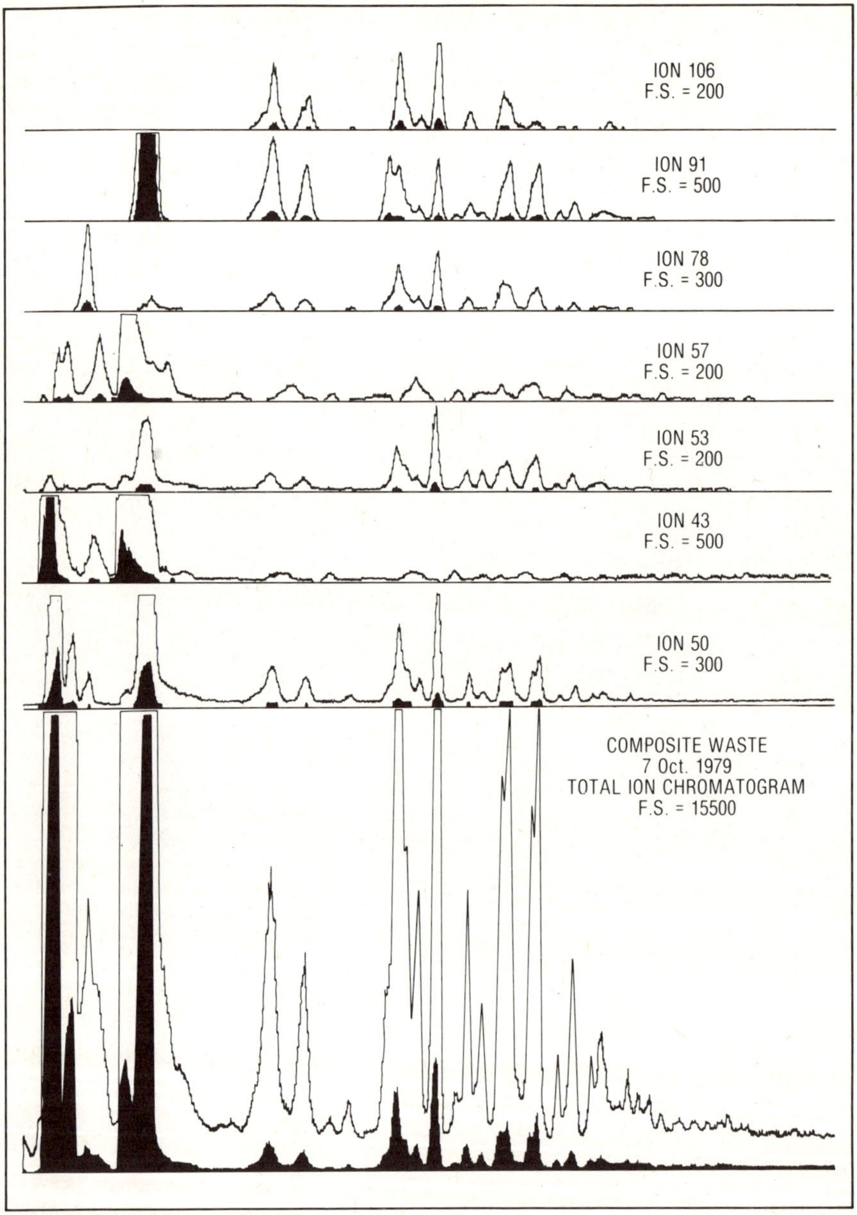

Figure 9.6. GC-MS total ion chromatogram with selected ion scans of composite waste sample collected on 7 October 1979 [column 0.3 cm o.d. × 305 cm SP-2100; carrier 20 ml min^{-1} He; 50°C (10 min) → 4°C min^{-1} → 180°C (20 min)].

Table 9.3. Concentration of Compounds Identified in the February 1980 Individual Waste Samples

Company[a]	Compound	Concentration (mg liter^{-1})	% of Volatiles
The Upjohn Mfg. Co. (200,000)	Acetone	300	4
	Ethylacetate	30	0.4
	Carbon tetrachloride	8,000	95
	Unknown	0.6	0.01
	Unknown ketone	8	0.1
	Other identified peaks (mostly ketones)	<1	<0.01
	Total	8,300	
Merck, Sharp & Dome Quimica de Puerto Rico (70,000)	Ethylmercaptan	200	87
	Unknown alcohol or ketone	<0.01	<0.004
	Methylisobutylketone	0.01	0.004
	Toluene	30	13
	Unknown alcohols or ketones	0.05	0.02
	Total	230	
Schering Corp. (62,000)	Acetone	15	42
	Chloroform	20	55
	Unknown	1	3
	Other unidentified peaks	<0.1	>1
	Total	36	
Pfizer Pharmaceutical Inc. (32,000)	Unknown	1	0.2
	Propylamine	400	85
	Unknown	40	9
	Unknown	7	1
	Toluene	10	2
	Ethylbenzene	1	0.2
	m,p-Xylene	5	1
	o-Xylene	0.5	0.1
	Other unidentified peaks	<0.1	<0.2
	Total	470	
Bristol-Alpha, Inc. (8,400)	Dichloromethane	100	5
	Unknown	1,000	45
	Trimethylsilyl	1,000	45
	Unknown	100	5
	Unknown	10	0.5
	Methylisobutylketone	4	0.2
	Toluene	1	0.05
	m,p-Xylene	0.02	0.001
	Dimethylaniline	5	0.2
	Total	2,200	
Squibb, Inc.	Acetone	60,000	95
	Methylisobutylketone	3,000	5
	Ethylbenzene	10	0.02
	m,p-Xylene	80	0.1
	o-Xylene	20	0.03

Table 9.3. ***(Continued)***

Company[a]	Compound	Concentration (mg liter^{-1})	% of Volatiles
	Cumene	0.06	–
	Unknown ketone	0.5	–
	Total	63,000	
Capri	Acetone	10,000	99
	Chloroform	1	0.01
	Benzene	0.3	0.003
	Toluene	100	1
	Ethylbenzene	0.1	0.001
	Xylene	0.4	0.004
	All others	<0.2	<0.002
	Total	10,000	

[a]Numbers in parentheses are the weight of waste dumped (metric tons y^{-1}) for that particular company.

the discharged plume immediately after passage of the barge. At that time marker buoys were deployed and sampling was initiated. An attempt was then made to follow the section of the discharged plume that was marked by buoys and dyed with rhodamine WT. This dye, which acted as a conservative tracer, was added to the barge waste before dumping.

Figure 9.4 shows a GC-MS ion spectrum from a sample (station DS/9) taken at 1 m depth, 30 min after the 9 February 1978 dump. This station contained the highest concentration of volatile organics and had the highest dye levels observed (1.05 μg liter^{-1}). However, the concentrations of both dye and volatile organics are three to four orders of magnitude less than the barge waste concentrations. A rapid dilution of the waste material occurs immediately after waste discharge in the wake of the barge (see Kullenberg, Chapter 4).

Compounds identified by GC-MS techniques in the waste material and at station DS/9 are presented in Table 9.1 (numbers correspond to numbered peaks in Figs. 9.4 and 9.7). Identified volatile organic compounds include hydrocarbons, chlorinated hydrocarbons, alcohols, esters, ketones, and nitrogen compounds. A comparison of the plume sample and barge sample (Fig. 9.4) shows that many compounds present in sample DS/9 are not present in the composite waste from the barge. The absence of these compounds in the composite waste is probably due to unrepresentative sampling of the barge wastes, which were sampled from the fill pipe as the barge was loaded.

Most of the compounds found in the DS/9 water sample, but not in the composite waste barge sample, are not found in unpolluted oceanic waters. They must therefore, result from the waste dumping. Over 25% of the organic content of the discharged waste is composed of volatile organic solvents. The remaining compounds are probably higher molecular weight compounds used as intermediates in pharmaceutical production.

Quantitative determinations were made on several volatile organic samples after

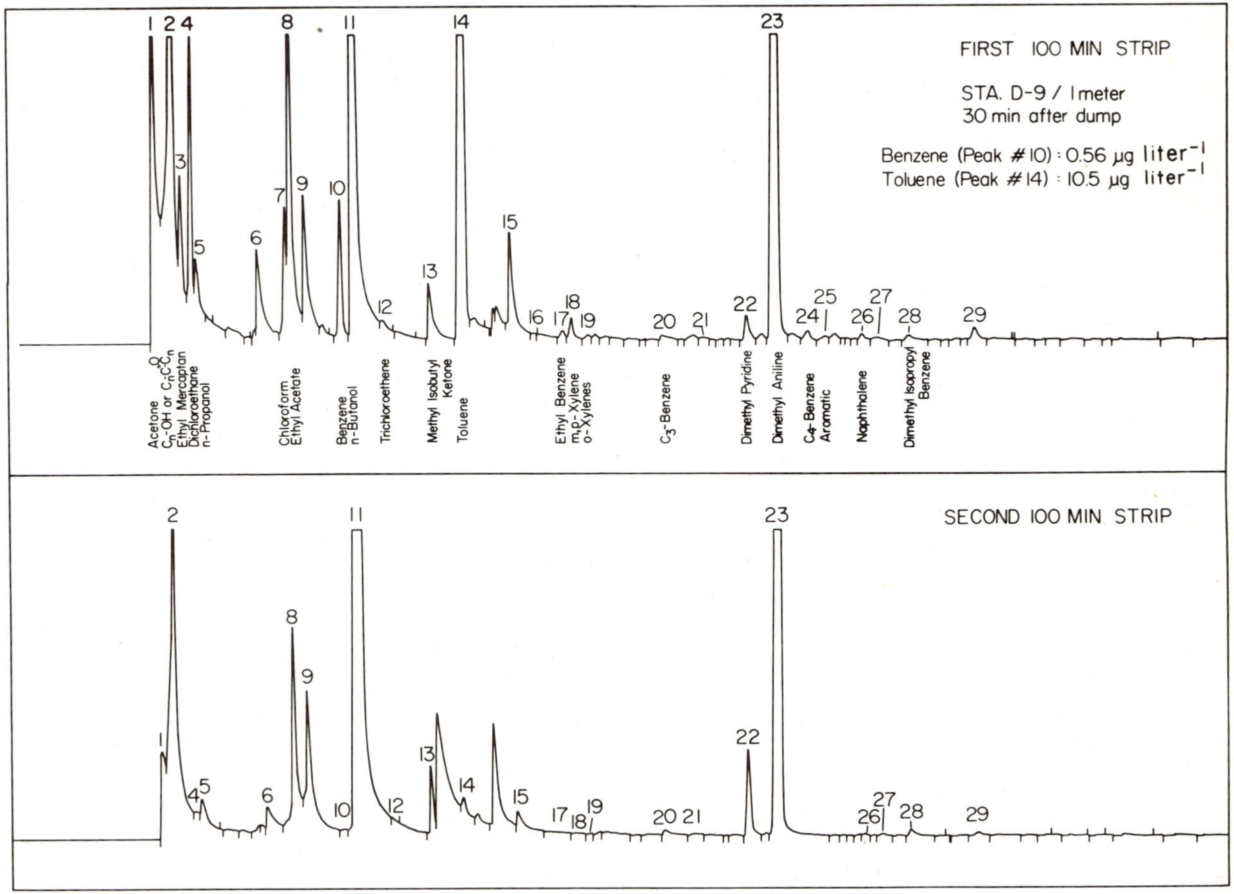

Figure 9.7. Chromatograms of surface water sample taken at station DS/9 (9 February 1978) 30 min after dumping. Peak numbers correspond to components in Table 9.1.

the February 1978 dump. Two gas chromatograms at the same attenuation, for two successive 100 min purgings of water sample DS/9, are shown in Fig. 9.7. The first-purge chromatogram displays all volatile organic compounds detectable. The 100 min purging time completely removes all volatile hydrocarbons and most volatile chlorinated hydrocarbons (Sauer et al., 1978). These compounds, such as benzene (peak 10) and toluene (peak 14), are not observed in the second (lower) chromatogram. Concentrations of benzene and toluene at station DS/9 (1 m depth) were 0.56 and 10.5 μg liter^{-1}, respectively. These values are two to three orders of magnitude greater than oceanic base line values, 10–40 ng liter^{-1} (Sauer, 1980). Many of the more soluble compounds with low vapor pressures, however, (for example, acetone, ethylacetate, *n*-butanol, methylisobutylketone, dimethylaniline) are observed in the second chromatogram. The purge-and-trap technique used does not completely remove these more soluble volatile components. Small percentages of these compounds are removed during each successive strip. Concentration calculations can be made, however, using the calculated partition coefficients (equations 1 to 3). Estimates for the concentrations of butanol and dimethylaniline in the plume are about 1–10 mg liter^{-1} in sample DS/9. In the February 1978 samples, approximately 70% of all organic solvents in the discharged waste consisted of these two compounds.

The persistence of selected volatile organics in the water column at the dumpsite was also studied following the 1978 dump. Station DS/9 samples had the highest concentrations of both volatiles and dye. Most volatile concentrations decreased gradually to approximately baseline values within 24 h after the dump (Table 9.4). However, many solvents could be observed in surface waters for a long period of time. The gradual decrease (dilution) in dye concentration was due principally to the entrainment of surface water with the waste plume. Since dye is a conservative tracer, concentration changes of dye indicate the physical effects of mixing.

Table 9.4. Concentrations of Volatile Hydrocarbons and Dye in Water Samples Taken during the Waste Dump of 9 February 1978[a]

Time (h) after Dump	Station	Depth (m)	Concentration (μg liter^{-1})			
			Dye	Benzene	Toluene	Xylenes
0.5	DS/9	1	1.05	0.56	10.5	0.15
		10	1.04	0.46	9.9	0.14
1.8	DS/10	1	0.43	0.22	4.7	–
		5	0.28	0.13	3.1	0.06
5.0	DS/11	1	0.27	0.11	2.8	0.05
		5	0.14	–	–	–
10.7	DS/12	1	0.01	0.008	0.030	0.07
		10	0.01	–	–	–
12.0	DS/13	1	0.06	0.008	0.044	–
Background		1	–	0.008	0.016	0.054

[a]Absence of data, indicated by dash in table, denotes no information available.

Volatile organics are not conservative, and undergo both dilution (mixing) and evaporation.

To illustrate the different processes that affect dye and volatiles, benzene concentrations were plotted against dye concentration (Fig. 9.8). Time after dumping is indicated for each point. Since the dye concentrations in samples DS/9 to DS/13 (0.5 to 12 h after dumping) covered a range of two orders of magnitude, the volatiles selected for these mixing (dilution) evaporation (venting) studies should have concentrations with at least this magnitude difference between maximum concentration and background levels. Benzene and toluene fit this prerequisite. Xylenes do not since only slightly higher than background concentrations are present even before the effect of evaporation becomes evident.

The dilution line in Fig. 9.8 represents the decrease of dye and volatile concentrations for a simple two-component mixing model (that is, no effect from evaporation of volatiles). The amount of a volatile component lost by evaporation is calculated as the difference between the concentration predicted from the dilution line and the background value. This difference could be due to separation of dissimilar phases in the water column; one phase being a separate water-insoluble solvent waste phase, and the other being a water-soluble phase containing the dye. Since

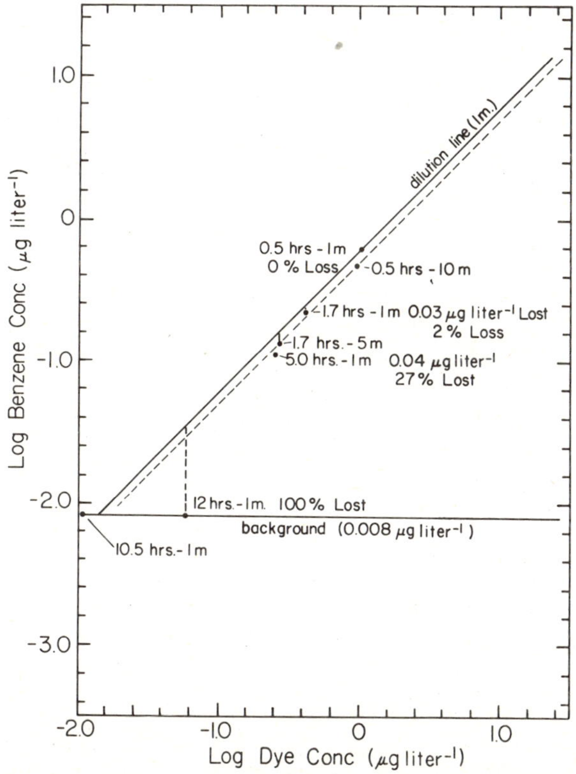

Figure 9.8. Plot of benzene versus dye concentrations in the waste plume following a dump on 9 February 1978.

dye is highly soluble in water, and solvents (volatile organics) are not very soluble, separation of these two components could give an impression of volatile loss due to evaporation. This possibility is remote, however, since volatile organic concentrations in the water are well below saturation.

Loss of neither benzene nor toluene due to evaporation was evident 2 h after the dump (Table 9.4). By 5.0 h, however, 27% of the benzene was lost, apparently through evaporation. After 12 h benzene had decreased to background levels. Toluene, on the other hand, did not show any discernible evaporation until 10.5 h after the dump, and did not reach background levels within 12 h, the duration of the experiment. The different rates of evaporation between the two volatiles result from different vapor pressures. Benzene, having a higher vapor pressure, has a higher rate of evaporation from the water column into the atmosphere. Evaporation rates for nonhydrocarbon volatiles, such as butanol and dimethylaniline, were not determined, since low purging and analysis efficiencies precluded very high precision in the results. A very low rate of evaporation is predicted for these heterocompounds because of their lower vapor pressure.

Evaporation rates are important in assessing the potential effects of toxic components found in the wastes on marine organisms. Since volatile hydrocarbons (for example, benzene and toluene) evaporate faster than the more soluble heterocompounds such as butanol and dimethylaniline, their potential for toxic interactions with organisms is reduced.

9.4.3. Spatial Distributions

A gas chromatogram has a specific pattern that is representative of the components contained in a sample. When a particular sample (such as a waste sample) has a characteristic, definable chromatographic pattern, this pattern can be described as a "fingerprint" of that sample. The volatile organic "fingerprint" in the waters north of Puerto Rico showed a pattern similar to the composition of the composite waste discharged during both the February 1978 and October 1979 cruises.

Comparison of a water sample taken from the dumpsite three days after the February 1978 dumping (Fig. 9.9) with a sample from the waste plume immediately after dumping (Fig. 9.7) illustrates this "fingerprint" method. The chromatogram of the sample taken three days after dumping at the site indicates that the dumpsite is not completely flushed of waste. To insure a contamination-free sample, this surface seawater was taken from a small rubber boat, 200 m upwind and upcurrent of the *R/V Knorr*. The sample was collected directly into a volatile-free, 2-liter glass bottle, by lowering the bottle below the surface over the bow of the boat and filling it at arm's length. The chromatogram of the day-three sample (Fig. 9.9) is very similar to the chromatogram of the sample taken in the waste plume (Fig. 9.7), but the attenuation is almost two orders of magnitude less than that of the waste plume chromatogram. The number of volatile components in the day-three sample is also not as great as from samples in the waste plume, and all volatile levels are considerably lower. The concentrations of benzene, toluene, and xylenes are 8, 16, and 54 ng liter^{-1}, respectively. These values are comparable to concentrations

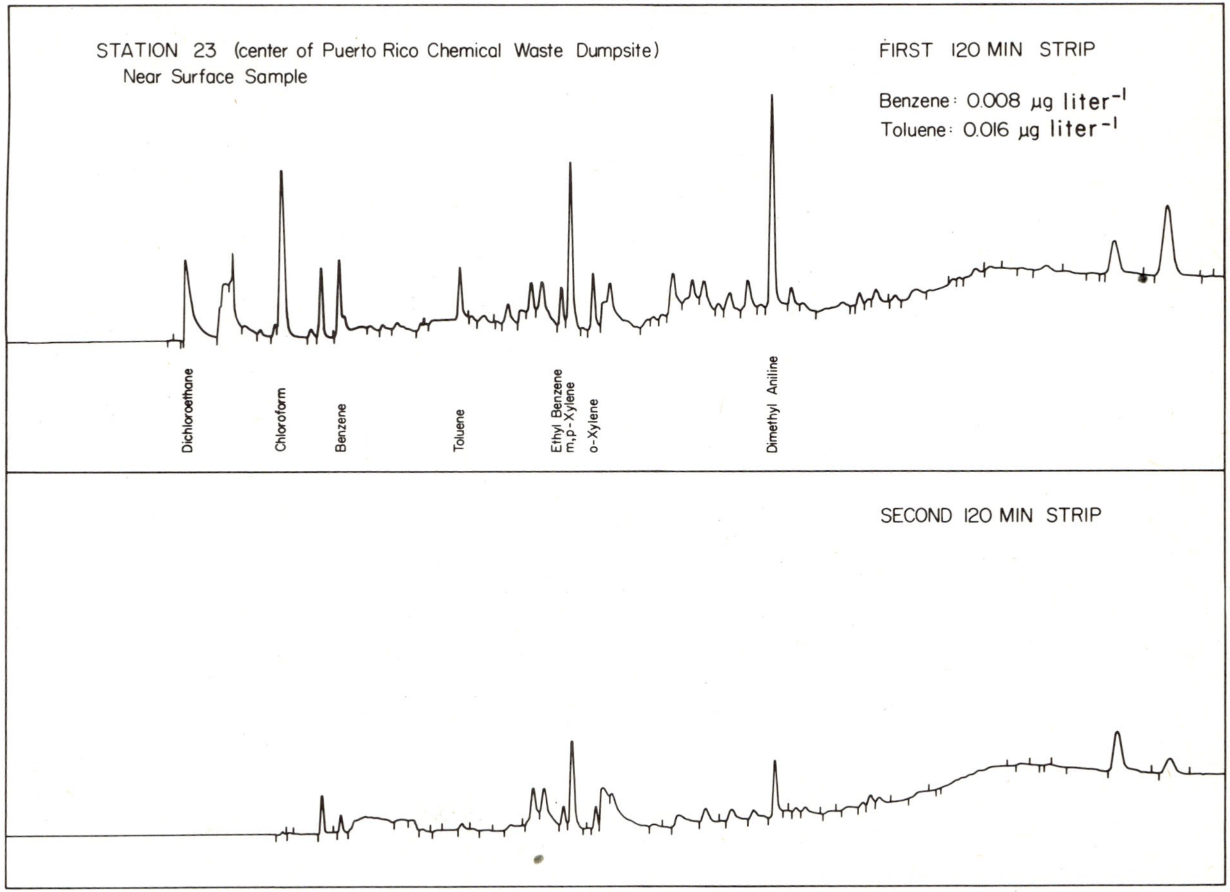

Figure 9.9. Chromatogram of surface water taken at the center of the Puerto Rico dumpsite three days after a dump, February 1978.

found in unpolluted coastal waters in the Gulf of Mexico (Sauer, 1980). Not typical in coastal waters, however, is the presence of dimethylaniline. This compound was apparently derived from the Bristol-Alpha waste during the 1978 sampling, but was not present in the 1979 waste (Table 9.3). In fact, dimethylaniline was not found even in polluted waters of the Mississippi Delta (Sauer et al., 1978) and coastal waters of Massachusetts (Schwarzenbach et al., 1978). Since this base line sample was taken three days after a dump (5 February 1978), it indicates that solvents such as dimethylaniline, with solubilities similar to chloroform and vapor pressures lower than the xylenes, persist in surface waters for extended periods of time. The baseline concentration of the dimethylaniline was approximately 0.1 μg liter^{-1}. This concentration is four to five orders of magnitude less than at station DS/9, taken 30 min after the dump.

Volatile organic "fingerprints" during the October 1979 cruise were found in 36 of the 37 stations occupied in the 23,000 km^2 area north of Puerto Rico (Fig. 9.1). These "fingerprints" were characterized by a series of alkyl-substituted, one-ring aromatic compounds which were readily apparent in the sample chromatograms (Figs. 9.10 to 9.12). The component patterns in these chromatograms were similar to the pattern observed in the composite waste discharged during this period (Figs. 9.5 and 9.6). Figure 9.10 illustrates a first and second purge of a station 7 sample. Most volatile organic components were completely removed with one purge.

The only station not showing a volatile "fingerprint" was station 1. All other stations contained substituted one-ring aromatics with total aromatic volatiles varying in concentration from approximately 0.1 μg liter^{-1} (station 27) to 8 μg liter^{-1} (station 5). Xylene (peak 7, Figs. 9.10 to 9.12) is the dominant peak in most chromatograms. It varied in concentration from 50 ng liter^{-1} at station 27 to 2 μg liter^{-1} at station 5. Although toluene (peak 4) was the major peak in the composite waste (Fig. 9.5), is was found at significantly lower concentrations, relative to other substituted benzenes, in the base line stations. These lower values were probably the result of different relative amounts of toluene being discharged over a several week period prior to the barge sampling of 7 October. However, if the relative concentrations of aromatics discharged at the site in the weeks prior to the base line survey were similar to the composite waste of 7 October, the data may represent selective alteration or loss of toluene from the water column.

The spatial distribution of the waste in the surface waters of the dumpsite area is illustrated in Fig. 9.13. Xylene, which is typical of all the aromatic components, showed a pronounced horizontal trend. Highest levels were found toward the east and northeast. Surface water with a low volatile "fingerprint" is apparent in the northwest section of the site. The volatile organic distributions suggest a gyre in the area, causing waste to move to the northeast away from the dumpsite, then south and west just north of the island. Note that, except for station 1, all stations show the same characteristic volatile "fingerprint." Obviously, the dumpsite is not swept clean of waste by a westward moving Antilles Current.

This interpretation of surface water current patterns, based on volatile organic data alone, is confirmed by the surface salinity and temperature field. Surface salinities for this area (Carnes et al., 1980) show an intrusion of low salinity waters

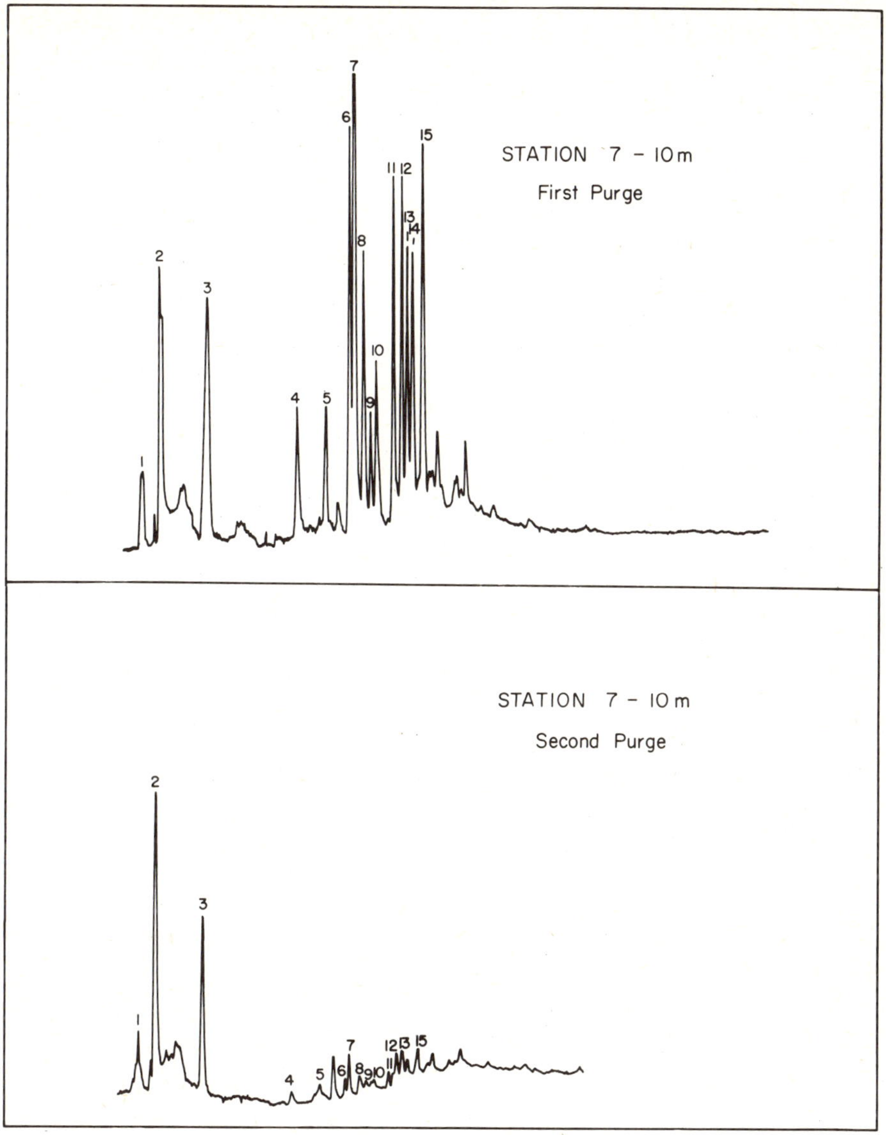

Figure 9.10. Chromatogram of volatile organics at station 7 (see Fig. 9.1 for location) taken in October 1979 [column 0.3 cm o.d. × 305 cm SP-2100; carrier 50 ml min^{-1} He; 50°C (10 min) → 10°C min^{-1} → 180°C (20 min)].

where the xylene and other aromatic concentrations are the lowest. Carnes et al. (1980) observed a strong meridional structure in the dynamic topography down to 100 m. They concluded that this feature was the result of either the presence of a mesoscale gyre or a seasonal or semipermanent meridional flow induced by the island topography. Our volatile organic data correspond well with the hydrographic description of this area.

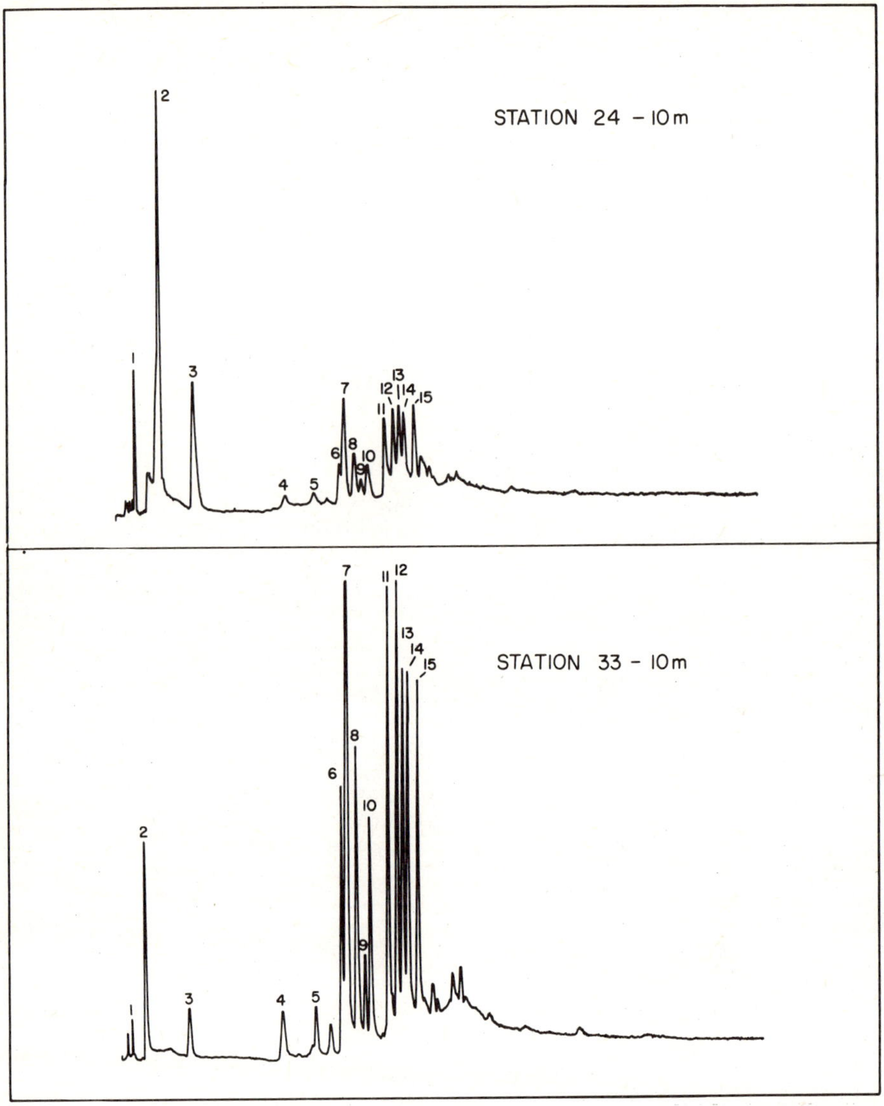

Figure 9.11. First purge chromatograms of volatile organics at stations 24 and 33 (see Fig. 9.1 for locations) taken in October 1979 [column 0.3 cm o.d. X 305 cm SP-2100; carrier 50 ml min^{-1} He; 50°C (10 min) → 10°C min^{-1} → 180°C (20 min)].

From analysis of samples taken at depth in the water column, we observe that waste penetrated the thermocline in the dumpsite region. Figure 9.12 shows a chromatogram of a sample taken from 50 m deep in the center of the dumpsite area. A chromatographic pattern similar to that of the surface waters was observed. Other waste samples taken down to a depth of 200 m also showed the characteristic aromatic "fingerprints." Since these samples were taken with Niskin bottles,

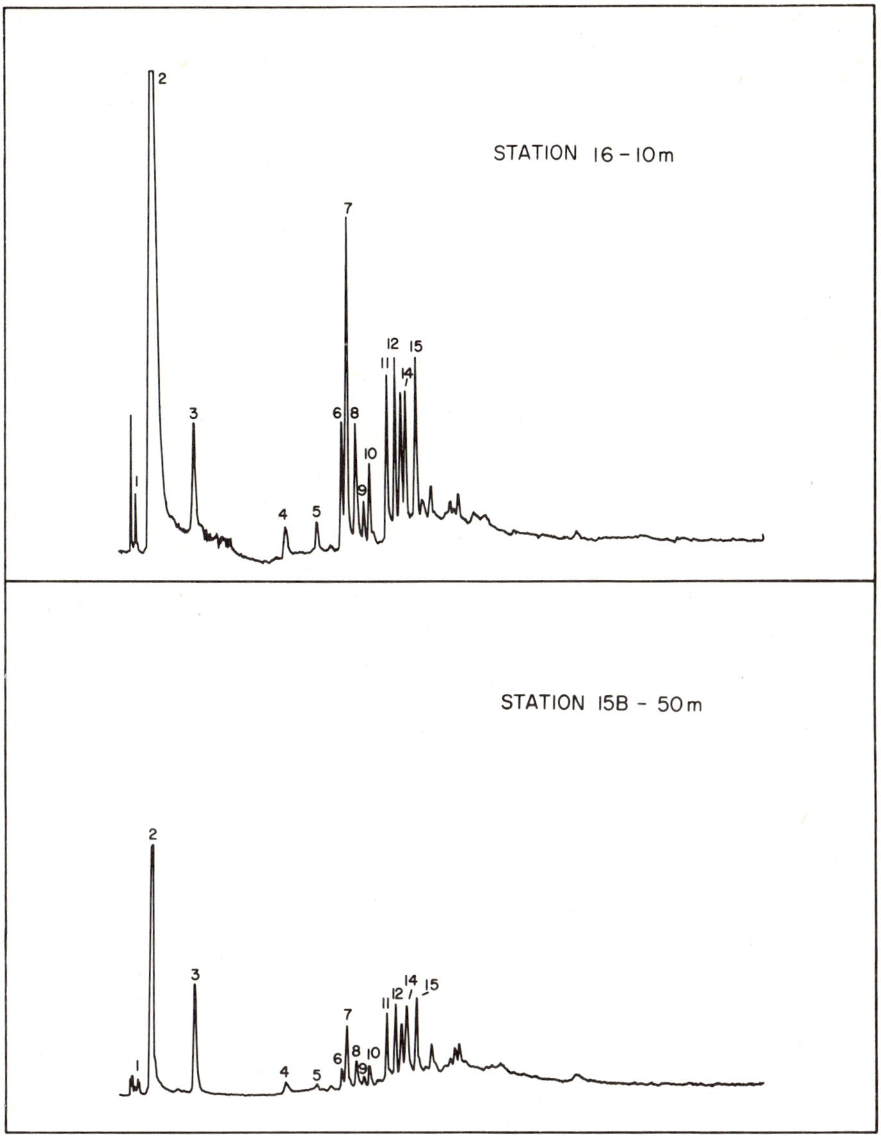

Figure 9.12. First purge chromatograms of volatile organics at stations 16 and 15B (see Fig. 9.1 for locations) taken in October 1979 [column 0.3 cm o.d. × 305 cm SP-2100; carrier 50 ml min^{1} He; 50°C (10 min) → 10°C min^{-1} → 180°C (20 min)].

which had to penetrate the mixed layer open, it is possible that we measured volatiles adsorbed on the walls of Niskin bottles. However, insignificant attachment to Niskin bottle walls would be expected, because these organic compounds are highly soluble. The mechanism by which the waste penetrates the thermocline is unknown, although migration of organisms or adsorption or inclusion into fecal pellets are possibilities.

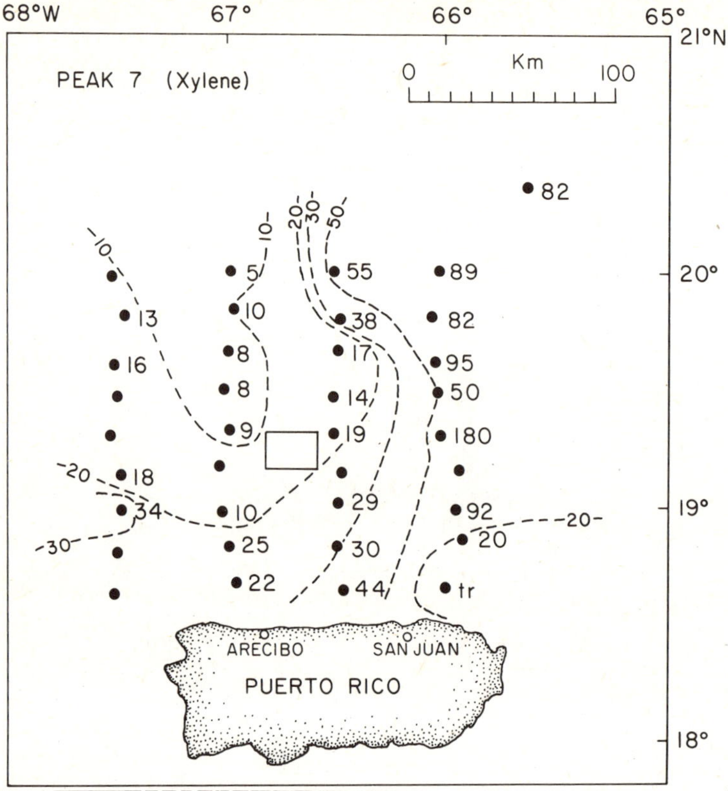

Figure 9.13. Spatial distribution of xylene (peak 7 in Figs. 9.10 to 9.12) taken in October 1979. Numbers are relative units; box denotes dumpsite.

9.5. CONCLUSIONS

Ocean dumping sites are typically chosen both for their proximity to land and for the ability of the area to be flushed quickly by dominant currents. The Puerto Rico dumpsite is located only 74 km offshore, and the major current maps indicate that the site is in a region of rapid westward current flow. However, mesoscale distributions of the volatile organics from the area north of Puerto Rico indicate that the dumped material is not rapidly flushed from the discharge area. Although the Antilles Current generally moves through the dumpsite in a westerly direction, volatile organic waste is dispersed over a wide area and was observed in stations as far as 130 km northeast of the dump location.

Characteristic volatile organic "fingerprints" were observed in sample depths down to 200 m. It appears that the discharged pharmaceutical wastes have penetrated the thermocline. Since little vertical mixing is apparent from the STD profiles (Fig. 9.2), transport of volatile organics to depth must be accomplished by attachment to more dense particulate material. Transport of volatiles by attachment to vertically migrating organisms or in zooplankton fecal pellets are two mechanisms that should be investigated.

This study has shown that dumped materials with a characteristic volatile organic "fingerprint" can be easily traced using gas chromatographic techniques now available. Mapping of the grid area and determination of degree of dispersion was facilitated by the presence of several alkyl-substituted aromatic compounds (for example, C_3 and C_4 benzenes), which were persistent in the study area. Other compounds, such as benzene and toluene, were not as widely distributed since they evaporate more rapidly than the less volatile alkyl-substituted aromatics and the more soluble heterocompounds. Thus, the potential toxic effects of the more volatile solvents are diminished, since their possible interaction time with marine organisms is reduced.

ACKNOWLEDGMENTS

The authors wish to thank Roger Burke, Jr., for his invaluable assistance in sampling and analysis of shipboard samples. Claude R. Schwab provided useful details of the 1978 cruise and stimulated discussion of the earlier data. Jack Becthold proved invaluable in operating the STD, and Masamichi Inoue kindly provided the digitized STD data. We wish to thank Dr. Thomas P. O'Connor, who served as Chief Scientist on the *Mt. Mitchell* and was responsible for much of the sample design during the October 1979 cruise. He also helped in sampling and provided waste samples for analysis. This work was supported by the National Oceanic and Atmospheric Administration under grant 04-8-M01-55.

REFERENCES

Bertsch, W., E. Anderson, and G. Holzer. 1975. Trace analysis of organic volatiles in water by gas chromatography–mass spectrometry with glass capillary columns. *Journal of Chromatography*, **112**, 701–718.

Butler, L. D., and M. F. Burke. 1976. Chromatographic characterization of porous polymers for use as adsorbents in sampling columns. *Journal of Chromatographic Science*, **14**, 117–122.

Carnes, M., M. Inoue, and T. Ichiye. 1980. Hydrographic data from the Puerto Rico waste dumping study cruises of 1978 and 1979. Technical Report 80-9-T, Department of Oceanography, Texas A&M University, College Station, Texas, 332 pp.

Heller, S. R., and C. W. A. Milne. 1978. EPA/NIH mass spectral data base, Volume 1. U.S. Government Printing Office, Washington, D.C., 988 pp.

May, W. E., S. N. Chesler, S. P. Cram, B. H. Gump, H. S. Hertz, D. P. Enagonio, and S. M. Dyszel. 1975. Chromatographic analysis of hydrocarbons in marine sediments and seawater. *Journal of Chromatographic Science*, **13** 535–540.

McLafferty, F. W. 1973. Interpretation of mass spectra. Benjamin, London, 278 pp.

Sauer, T. C., Jr. 1978. Volatile liquid hydrocarbons in the marine environment. Ph.D. Thesis, Texas A&M University, College Station, Texas, 317 pp.

Sauer, T. C., Jr. 1980. Volatile liquid hydrocarbons in waters of the Gulf of Mexico and Caribbean Sea. *Limnology and Oceanography*, **25**, 338–351.

Sauer, T. C., Jr., W. M. Sackett, and L. M. Jeffrey. 1978. Volatile liquid hydrocarbons in the surface coastal waters of the Gulf of Mexico. *Marine Chemistry*, **7**, 1–16.

Schwab, C. R., T. C. Sauer, G. F. Hubbard, H. Abdel-Reheim, and J. M. Brooks. 1981. Chemical and biological aspects of ocean dumping at the Puerto Rico dumpsite. *In*: Ocean Dumping of Industrial Waste, B. H. Ketchum, D. R. Kester, and P. K. Park (Eds.). Plenum Press, New York, pp. 247–273.

Schwarzenbach, R. P., R. H. Bromund, P. M. Gschwend, and O. C. Zafiriou. 1978. Volatile organic compounds in coastal seawater: Preliminary results. *Journal of Organic Geochemistry*, **1**, 93–107.

Wust, G. 1964. Stratification and Circulation in the Antillean-Caribbean Basin, Part 1., Spreading and Mixing of the Water Types with an Oceanographic Atlas. Columbia University Press, New York, 201 pp.

PART IV: BIOLOGICAL AND TOXIC EFFECTS

Pharmaceutical Wastes

10

MICROBIAL COMMUNITIES IN SURFACE WATERS AT THE PUERTO RICO DUMPSITE

Fred L. Singleton, Jody W. Deming, Emily R. Peele, Benzion Cavari, Bruce Gunn, and Rita R. Colwell

Department of Microbiology
University of Maryland
College Park, Maryland

ABSTRACT

A variety of microbiological parameters were determined for surface waters in and surrounding the Puerto Rico dumpsite which is used for disposal of pharmaceutical wastes. Specific activities of microbial populations were derived from comparisons of activity measurements (uptake of radiolabeled substrates or substrate-responsive cell numbers determined by epifluorescent microscopy) and total cell numbers. Highest values were observed in samples from stations in or near the dumpsite. Similarly, largest numbers of colony-forming bacteria, enumerated on marine agar, were obtained in the vicinity of the dumpsite.

Total colony-forming bacteria were enumerated at all stations, with several different culture media, and randomly selected isolates were identified to develop diversity indices for the culturable bacterial community. Bacteria isolated on marine agar were found to be predominantly members of the *Vibrio/Aeromonas* group, with the more typical marine pseudomonads comprising less than 9% of the community. In the vicinity of the dumpsite, large numbers of Gram-positive bacteria, that is, micrococci, staphylococci, and bacilli, were recovered from water samples plated on marine agar, as well as from those plated on plate count agar, which selects against bacteria requiring sea salts for growth.

Results obtained, in particular diversity index measurements and the persistence of culturable, waste-specific organisms at the dumpsite, suggest that alterations in the natural microbial populations of surface waters of the Puerto Rico dumpsite and environs have occurred.

10.1. INTRODUCTION

Selected groups of bacteria have been employed as indicators of environmental alterations to monitor the presence of domestic wastes (American Public Health Association, 1976) and more recently to evaluate effects arising from massive oil spills and other catastrophic events (Colwell and Walker, 1977). The usefulness of bacteria as indicators of environmental perturbations, resulting from chronic or low-level input of allochthonous materials in an aquatic environment, has not been investigated extensively. In view of their ubiquitous distribution, short generation times, and sensitivity to changes in the physical and chemical environment, bacteria may be the first component of the biota to reflect environmental perturbations.

Problems of waste disposal have increased in severity in recent years, becoming of crisis proportion in some cases. The need for an immediate solution has made ocean dumping a seemingly attractive alternative to land disposal. However, there is little information on potential acute or chronic biological effects resulting from waste disposal activities, especially at deep ocean dumpsites (Ketchum et al., 1981).

The waters of the Puerto Rico Trench, with recorded depths in the range 6000-

8000 m, presently serve as a dumpsite for pharmaceutical wastes (Chapter 9) and will continue to do so until Puerto Rico completes a regional wastewater treatment system at Barceloneta, scheduled to begin operation in 1981. Swan (1979) has reported on the composition of the waste from pharmaceutical industries in Puerto Rico, as well as on the contributions of the various industries to the waste. Results of analyses of composite waste samples, collected from storage tanks prior to barge disposal, indicate metal concentrations of $<1\ \mu g\ g^{-1}$, and a relatively high organic content, about 1-4% by weight (T. O'Connor, personal communication). However, because industries contributing to the waste employ different batch processes throughout the year, the wastes destined for ocean disposal are not of uniform composition. This variability in the composition of the waste confounds laboratory attempts to assess biological impacts of ocean disposal of these pharmaceutical wastes.

Earlier studies of the effects of pharmaceutical waste disposal on macroorganisms at the Puerto Rico site were inconclusive (Murphy et al., 1981; Wohlschlag et al., 1981). Recent work, however, shows this waste does affect fish and plankton (Chapters 11-13). The objective of our study was to apply several microbiological methods, including use of various culturing media, microscopic examination of total and active bacteria, uptake of radiolabeled substrates, and ATP measurements, to evaluate what changes, if any, in the naturally occurring bacterial communities of surface waters at the Puerto Rico dumpsite result from ocean disposal of pharmaceutical wastes.

10.2. EXPERIMENTAL

10.2.1. Research Site

The Puerto Rico dumpsite is an area of 500 km^2, approximately 64 km north of Arecibo, Puerto Rico (Fig. 10.1). The amount of material dumped annually is estimated to be $3.2 \times 10^5\ m^3$ and includes wastes from seven pharmaceutical industries. The wastes are collected in a common holding tank and, at intervals of two or three days, are transferred to barges which are towed to the dumpsite. The composite waste is released over an area of several square kilometers at a rate of 70 $m^3\ km^{-1}$.

10.2.2. Sample Collection

Water samples were collected aseptically in sterile Niskin bags (General Oceanics, Miami, Florida). All water samples were processed on board ship within 15 min of collection. Unless otherwise indicated, samples were collected from a depth of 1 m.

10.2.3. Bacterial Enumeration

Total viable, aerobic, heterotrophic bacteria were enumberated by culturing on Marine Agar 2216 (MA) (Difco Laboratories, Detroit, Michigan). Appropriate

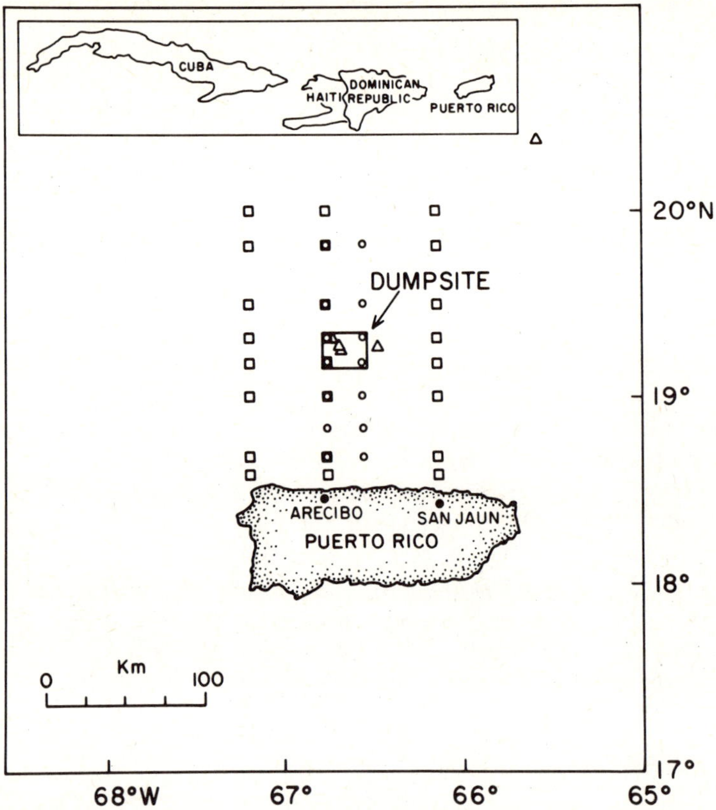

Figure 10.1. The pharmaceutical waste dumpsite 64 km north of Puerto Rico, and stations sampled in October 1979 (△), December 1979 (○), and February 1980 (□).

volumes of water (usually 1, 10, and 100 ml) were filtered in triplicate through 0.2-μm nitrocellulose membrane filters (Schleicher and Schuell, Inc., Keene, New Hampshire) and the filters placed onto MA culture plates. Numbers of heterotrophic bacteria in the surface water samples that did not require sea salts or media containing sea salts were enumerated using membrane filters and place count agar (PCA) (Difco Laboratories, Detroit, Michigan).

Total bacterial numbers were determined by acridine orange staining and epifluorescent microscopy (Francisco et al., 1973; Daley and Hobbie, 1975; Hobbie et al., 1977). Samples of surface water were preserved with formalin (final concentration 2% v/v). After staining, samples were filtered through a 0.2-μm membrane filter (Nuclepore Corp., Pleasanton, California) prestained with Irgalan black, and the total number of cells counted using a Zeiss Standard 18 microscope, equipped with an IV FL epifluorescence condensor, 100-watt halogen lamp, BP450-490 band-pass filter, FT 510 beam splitter, and LP520 barrier filter (Carl Zeiss, Inc., New York, New York). For each sample, a minimum of 10 randomly selected fields, with at least 50 cells per field, were counted to obtain acridine orange direct counts (AODC).

Numbers of viable substrate-responsive bacteria were determined by acridine orange staining and epifluorescent microscopy, according to the procedure of Kogure et al. (1979). Twenty-cm^3 seawater samples were supplemented with 0.025% yeast extract (final concentration) and 0.002% nalidixic acid (final concentration) (Sigma Chemical Co., St. Louis, Missouri). Nalidixic acid is an antibiotic that inhibits cell division in Gram-negative bacteria by blocking DNA synthesis. All samples were incubated for 6 h in darkness at *in situ* temperatures and fixed with formalin. The effect of nalidixic acid on actively metabolizing cells capable of utilizing yeast extract, that is, substrate-responsive cells, can be visualized by acridine orange staining and epifluorescent microscopy. The cells become enlarged or elongated, and bright orange in color. Inactive or non-substrate-responsive cells appear as small green rods and cocci. Both total cell numbers and direct viable counts (DVC), that is, number of substrate-responsive cells, were determined for each sample.

10.2.4. Bacterial Identification

Ten colonies from one MA culture plate per sample per station were randomly selected, purified, and identified to genus. Biochemical and morphological characteristics described in *Bergey's Manual of Determinative Bacteriology* (Buchanan and Gibbons, 1974) and the identification schemes of Bain and Shewan (1968) and Gibson et al. (1977) were used for identification of the isolates.

10.2.5. Heterotrophic Activity

Using the method of Hobbie and Crawford (1969), heterotrophic activity of the microbial flora of samples of surface water was determined. Fifty-ml water samples, aseptically collected, were placed into sterile 125-ml Erlenmeyer flasks and supplemented with a ^{14}C-labeled amino acid mixture (57 mCi mg-atom^{-1} carbon, Amersham Corp., Arlington Heights, Illinois). Final concentrations of amino acids were 25, 50, 75, and 100 μg liter^{-1}. All systems supplemented with amino acids, including controls, which were samples fixed with 0.5 ml of a solution of 30% trichloracetic acid (weight/volume, w/v) and 0.05% mercuric chloride (w/v) in 40% formalin (v/v), were duplicated for each sample site. After addition of the labeled amino acids, flasks were sealed with serum stoppers fitted with a plastic cup containing a precombusted (450°C, 3 h) glass fiber filter. All flasks were incubated at 25°C for 3 h and the reaction terminated by injection of the trichloracetic acid-mercuric chloride-formalin mixture. Filter wicks were saturated with 0.2 ml phenothylamine to absorb $^{14}CO_2$, and the flasks were incubated for an additional hour with shaking.

After the second incubation period, the filter wicks were placed in scintillation vials, and the water samples were filtered through a 0.2-μm membrane filter (Millipore Corp., New Bedford, Massachusetts). After rinsing with 10 ml filtered (0.2-μm) seawater, the filters were placed in scintillation vials for transport to the laboratory at College Park, Maryland. In the laboratory, a 10-ml volume of scintillation cocktail (Amersham Corp., Arlington Heights, Illinois) was added to each vial, and radioactive counts were made using a Beckman liquid scintillation spectrometer Model No. LS-3155T. Calculations of V_{max}, maximum uptake velocity for the substrate used, and of the microbial population of the water sample, were made according to the procedure of Hobbie and Crawford (1969).

10.2.6. Specific Activity Indices

Based on concepts discussed by Wright (1978) and a method developed by Orndorff and Colwell (1980), indices were calculated to depict the relationship between microbial heterotrophic activity and number of bacteria present in a given sample. The index V_{max}/AODC utilized V_{max} as a measure of potential heterotrophic activity in relation to the number of bacteria determined by AODC. A second index, DVC/AODC, compared the number of viable bacteria capable of utilizing yeast extract (Kogure et al., 1979), and the total number of bacterial cells.

10.2.7. ATP Measurements

Particulate ATP concentrations were determined for surface water samples collected in February 1980 by the method of Holm-Hansen and Booth (1966). ATP from the particulate fractions of 100-ml seawater samples, collected on 0.2-μm membrane filters (Millipore Corp., New Bedford, Massachusetts), was extracted in boiling 0.05*M* Tris buffer, frozen, and returned to the laboratory for analysis. Purified firefly luciferase (E. I. du Pont de Nemours & Co., Wilmington, Delaware) and a model 3000 ATP Photometer (SAI Technology Company, San Diego, California) were used to determine ATP concentrations.

10.3. RESULTS

10.3.1. Preliminary Cruise

A preliminary cruise was carried out in May 1979 to test methods and examine microbial populations in surface waters. Samples were collected in the waste plume of an actively discharging barge and at locations assumed to be sufficiently away from the dumpsite to be uninfluenced by dumping activities. However, the latter assumption was not achieved, as samples collected were subsequently found to contain waste materials, shown by the presence of waste-specific volatile organic compounds detected by gas chromatographic analysis.

The cruise plan for October 1979 included shipboard chromatographic analyses of samples immediately after collection so that control (waste-free) sites could be established. Surface water samples were collected for chemical analyses from a grid of 36 stations encompassing the dumpsite and a surrounding region of approximately 2.4×10^4 km^2. Detectable quantities of volatile organic waste-specific compounds were present in all water samples collected, including those from

Table 10.1. Bacteriological Analysis of Surface Waters East of the Dumpsite and in the Waste Plume (October 1979)

Station		Specific Activity[a]	Marine Agar 2216	Plate Count Agar
9B-5m	144 km NE of dumpsite	75.0	47[b]	10
15B-5m	16 km E of dumpsite	9.3	>100	4
D1-1m	Waste plume (T_0 h)[c]	N.A.[d]	7	19
D2-5m	Waste plume (T_3 h)	<0.5	6	27
D4-5m	Waste plume (T_{13} h)	N.A.[d]	6	102

[a](Direct Viable Count/Acridine Orange Direct Count) $\times 10^{-3}$.
[b]Colony-forming units per milliliter.
[c]Signifies time in hours after discharge of the wastes.
[d]Not available.

144 km northeast of the dumpsite (Brooks et al., 1980). Therefore, the boundaries of the waste-impacted area could not be clearly defined. Surface water samples for microbiological analyses were collected from stations with relatively low quantities of volatile organic waste-specific compounds and from within a waste plume tracked to monitor acute effects of the dumped waste on the microbial community (Table 10.1). Total numbers of culturable, aerobic heterotrophic bacteria, enumerated on marine agar (MA), were consistently low in the waste plume, compared to stations east and northeast of the dumpsite itself. In contrasts, number of bacteria cultured on plate count agar (PCA) increased markedly during the timed sampling sequence.

In light of the apparent ubiquitous distribution of the waste and its variable composition, we abandoned the earlier cruise plans designed to find waste-free control sites within a workable distance from the dumpsite (<150 km). Instead, sampling sites along transects extending from coastal waters, through the dumpsite, and into areas of the ocean 64 km beyond the dumpsite, were established for the remaining cruises.

10.3.2. Enumeration and Identification of Culturable Bacteria from Transect Samples

Total numbers of culturable, aerobic heterotrophic bacteria on MA were relatively stable for samples collected along transects through the dumpsite region (Fig. 10.2). However, during December 1979, our results revealed significant increases ($P \leq 0.05$) in numbers of culturable heterotrophs at stations due north (16 km) of the dumpsite. Counts for samples collected at stations 32 km north of the dumpsite were similar to those obtained in surface waters within, and south of, the designated dumping area.

During February 1980, large numbers of heterotrophic, colony-forming bacteria obtained on MA were again recorded at stations north and northwest of the dump-

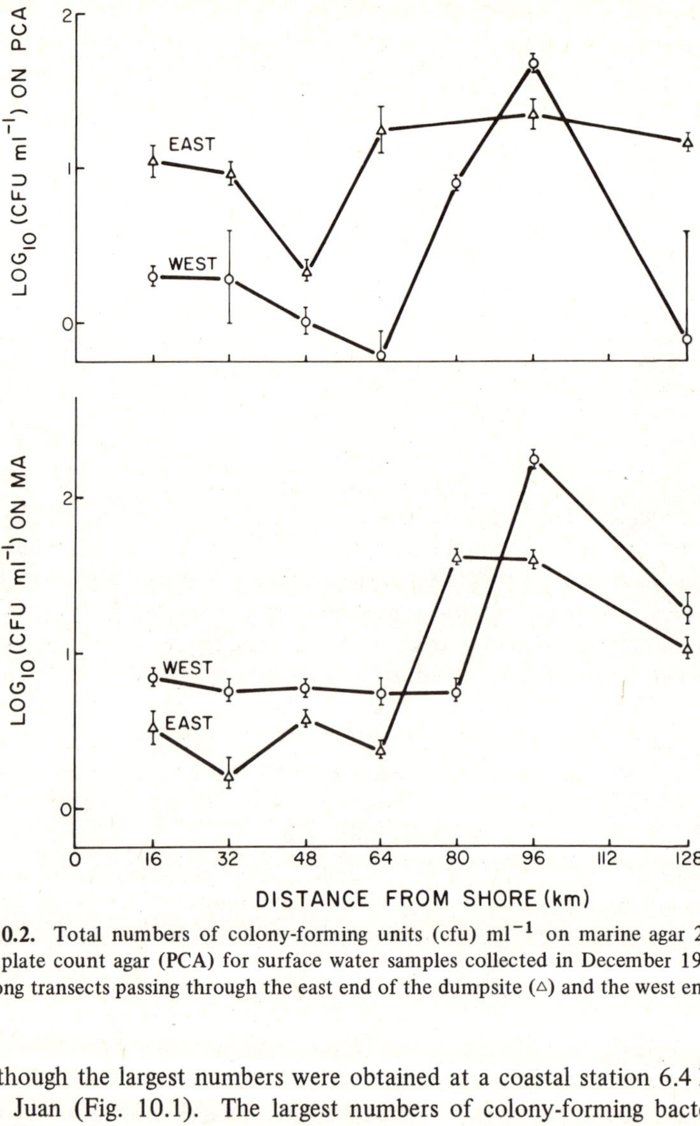

Figure 10.2. Total numbers of colony-forming units (cfu) ml^{-1} on marine agar 2216 (MA) and on plate count agar (PCA) for surface water samples collected in December 1979, at stations along transects passing through the east end of the dumpsite (△) and the west end (○).

site, although the largest numbers were obtained at a coastal station 6.4 km north of San Juan (Fig. 10.1). The largest numbers of colony-forming bacteria were usually found at stations north of the dumpsite.

A significant portion of the culturable bacterial community did not require seawater, or media prepared with the major salts of seawater, for growth, as shown by the numbers of microorganisms growing on PCA (Fig. 10.2). The numbers of bacteria that grew on PCA were found to vary within and between transects, as well as between sampling times. In general, the largest numbers of bacteria growing on PCA were recovered from water samples collected at stations north of the dumpsite during December 1979. In February 1980, fewer bacteria were recovered on PCA than during the December cruise.

The culturable bacterial community of the surface waters in and around the

Table 10.2. Taxonomic Distribution of Bacterial Strains Isolated from Marine Agar 2216 for Three Cruises to the Puerto Rico Trench during 1979

	Time of Sampling		
Genus	May	October	December
Pseudomonas	1.3[a]	8.1	2.2
Vibrio/Aeromonas	86.3	81.8	74.3
Photobacterium	3.1	0	1.4
Lucibacterium	0	4.0	0
Flavobacterium	0.9	5.1	3.5
Acinetobacter/Moraxella	3.5	1.0	2.2
Unknown Gram-negative rods	0	0	1.4
Unidentified Gram-positive cocci	0	0	15.0
Total number of strains	227	99	140

[a]Percent of total number of strains examined.

dumpsite, as recovered on MA, was found to be dominated by *Vibrio* and *Aeromonas* spp. (Table 10.2), an unexpected finding, since it is generally accepted that the bacterial community of open ocean waters contains a predominance of pseudomonads. Instead, *Pseudomonas* spp. were found to comprise less than 9% of the culturable community. During December 1979, unexpectedly large numbers of culturable organisms were obtained from stations 16 km north of the dumpsite and were found to be Gram-positive (Fig. 10.2, Table 10.3). As the incidence of Gram-positive bacteria increased, Gram negatives, notably *Vibrio* and *Aeromonas* spp., decreased.

10.3.3. Heterotrophic Activity

In February 1980, a decline in heterotrophic bacterial activity correlated with distance from land was observed, with highest uptake velocities detected for a coastal station 6.4 km north of San Juan (Fig. 10.3). Uptake velocities were lower for samples collected at stations along a transect 40 km west of the dumpsite than at stations on a parallel transect 40 km east of the dumping area. Along the eastern transect, stations far enough from land to be free from terrestrial influence yielded heterotrophic activity measurements showing little variation. In general, rates of activity were lowest within the dumpsite or at stations of similar latitude.

10.3.4. ATP Measurements

Concentrations of particulate ATP were found to vary considerably within and between transects (Fig. 10.3). Values did not correlate with other microbiological measurements, and no clear patterns emerged which aided in evaluating effects of waste dumping activities in the dumpsite region.

Table 10.3. Distribution of Gram-Negative and Gram-Positive Bacteria Isolated on Marine Agar 2216 from Surface Water Sample from the Puerto Rico Trench (December 1979)

Distance from Shore (km)	%Gram-Positive Organisms	%Gram-Negative Organisms	%*Vibrio/Aeromonas*
		Eastern Transect	
16	0	100	90
32	0	100	80
48	0	100	90
64	0	100	100
80	0	100	100
96	60	40	40
128	70	30	30
		Western Transect	
16	0	100	90
32	0	100	100
48	0	100	70
64	0	100	90
80	0	100	90
96	60	40	30
128	0	100	40

10.3.5. Total Counts, Direct Viable Counts, and Specific Activity of Bacterial Populations

Total numbers of bacteria (AODC) in water samples collected during the December 1979 cruise were relatively stable, ranging from 1.93×10^5 to 3.24×10^5 per ml (Fig. 10.4). On the average, largest numbers of bacteria were found in surface water samples collected along the western side of the dumpsite, with differences in counts between the parallel east-west transects ranging from 10 to 19%. The proportion of the total bacterial population which was responsive to substrate addition, that is, the direct viable count (DVC) (Kogure et al., 1979) ranged from 0.2 to 13%, depending on location.

When the direct viable count was related to total numbers of bacteria determined by AODC, the resulting activity index (DVC/AODC) suggested marked differences in specific activity among various water samples. In October 1979, the specific activity index determined for a seawater sample collected directly in the waste plume was considerably lower than the indices calculated for samples taken 16 and 144 km away from the dumpsite (Table 10.1). For transect stations in December 1979, highest specific activities were recorded for samples collected at eastern stations within the dumpsite (Fig. 10.4).

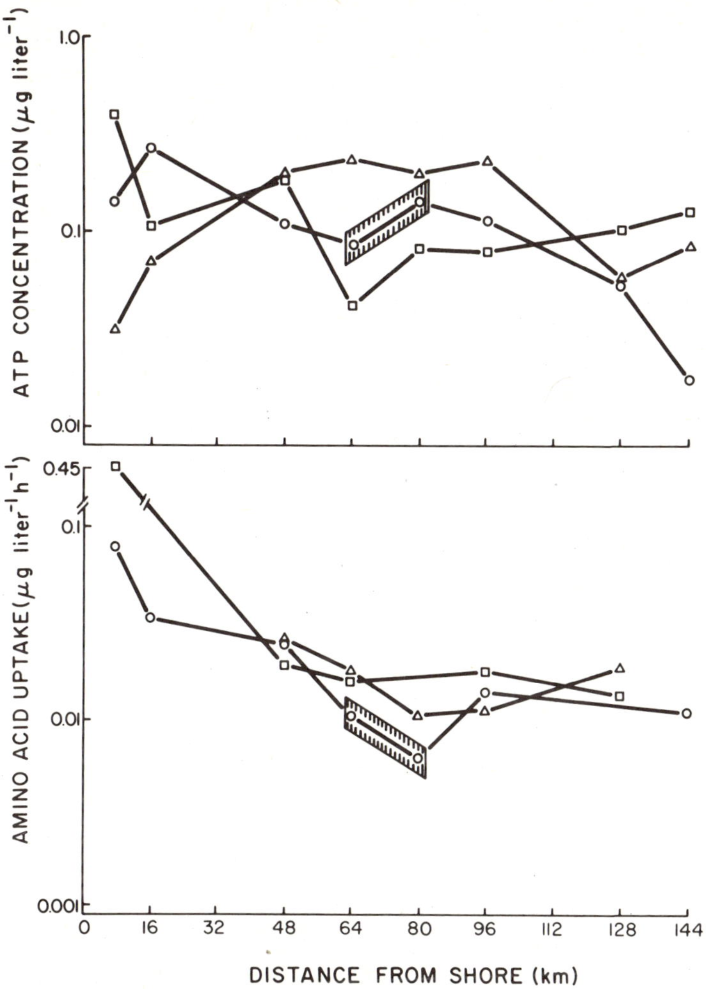

Figure 10.3. ATP concentrations and heterotrophic bacterial activity, measred by uptake of [^{14}C]amino acids, for surface water samples collected in February 1980, at stations along transects 40 km each of the dumpsite (■), passing through the western end of the dumpsite (○), and 40 km west of the dumpsite (△). Shaded area indicates station within the dumpsite.

A comparison of total numbers of bacteria (AODC) and specific activities (DVC/AODC) for water samples collected in December 1979 with data from the same transect sampled in February 1980 indicated little seasonal variation between December and February (Fig. 10.5). In the absence of marked seasonal variations, specific activities based on V_{max}/AODC, obtained for two stations on the eastern transect sampled in December 1979, were compared to similar indices obtained for stations along the western transect sampled in February 1980 (Fig. 10.6). Although total numbers of bacteria were consistently higher for samples collected on the western transect, specific activity indices based on V_{max}/AODC for the two

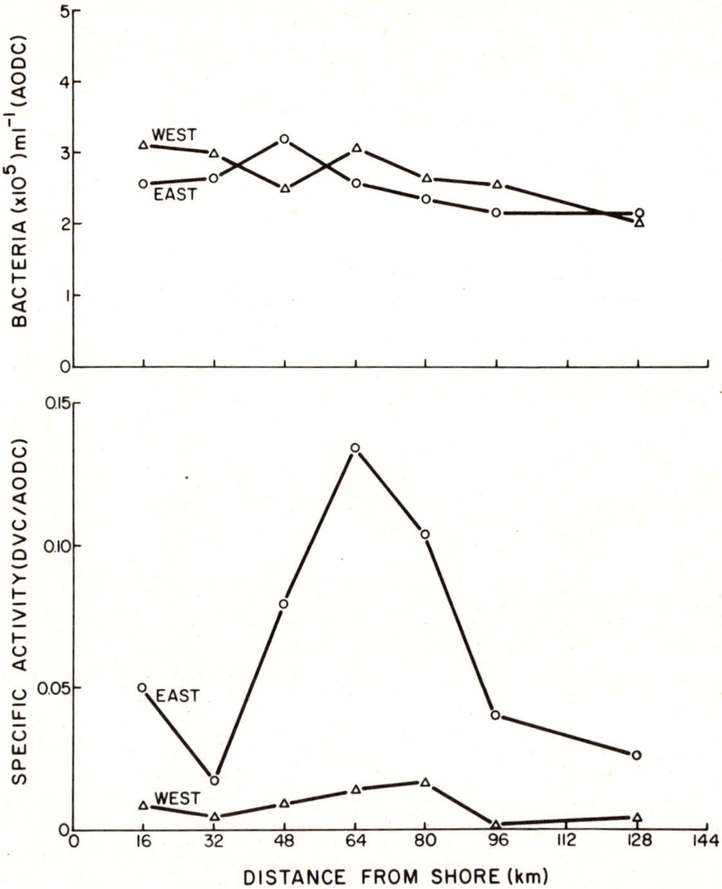

Figure 10.4. Total numbers of bacteria per ml, determined by acridine orange direct counts (AODC), and specific activities, calculated as the ratio of direct viable counts (DVC) and total counts (AODC), for surface water samples collected in December 1979, at stations along transects passing through the eastern end of the dumpsite (○) and the western end (△).

samples collected on the eastern transect were highest, in agreement with indices based on DVC/AODC (Figs. 10.4 and 10.6).

10.4. DISCUSSION

Because of their short generation times, microorganisms are predictably the first component of the biota in an ecosystem to demonstrate effects of environmental stress. For the same reason, they may be the first to recover, if only in numbers, from acute effects, such as those that occur immediately in the wake of a waste disposal barge. Accordingly, only gross effects of a pollutant on natural microbial populations can be detected by techniques that measure changes in population size

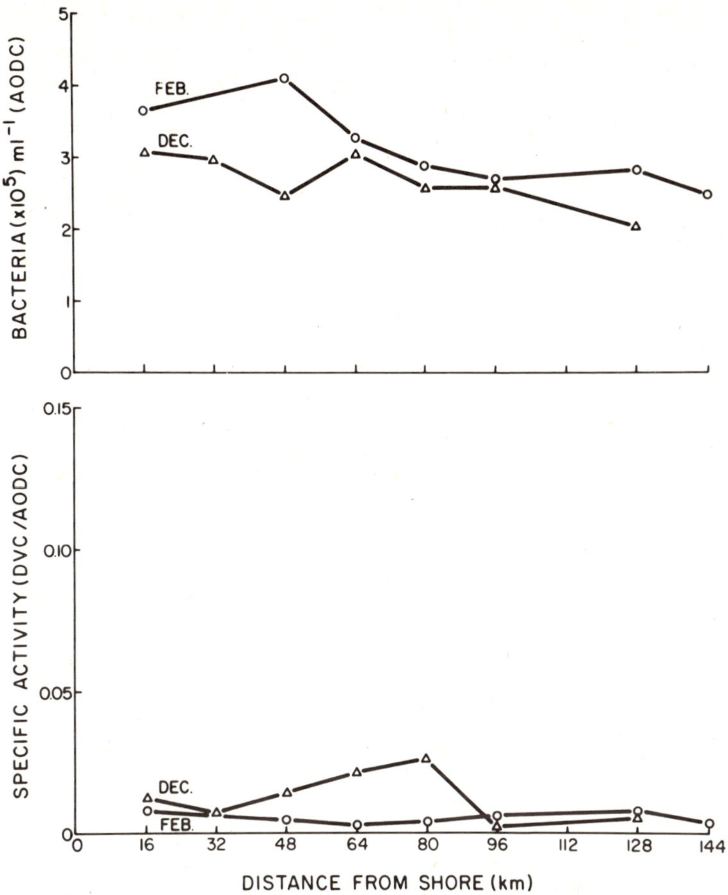

Figure 10.5. Total numbers of bacteria per ml, determined by acridine orange direct counts (AODC), and specific activities, calculated as the ratio of direct viable counts (DVC) to total counts (AODC), for surface water samples collected in December 1979 (△) and February 1980 (○) at the same stations along a transect passing through the western end of the dumpsite.

or overall levels of microbial activity. Recognition of more subtle, long-term effects, which are likely to be expressed as changes in community structure or function rather than size or activity, generally requires tedious methodologies that are limited to measurements of small or selective portions of the microbial community. For example in open ocean waters, including the Puerto Rico dumpsite, culturable bacteria, for which subsequent taxonomic analysis can be performed, represent only a small fraction of the total bacterial community as determined by acridine orange direct counts. For this study, microbial communities in surface waters of the Puerto Rico dumpsite and environs were examined using a variety of techniques, so that the effects, if any, of chronic exposure to pharmaceutical wastes could be evaluated from a broad basis of microbial data.

Although bacterial populations, measured by direct microscopic counts, ap-

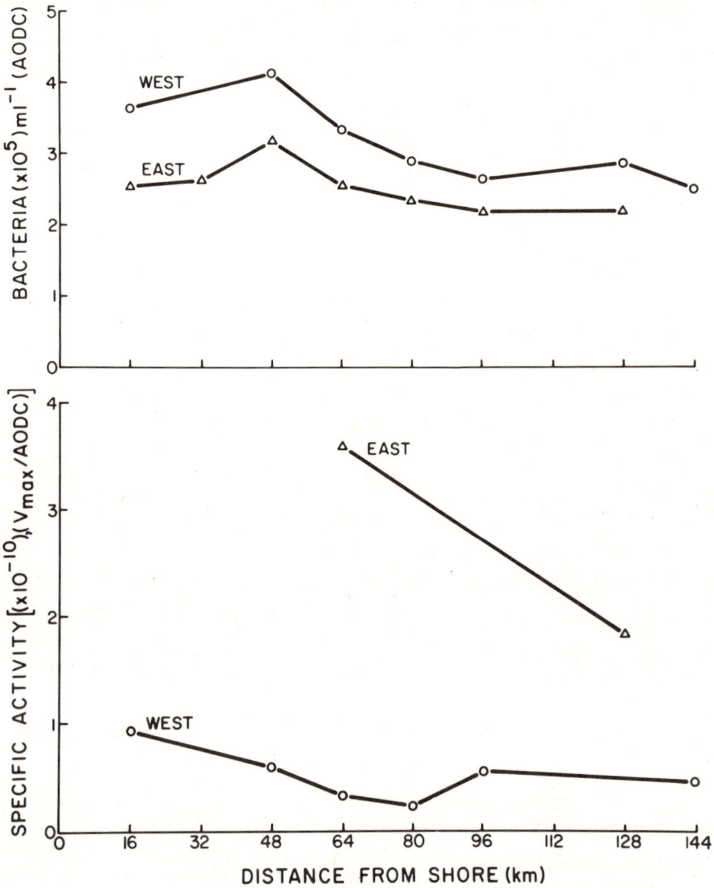

Figure 10.6. Total numbers of bacteria per ml, determined by acridine orange direct counts (AODC), and specific activities, calculated as the ratio of maximum uptake velocity of [^{14}C] amino acids (V_{max}) to total counts (AODC), for surface water samples collected in December 1979 at stations along a transect passing through the western end of the dumpsite (○), and collected in February 1980 at two stations along a transect passing through the eastern end of the dumpsite (△).

peared to be relatively stable throughout the region, changes in the specific activity indices DVC/AODC and V_{max}/AODC occurred which indicated variations in bacterial heterotrophic activity that may be related to location and dispersion of the pharmaceutical wastes (Figs. 10.4 to 10.6). The lowest specific activity (DVC/AODC) was recorded for a water sample collected in the waste plume 5 h after initial entry of the waste into surface waters (Table 10.1). Specific activity determined by DVC/AODC related only to populations of Gram-negative organisms, since Gram-positive bacteria are generally resistant to nalidixic acid used in the DVC assay. Small numbers of bacteria cultured on marine agar, all of which were found to be Gram-negative organisms, were also measured for those samples collected in the waste plume. During December 1979 and February 1980, when samples were collected at established transect stations and not within a waste plume, or in association with any (known) recent dumping activities, specific activities based on DVC/AODC and V_{max}/AODC were highest for samples collected within the dumpsite and at similar latitudes.

Low activity in the waste plume itself can be explained if there is an immediate, but temporary, inhibitory effect on the indigenous microflora arising from abrupt changes in exogenous substrate levels. It is also possible that the observed rate of uptake of amino acids underestimated microbial activity because of competition with other substrates introduced with the pharmaceutical waste material. After an acclimation period, the activity of indigenous bacteria may be expected to recover to previous or new levels, an explanation compatible with the higher specific activities recorded for transect stations sampled in the dumpsite, and particularly for those near its eastern boundary. Preliminary laboratory studies indicate that low levels of composite waste can stimulate bacterial activity in aquatic microcosms, causing significant increases in selected species of the bacterial community (F. L. Singleton and R. R. Colwell, unpublished results). Effects of the waste on heterotrophic activity, measured by changes in the uptake of radiolabeled substrates, in similar laboratory studies, have not yet been determined.

Taxonomic analysis indicates an apparent restructuring of the culturable segment of the bacterial community in these waters (Table 10.2). Such a dominance of the culturable community by members of the *Vibrio/Aeromonas* group has not been previously reported for waters of the open ocean. Sieburth (1971) conducted a survey of 381 samples collected to depths of 1200 m at 47 offshore stations in the Atlantic, Caribbean, and Pacific. His results showed that *Pseudomonas* spp. were dominant in the Atlantic and Caribbean samples. Similarly, Murchelano and Brown (1971) found that, in Long Island Sound, *Pseudomonas* spp. were dominant, accounting for about 41% of the culturable bacteria. The most common genera of bacteria in ocean waters have been reported to be *Pseudomonas*, *Vibrio*, *Flavobacterium*, *Arthrobacter*, *Caulobacter*, *Hyphomicrobium*, *Cytophaga*, *Acinetobacter*, and *Photobacterium* (Jones and Colwell, 1979). Within the dumpsite area, however, *Pseudomonas* and *Flavobacterium* spp. were found to occur at very low frequencies (<9%) (Table 10.2).

It is generally accepted that the majority of bacteria found in open ocean waters are Gram-negative organisms (ZoBell, 1946). Kriss (1963, 1971) reported isolation of Gram-positive organisms, (*Micrococcus* spp.) from water and sediment samples collected in the Black Sea, but did not conclude that they comprised a significant portion of the natural flora. In this study, large numbers of Gram-positive organisms were detected and have subsequently been identified as *Microcossus* spp. and *Staphylococcus* spp., including *Staphylococcus aureus* and *Staphylococcus epidermidis* (Gunn et al., 1982). Staphylococci accounted for the majority (90%) of the strains isolated on PCA during December 1979. It is possible that some of the Gram-positive cocci were derived originally from the pharmaceutical wastes disposed at the dumpsite. The presence of unusually large numbers of Gram-positive cocci, isolated on both MA and PCA, near the dumpsite, the increase in viable counts on PCA observed in waste plume studies in October 1979 (Table

10.1), and the presence of viable, Gram-positive cocci in all waste samples examined to date (data not presented), provide evidence in support of such an hypothesis. It is also possible that the persistence of pharmaceutical wastes in these waters gives a competitive advantage to the otherwise small population of indigenous Gram-positive bacteria. Taxonomic studies are in progress to trace the source of selected Gram-positive isolates, and to measure their ability to survive and grow in seawater containing varying concentrations of waste.

10.5. CONCLUSIONS

The widespread occurrence of waste-specific organic compounds throughout the study region illustrates that the potential impacts of ocean disposal of pharmaceutical wastes are not limited to the dumpsite itself. The results of the microbiological studies reported here suggest that ocean disposal of pharmaceutical wastes has immediate, though transient, effects on the size and specific activity of the microbial community in surface waters of the dumpsite. However, the recovery of the microbial populations from the acute effects of the pharmaceutical waste is accompanied by detectable, persistent alterations in the taxonomic composition of the culturable bacterial community of the receiving waters.

ACKNOWLEDGMENTS

The authors thank the officers and crew of *R/V Mt. Mitchell* and *R/V Eastward* for assistance in the collection of samples. We also thank Dr. George Harvey of the Atlantic Oceanographic and Meterological Laboratories, Miami, Florida, for gas chromatographic analysis of water samples.

This research was performed under National Oceanographic and Atmospheric Administration grant NA 79AA-D-00062. Support for bacterial identification and classification was provided, in part, by National Science Foundation grant DEB 77-14646,A02.

REFERENCES

American Public Health Association. 1976. Standard Methods for the Examination of Water and Wastewater, 14th ed. American Public Health Association, New York, 1193 pp.

Bain, N., and J. M. Shewan. 1968. Identification of *Aeromonas*, *Vibrio* and related organisms. *In*: Identification Methods for Microbiologists, Part B, B. M. Gibbs and D. A. Shapton (Eds.). Academic Press, New York, pp. 79–84.

Brooks, J. M., D. A. Wiesenburg, G. Bodennec, and C. R. Schwab. 1980. Dispersion and background measurements of volatile organics at Puerto Rico Deep Water Industrial Waste Dumpsite. Abstracts of 43rd Annual Meeting of the American Society of Limnology and Oceanography, Knoxville, Tennessee.

Buchanan, R. E., and N. E. Gibbons (Eds.). 1974. Bergey's Manual of Determinative Bacteriology, 8th ed. Williams and Wilkins, Baltimore.

Colwell, R. R., and J. D. Walker. 1977. Ecological aspects of microbial degradation of petroleum in the marine environment. *Critical Reviews in Microbiology*, **5**, 423–445.

Daley, R. J., and J. E. Hobbie. 1975. Direct counts of aquatic bacteria by a modified epifluorescence technique. *Limnology and Oceanography*, **20**, 875–882.

Francisco, D. E., R. A. Mah, and A. C. Rabin. 1973. Acridine orange epifluorescence technique for counting bacteria in natural waters. *Transactions of American Microscopical Society*, **92**, 416–421.

Gibson, D. W., M. S. Hendrie, N. C. Houston, and G. Hobbs. 1977. The identification of some Gram negative heterotrophic aquatic bacteria. *In*: Aquatic Microbiology, F. A. Skinner and J. M. Shewan (Eds.). Academic Press, New York, pp. 135–159.

Gunn, B. A., F. L. Singleton, E. R. Peele, and R. R. Colwell. 1982. Isolation and enumeration of Gram-positive cocci in marine and estuarine waters. *Journal of Applied Bacteriology*, in press.

Hobbie, J. E., and C. C. Crawford. 1969. Respiration corrections for bacterial uptake of dissolved organic compounds in natural water. *Limnology and Oceanography*, **14**, 528–532.

Hobbie, J. E., R. J. Daley, and S. Jasper. 1977. Use of Nuclepore filters for counting bacteria by fluorescence microscopy. *Applied and Environmental Microbiology*, **33**, 1225–1228.

Holm-Hansen, O., and C. R. Booth. 1966. The measurement of adenosine triphosphate in the ocean and its ecological significance. *Limnology and Oceanography*, **11**, 510–519.

Jones, G. E., and R. R. Colwell. 1979. Microbiology. A summary and analysis of environmental information on the continental shelf from the Bay of Fundy to Cape Hatteras (1977) Vol. 1, Book 3, Chapter 17, pp. 1–95. Bureau of Land Management Report, South Gardiner, Maine.

Ketchum, B. H., D. R. Kester, and P. K. Park (Eds.). 1981. Ocean Dumping of Industrial Wastes. Plenum Press, New York, 525 pp.

Kogure, K., U. Simidu, and N. Taga. 1979. A tentative direct microscopic method for counting living marine bacteria. *Canadian Journal of Microbiology*, **25**, 415–420.

Kriss, A. E. 1963. Marine Microbiology (Deep Sea), J. M. Shewan and Z. Kabata (Trans.). Wiley-Interscience, New York, 536 pp.

Kriss, A. E. 1971. Oceanic microbiology: Ecology and geography of microorganisms. *Mikrobiologia*, **40**, 904–911.

Murchelano, R. A., and C. Brown. 1970. Heterotrophic bacteria in Long Island Sound. *Marine Biology*, **7**, 1–6.

Murphy, L. S., P. R. Hoar, and R. A. Belastock. 1981. The effects of industrial wastes on marine phytoplankton. *In*: Ocean Dumping of Industrial Wastes, B. H. Ketchum, D. R. Kester, and P. K. Park (Eds.). Plenum Press, New York, pp. 399–410.

Orndorff, S. A., and R. R. Colwell. 1980. Effect of Kepone on estuarine microbial activity. *Microbial Ecology*, **6**, 357–368.

Sieburth, J. M. 1971. Distribution and activity of oceanic bacteria. *Deep-Sea Research*, **18**, 1111–1121.

Swan, R. 1979. Pharmaceutical industry sludge: drug makers face waste management headache. *Sludge Magazine* **2**(4), 21–25.

Wright, R. T. 1978. Measurement and significance of specific activity in the heterotrophic bacteria of natural waters. *Applied and Environmental Microbiology*, **36**, 297–305.

Wohlschlag, D. E., F. R. Parker, E. D. Findlay, and J. A. Kinney. 1981. Metabolic sensitivity of fish to ocean dumping of industrial wastes. *In*: Ocean Dumping of Industrial Wastes, B. H. Ketchum, D. R. Kester, and P. K. Park (Eds.). Plenum Press, New York, pp. 455–471.

ZoBell, C. E. 1946. Marine Microbiology. Chronica Botanica, Waltham, Massachusetts, 240 pp.

11

PHYTOPLANKTON: COMPARISON OF LABORATORY BIOASSAY AND FIELD MEASUREMENTS

Lynda S. Murphy, Elin M. Haugen, and Jeffrey F. Brown

Bigelow Laboratory for Ocean Sciences
West Boothbay Harbor, Maine

ABSTRACT

A laboratory bioassay was developed to predict the effects on the phytoplankton community of the several wastes disposed of at deep-ocean dumpsites. The bioassay measured change in chlorophyll fluorescence and cell number of specified clones in a controlled environment. For all phytoplankton clones tested, the pharmaceutical and the American Cyanamid wastes were more toxic than the DuPont-Edge Moor waste, which was more toxic than the DuPont-Grasselli waste.

Centric diatoms and dinoflagellates were more sensitive to the wastes than were pennate diatoms and some representatives of the monad classes, but some degree of resistance was shown in isolates of other classes established from polluted estuaries.

These bioassays predict that changes in community structure should occur at the concentration existing in the wake of the barge during dumps. Initial studies at the Puerto Rico dumpsite showed an immediate, short-term pulse effect on the community structure, with dinoflagellates decreasing and monads increasing.

11.1. INTRODUCTION

The oceans have always been a repository for man's wastes, but the ever-increasing volume (Chapter 1) of these wastes now raises the question of damage to the marine biota. It is a difficult question to evaluate. We know that initial dilution of most industrial wastes occurs rapidly in the wake of the dumping barge, but that further dilution is slow, and potentially toxic levels of wastes can remain in the environment for at least several days after a dump (Csanady, 1981). We know also that the response of specific phytoplankton strains to the various wastes is a function of the previous pollution exposure of that strain (Murphy and Belastock, 1980). But we have learned that the dumpsites are hydrographically complex (Chapters 4 and 5), and that this results in an enormous natural variation in phytoplankton biomass and species composition (Ortner and Murphy, 1981).

When a marine environment becomes polluted, the different sensitivities of the various exposed phytoplankters could result in an altered community structure, even in the absence of biomass changes. Such an alteration would affect the structure of the food web and could result in shifts in dominance. Since other environmental changes also alter phytoplankton biomass and community structure, demonstration of a clear cause-and-effect relationship between environmental contamination and phytoplankton damage has been difficult. The large natural variation in physical parameters at the deepwater dumpsites and the resulting changes in phytoplankton dominance, diversity, and biomass (Ortner and Murphy, 1981) have made it difficult to assign any specific effect to a single cause.

Bioassay studies using natural communities (Parsons et al., 1976; Thomas and Seibert, 1977; O'Connors et al., 1978) have been complicated by the effect of the container itself; changes in species composition caused by containerization were often as large as those caused by the stress being evaluated. Also, communities indigenous to the deep-ocean dumpsites are difficult to obtain and impossible to maintain as communities. Thus, we have based our system in the present work on measured responses of clonal cultures, using a variety of phytoplankton species from specified (and environmentally realistic) origins. Our goals were (1) to rank a wide range of clones by order of growth-rate depression caused by the several wastes being tested, (2) to determine ecological and/or phylogenetic trends in the rank order, and (3) on this basis, to predict what should happen, *in situ*. The industrial wastes tested originated from E. I. du Pont de Nemours & Co.'s Grasselli (Linden, New Jersey), and Edge Moor (Edge Moor, Delaware) plants, American Cyanamid (Linden, New Jersey), and the pharmaceutical industries in Puerto Rico.

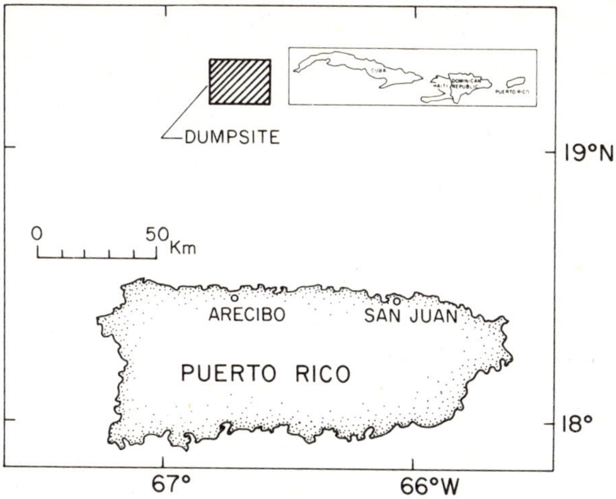

Figure 11.1. Chart of the location of the Puerto Rico dumpsite.

The predictions of the bioassay system were tested by subsequent on-site studies at the Puerto Rico dumpsite, where the hydrographic regime is less complex than at Deepwater Dumpsite-106 (DWD-106) in the New York Bight, and where only the pharmaceutical waste is dumped. The dumpsite lies about 74 km north of Arecibo, Puerto Rico (Fig. 11.1), and overlies the Puerto Rico Trench. The waste is barged by PCI International, Inc. (Anderson and Dewling, 1981), to the dumpsite and discharged directly in 2.5×10^3 m^3 lots, with total annual input of about 3.0×10^5 m^3. The pharmaceutical waste consisted of a composite of aqueous wastes from seven pharmaceutical firms located in Puerto Rico. They contain many organic and inorganic chemicals used for drug synthesis and purification (Atlas et al., 1981; Presley et al., 1981; Schwab et al., 1981). The actual composition varies from lot to lot, but in general, the trace metal concentrations are relatively low (< 1 mg liter^{-1}) and total organic carbon content is high (1–4%). A more detailed description of the organic chemistry of these wastes is given by Brooks et al. (Chapter 9).

11.2. THE BIOASSAY SYSTEM

11.2.1. Methods

The bioassay system described below is diagrammed in Fig. 11.2. The growth medium was a modification of the f/2 medium of Guillard and Ryther (1962), as follows: aged Sargasso seawater (stored in polyethylene carboys) was filtered through a 0.8-μm filter, then autoclaved in 2-liter Teflon containers. Nutrients, each previously autoclaved separately in Teflon containers, were aseptically added to the final concentration of f/2, except that EDTA was omitted. The resulting medium was aseptically dispensed into 25×150 mm sterile, capped culture tubes (25 ml per tube). Then 0.1 ml of waste was added as a solution (in sterile distilled water) 250 times the final concentration desired. This presolution of waste per-

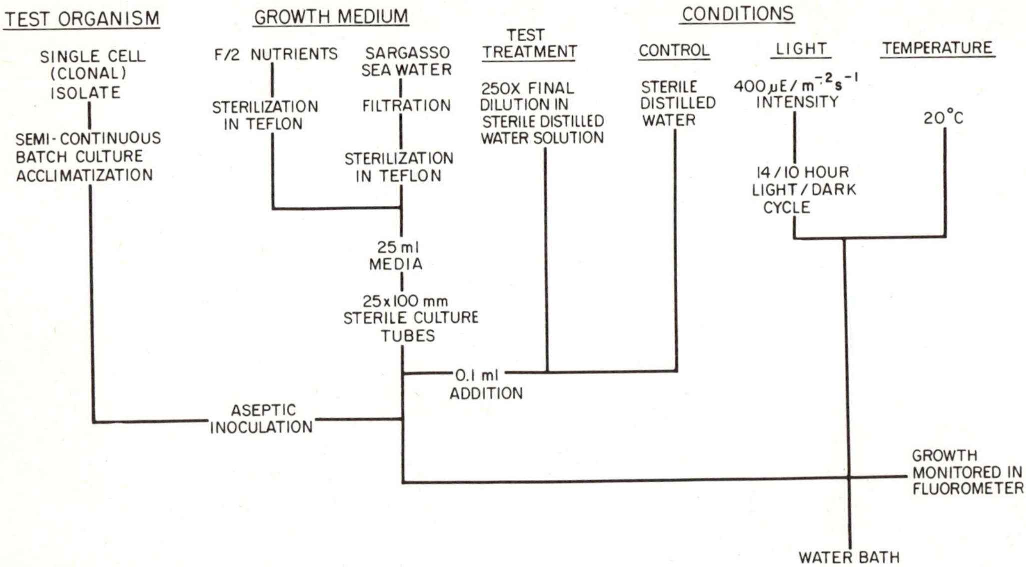

Figure 11.2. Flow diagram of the bioassay system.

mitted more accurate concentrations in the final dilution and did not affect average toxicity when compared to direct addition (Murphy et al., 1981a). Control tubes received 0.1 ml of sterile, distilled water.

The clones used in this study are all maintained in the culture collection of the Bigelow Laboratory for Ocean Sciences (Table 11.1). Before a bioassay, the clones were acclimated for several weeks under control test conditions (minus waste) in semicontinuous batch culture (Brand et al., 1981). At the onset of a bioassay, a specific clone was inoculated into each tube at a concentration calculated to give an initial reading of 1.0 fluorescence units in a Turner Design (Mountain View, California) Model 10 fluorometer. This corresponded to 10^3–10^4 cells ml^{-1}, depending on the clone.

Each bioassay consisted of several clones and several concentrations of waste and each set of conditions (clone and concentration) was repeated in triplicate. Thus, a single bioassay consisted of 100–300 tubes run simultaneously.

The tubes were placed in an illuminated water bath (Brand, 1980) at 20°C and 400 μE m^{-2} s^{-1} (E = einsteins), on a 14:10 h light:dark cycle. Each tube was removed from the water bath daily (at the same time each day), allowed to equilibrate 20 min to reduce the Kautsky induction effect (Loftus and Seliger, 1975), read in the fluorometer directly in the tube, and returned to the water bath. Readings continued until the control tube (0.1 ml sterile distilled water added in place of waste) reached the end of log phase growth (usually 72–120 h, depending on the clone). The growth rate, μ, defined as divisions d^{-1}, was calculated from the logarithm of the fluorescence during the log phase of growth, using linear regression. The depression in growth rate caused by the waste was expressed as $\mu_w:\mu_c$ (the ratio of growth rate in the treated tubes to growth in the control tubes). Each response was the average of the triplicates.

Table 11.1. Origin of Clones Used in This Study

Species	Clone	Origin	Date	Isolator
BACILLARIOPHYCEAE				
Thalassiosira pseudonana	3H	Great South Bay, Long Island, New York	1958	R. Guillard
	W	Wümme River Estuary, Bremen, Germany	1973	E. Paasche
	Swan-1	Swan River Estuary, Perth, W. Australia	1965	R. Davis
	13–1	33° 11′N, 65° 15′W, Sargasso Sea	1958	R. Guillard
	58–102	39° 05′N, 71° 56′W, Continental Slope	1976	L. Murphy
Phaeodactylum tricornutum	Phaeo	Great South Bay, Long Island, New York	1952	J. Ryther
Nitzschia breviostus	0–1	Oyster Pond, Martha's Vineyard, Massachusetts	1956	R. Guillard
Fragilaria pinnata	13–3	Sargasso Sea	1958	R. Guillard
DINOPHYCEAE				
Scrippsiella trochoidea	Peri	British coastal waters, exact origin unknown	Unknown	M. Parke
CHLOROPHYCEAE				
Dunaliella tertiolecta	Dun	Plymouth, England	Unknown	From the Cambridge Culture Collection
EUSTIGMATOPHYCEAE				
Nannochloropsis salina	GSB Sticho	Great South Bay, Long Island, New York	1952	J. Ryther
	Say 3	Great South Bay, Long Island, New York	1965	R. Guillard
PRYMNESIOPHYCEAE				
Isochrysis galbana	Iso	Plymouth, England	1956	M. Parke
Pavlova (*Monochrysis*) *lutheri*	Mono	Finnish tidal rock pool	1956	M. Droop
Pavlova sp.	Nep	Gulf Stream	1958	R. Guillard

At the end of the fluorescence determinations, an aliquot from each tube was preserved with Lugol's iodine solution for subsequent cell counts. The increase in chlorophyll fluorescence was found to be proportional to the increase in cell numbers during the logarithmic phase of growth.

11.2.2. Results

The toxicities of the four major wastes disposed of at deep-ocean dumpsites are compared in Table 11.2. The DuPont-Grasselli waste was clearly the least toxic to the *Thalassiosira pseudonana* clones tested. At 0.1 ml liter^{-1}, the highest concentration of this waste likely to persist in the wake of a barge (Kohn and Rowe,

Table 11.2. Relative Growth Rates ($\mu_W:\mu_c \pm \sigma$) of Four Clones of *Thalassiosira pseudonana* in Different Waste Concentrations

Wastes and Concentrations (ml liter^{-1})	Neritic Clones		Oceanic Clones	
	3H	W	13-1	58-102
Dupont-Grasselli				
0.1	1.01 ± 0.02	1.01 ± 0.01	0.95 ± 0.02	0.82 ± 0.02
1.0	0.98 ± 0.02	1.01 ± 0.02	0.73 ± 0.05	0.62 ± 0.04
Dupont-Edge Moor				
0.1 (pH 7.8)	1.13 ± 0.06	1.08 ± 0.03	1.13 ± 0.12	1.16 ± 0.05
1.0 (pH 4.1)	Died	Died	Died	Died
1.0 (pH 8.1)	0.68 ± 0.03	0.63 ± 0.11	0.45 ± 0.09	0.22 ± 0.12
None[a] (pH 4.1)	Died	Died	Died	Died
American Cyanamid				
0.1	0.97 ± 0.01	0.65 ± 0.01	0.76 ± 0.05	0.82 ± 0.04
1.0	0.44 ± 0.02	0.01 ± 0.02	Died	Died
Puerto Rico Pharmaceutical (Lot: February, 1979)				
0.1	0.98 ± 0.04	0.97 ± 0.02	0.80 ± 0.02	0.68 ± 0.05
1.0	Died	Died	Died	Died

[a]This test system contained no waste; the pH of the medium was reduced to 4.1 by the addition of HCl.

1981), all clones except the oceanic 58-102 showed normal growth. At 1.0 ml liter^{-1}, there was significant difference at the 5% level between the growth-rate responses of the neritic and the more sensitive oceanic clones.

The DuPont-Edge Moor waste was slightly stimulatory to growth at 0.1 ml liter^{-1}. Low-dose stimulation by toxic substances has been reported by Dunstan et al. (1975), Parsons et al. (1976), and Murphy et al. (1981a). At 1.0 ml liter^{-1}, the pH of the medium was 4.1 and none of the clones survived. An equivalent toxic effect was produced by reducing the pH of the medium to 4.1 by addition of HCl, but medium containing 1.0 ml liter^{-1} waste whose pH was increased to pH 8.1 with NaOH, was itself clearly toxic, significantly more so (at the 5% level) to oceanic than to neritic clones.

The American Cyanamid waste was highly toxic at 1.0 ml liter^{-1}. Only clone 3H showed any growth, and that was reduced to 44% of control growth. At 0.1 ml liter^{-1}, the resistant clone 3H was not affected by the waste, but the remaining clones showed significant (at the 5% level) reduction of growth.

The pharmaceutical waste was the most toxic of the wastes tested (on a per volume basis). It alone produced significant (at the 5% level) growth-rate depression at 0.1 ml liter^{-1}, the highest concentration likely to persist at the Puerto Rico dumpsite (0.01 to 0.1 ml liter^{-1}), 10 h after dumping (T. P. O'Connor, personal communication). Again, neritic clones were significantly (at the 5% level) less affected than oceanic clones. At 1.0 ml liter^{-1}, both the neritic and the oceanic clones were killed.

Thus, growth rate was found to be a function of the type and level of waste, and of the origin of the clone tested. These observed differences were all significant at the 5% level.

Table 11.3. Relative Growth Rates ($\mu_w:\mu_c \pm \sigma$) of Taxonomically Diverse Clones in Diluted Pharmaceutical Waste (Lot: October, 1979)

Clone	Origin	Concentration (ml liter^{-1}) 0.1	1.0
BACILLARIOPHYCEAE (centric)			
3H	Neritic, polluted	0.89 ± 0.02	Died
Swan-1	Neritic, clean	0.78 ± 0.06	Died
13–1	Oceanic	0.28 ± 0.03	Died
BACILLARIOPHYCEAE (pennate)			
Phaeo	Neritic, polluted	1.08 ± 0.03	0.91 ± 0.05
0–1	Neritic, clean	1.02 ± 0.09	0.15 ± 0.15
13–3	Oceanic	1.10 ± 0.08	Died
DINOPHYCEAE			
Peri	Neritic, polluted	0.80 ± 0.12	Died
CHLOROPHYCEAE			
Dun	Neritic, polluted	1.02 ± 0.03	1.10 ± 0.02
EUSTIGMATOPHYCEAE			
GSB Sticho	Neritic, polluted	1.02 ± 0.07	Died
Say 3	Neritic, polluted	1.03 ± 0.03	Died
PRYMNESIOPHYCEAE			
Iso	Neritic, polluted	0.87 ± 0.05	0.05 ± 0.09
Mono	Neritic, clean	0.62 ± 0.02	Died
Nep	Oceanic	0.63 ± 0.06	Died

The responses of all thirteen clones, representing five classes of marine phytoplankton, were measured against a single waste (pharmaceutical, October 1979 lot) (Table 11.3). At 1.0 ml liter^{-1}, only the pennate diatom clone Phaeo and the chlorophyte clone Dun showed any significant growth. At 0.1 ml liter^{-1}, neither the pennate diatoms nor certain monads (the chlorophyte and the eustigmatophytes) were affected by the waste; the representatives of the other classes all showed at least some growth-rate depression.

In general, the extent of growth-rate depression within any one class was a function of origin of the tested clone. Clones from each of the types tested were not available for all classes. But for those tested, there was a tendency for clones from polluted estuaries to be less affected by the waste than clones from clean estuaries, and for clean-estuary clones to show less effect than clones of oceanic origin.

The highly resistant clone Dun was allowed to grow in the presence and absence of the pharmaceutical waste (0.1 ml liter^{-1}), both in the light and in the dark.

Table 11.4. Effect of Pharmaceutical Waste (Lot: October, 1979) at 0.1 ml liter^{-1} on the Growth Rate of Light- and Dark-Grown *Dunaliella tertiolecta*

	Treatment: Light–Control	Light–Waste	Dark–Control	Dark–Waste
Growth rate (μ) (divisions day^{-1} $\pm \sigma$)	1.78 ± 0.03	1.79 ± 0.02	0.33 ± 0.02	1.45 ± 0.42
$\mu_w:\mu_c \pm \sigma$	–	1.01 ± 0.02	–	4.39 ± 0.90

Growth rates in the light were monitored by fluorescence, and in the dark by direct cell count. Table 11.4 shows the effect of the waste on growth rate under these four sets of conditions. Although the difference between waste-treated and control growth rates was not significant when they were grown in the light, dark-grown cultures responded to the waste by a fourfold increase in growth rate, resulting in growth similar to light-grown cultures. The large standard deviation (Table 11.4) for dark-grown waste-exposed cultures was caused by a tendency for the cells to clump around waste particles, which made counting difficult. This tendency was similar to the behavior of the phagotrophically feeding *Gymnodinium* described by Spero and Morée (1981). The waste contains bacteria (Singleton et al., Chapter 10) which may accumulate on these particle surfaces, but to our knowledge *Dunaliella* has never been reported to feed on bacteria. Clone Dun may be able to utilize some unidentified organic component of the waste for heterotrophic growth.

11.3. DUMPSITE STUDIES

11.3.1. Methods

Phytoplankton samples were collected in May and October of 1979 at the Puerto Rico dumpsite (Fig. 11.3). On each occasion, the pharmaceutical waste was dyed with rhodamine and tracked by drogue and fluorometry. Water samples were drawn from Niskin (General Oceanics, Inc., Miami, Florida) bottle casts at specified depths at predump stations (background or control stations) and at stations in the tracked waste plume for 56 and 36 h after the dump, respectively. Each water sample collected (280 ml) was immediately preserved with Lugol's solution on shipboard and stored in reduced light.

In the laboratory, samples were concentrated by sedimentation to 0.7 ml and subsamples of 0.1 ml were scanned in a Palmer–Maloney cell using a trinocular light microscope (Carl Zeiss, Inc., New York). Enumerations of the major divisions of pennate and centric diatoms, dinoflagellates, cyanophytes, and monads (*sensu* Hasle, 1978) were made under 400X magnification; at least 80–100 cells per chamber were counted. Storage in acidic Lugol's preservative eliminated coccolithophores from the samples; they are not included in monad counts.

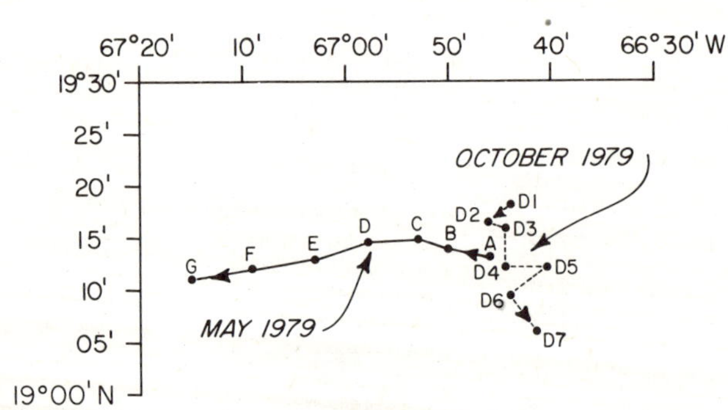

Figure 11.3. Locations of stations at the Puerto Rico dumpsite, May and October 1979.

11.3.2. Results

Total cell counts of the preserved samples from all stations varied from 2×10^3 to 9×10^3 cells liter^{-1}. A decrease in total cell number was observed after both the May 1979 and October 1979 dumps. At surface, 5-m, and 10-m depths, the decrease occurred at the first postdump stations, 3 h after the May dump and 3.5 h after the October dump. At 25 m, the decrease was seen at the second postdump stations, 9.7 h after the May dump and 8.3 h after the October dump. Since samples were not taken between these two sets of stations, neither exact peaks nor time courses can be determined. It seems likely, however, that the lag observed at the 25-m depths was due to the time required for the waste material to penetrate to that depth. Hulburt and Jones (1977) also noted a decrease in total cell number following a waste dump of American Cyanamid waste at DWD-106 (in the New York Bight) in August 1976.

Figures 11.4 and 11.5 shows the percent composition of major cell types observed at the dumpsite. In both May 1979 and October 1979, dinoflagellates tended to dominate at all depths sampled. Monads were the second largest group, and occasionally occurred in larger numbers than dinoflagellates. Pennate diatoms were important in October (Fig. 11.5), but not in May (Fig. 11.4). Centric diatoms and cyanophytes (*Oscillatoria erythraea*) occurred rarely, both in May and in October.

In the wake of the barge, a change to a monad-dominated community occurred (see "waste dump" in Figs. 11.4 and 11.5). This was seen at the first postdump station (3.0–3.5 h after the dumps) at the shallower depths, and at the second postdump stations (8.3–9.7 h after the dumps) at the 25-m depths. Figures 11.4 and 11.5 show actual cell counts for the major cell types observed. It can be seen that, with the exception of the 5-m October series of stations (Fig. 11.5a), the change from a dinoflagellate-dominated community to one dominated by monads was caused not only by a rapid decrease in dinoflagellates, but also by an actual increase in monads. At the 5-m October series of stations, both monads and dinoflagellates decreased, but dinoflagellates more rapidly than monads, so that this depth also resulted in a monad-dominated community at the first postdump station.

11.4. DISCUSSION

Neritic diatoms have often been observed to be more resistant to stress than their oceanic counterparts (Guillard and Ryther, 1962; Fisher, 1977; Murphy and Belastock, 1980; Murphy et al., 1981b). Murphy and Belastock (1980) and Murphy et al. (1981b) concluded that much of this resistance was due to previous exposure, rather than to a generalized resiliency of neritic forms. Our bioassay results have shown the generalization of both of these observations to groups other than centric diatoms: that neritic phytoplankton clones are more resistant than closely related oceanic clones, and that neritic isolates from polluted estuaries are more resistant than closely related isolates from relatively clean estuaries.

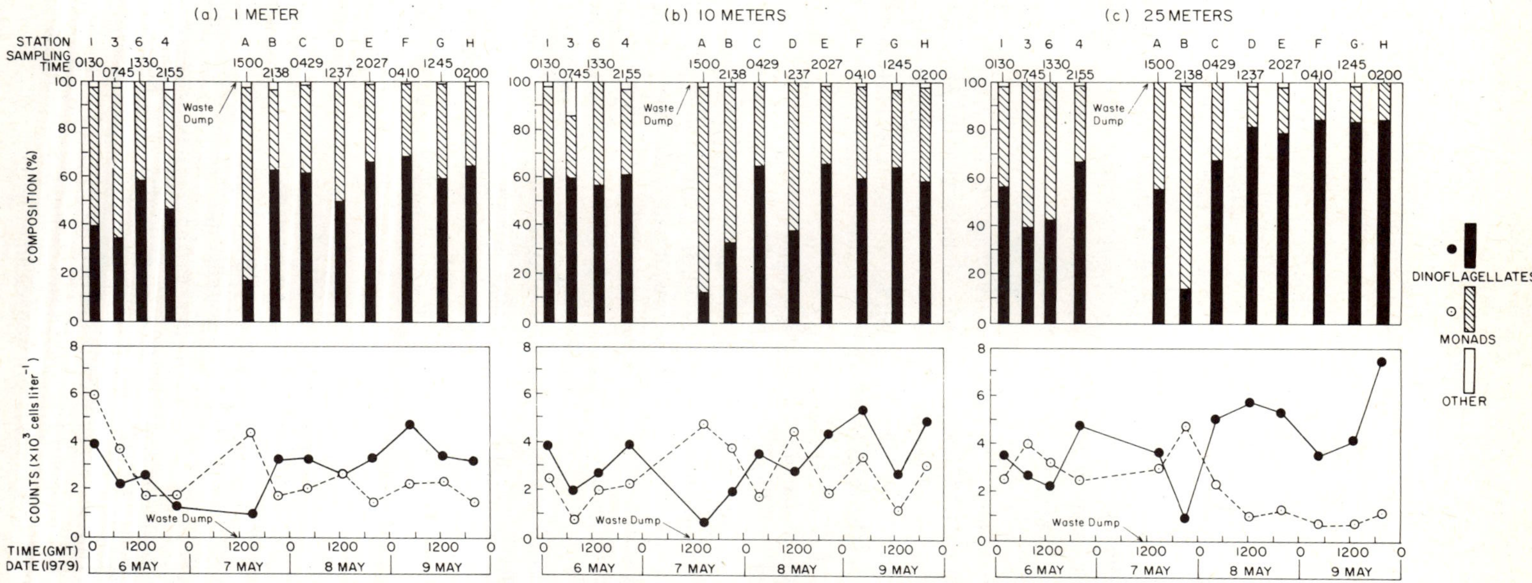

Figure 11.4. Changes in percent composition and actual cell counts of phytoplankton as a result of a pharmaceutical waste dump in May of 1979: (a) 1 m; (b) 10 m; and (c) 25 m.

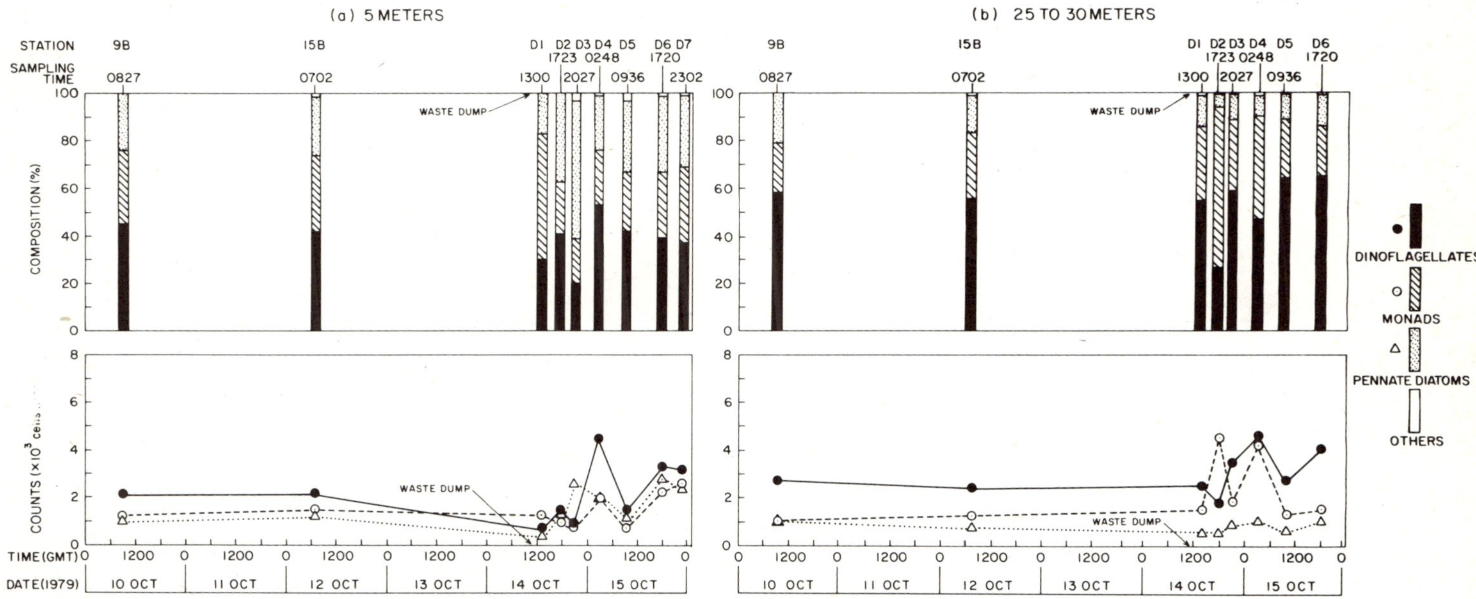

Figure 11.5. Changes in percent composition and actual cell counts of phytoplankton as a result of a pharmaceutical waste dump in October of 1979: (a) 5 m; and (b) 25–30 m.

The chlorophyte clone Dun and the pennate diatom clone Phaeo were significantly more resistant (at the 5% level) than any other neritic isolate. Other workers have demonstrated similar differential sensitivities in these clones. Dunstan et al. (1975) found clone Dun less sensitive to aromatic hydrocarbons than the centric diatom *Skeletonema costatum* (clone Skel, from Long Island Sound) or the dinoflagellate *Amphidinium carterae* (clone Amphi from Falmouth, Massachusetts). Menzel et al. (1970) found clone 13-1 (an oceanic isolate of the centric diatom *Thalassiosira pseudonana*; see Tables 11.1 to 11.3) far more sensitive than clone Dun to chlorinated hydrocarbons. Mosser et al. (1972) tested the effects of chlorinated hydrocarbons on mixed cultures of Dun and clone 3H (conspecific to clone 13-1, but neritic in origin; see Tables 11.1 to 11.3). Although the diatom outgrew the chlorophyte in the control cultures, the latter predominated in treated cultures, even at concentrations that did not inhibit 3H in pure culture.

Our dumpsite enumerations have shown a postdump increase in monads. Hasle (1978) defined a monad as "a small cell (about 10 μm or less) with a more or less spherical outline" and included unidentifiable flagellates in the group. Thus, the term monad shares a certain taxonomic vagueness with the term worm, which has been defined (Chandler and Read, 1930) as "any elongated creeping thing that is not obviously something else." Nonetheless, when enumerating preserved samples with a light microscope, it is often impossible to be more specific. Even with the most gentle filtration or with centrifugation, flagellae become lost, and class designations within the monads are based more on pigment composition than on morphology. The point is that chlorophytes like Dun would be counted with the monads.

Correlations between species composition changes and waste addition are difficult to document. Observed changes in population structure may be caused by a number of factors. In order to account for the changed structure solely on the basis of differential resistance to toxic wastes, it would be necessary to assume a very rapid growth rate. Division times in the open ocean have been estimated at from 3 h (Sheldon and Sutcliffe, 1978) to 5 d (Sharp et al., 1980). The observed ability of clone Dun to utilize the waste for growth could lead to higher growth rates.

In addition to these direct differential responses, an indirect effect could be caused or augmented by a waste-induced change in predator:prey ratios. Many of the dinoflagellates occurring at the dumpsite were colorless, and these nonphotosynthesizing plankters have been observed to feed, phagotrophically, on *Dunaliella* (Spero and Morée, 1981). Thus, any factor limiting the growth or feeding behavior of the dinoflagellates would permit an absolute increase in monads by reduction of grazing pressure on that group.

Except for the monad pulse observed in the wake of the barge, dinoflagellates dominated at 87% of the dumpsite stations. Monad domination was characteristic of the wake of the barge and occurred at the first postdump stations at 0, 5 and 10 m, or at the second postdump stations at 25 m. But monad-dominated samples occurred at 13% of other stations within the dumpsite. Further analyses may show these samples to be correlated with areas of residual waste; at present we must assume that such monad-dominated communities can occur naturally. The probability of such an occurrence would be approximately 0.13. The probability of every observed postdump monad pulse being a random, or naturally occurring event, would be the product of the separate probabilities, or $P \cong (0.13)^5 < 0.01$.

Although a direct cause-and-effect relationship cannot be rigorously proved, it is highly probable. The bioassay results predicted differential survival of monads and pennate diatoms, at the expense of dinoflagellates. Feeding habits of the waste-sensitive dinoflagellates and heterotrophic potential of the waste-resistant monads would augment this effect. Further, the concentration of waste persisting in the wake of the barge (0.01–0.1 ml liter^{-1}, 10 h after dumping) is within the range where such an effect should be anticipated. We believe that the observed short-term change in community structure is most likely caused by the waste disposal. We have, at present, no bioassay or field results demonstrating long-term, persistent changes in phytoplankton community structure.

11.5. CONCLUSIONS

Bioassay results showed that, although resistance to the pharmaceutical waste was a function of origin in any one group, the pennate diatoms and certain monads were more resistant than other groups.

In situ dumpsite results showed an immediate, short-term change in phytoplankton community structure, with monads increasing and dinoflagellates decreasing.

Although a direct cause-and-effect relationship cannot be rigorously proved, it is highly probable ($P > 0.99$) that the observed pulse was the result of waste-induced differential growth augmented by heterotrophic use of the waste in the monads, and by decreased feeding on monads by the dinoflagellates.

No persistent, long-term changes in phytoplankton community structure were observed.

ACKNOWLEDGMENTS

D. Sampson helped us with statistical analyses and J. Rollins prepared the figures; we are most grateful. R. Belastock and T. O'Connor obtained and preserved samples for us on the May and October 1979 cruises, respectively. Our thanks to P. Beane, S. Davis, L. Janks, M. Marston, and R. Wahlstrom, who cleaned glassware and counted cells. This study was supported by the National Oceanographic and Atmospheric Administration, grant number NA 80 ADD-00033.

REFERENCES

Anderson, P. W., and R. T. Dewling. 1981. Industrial ocean dumping in EPA Region II—Regulatory aspects. *In*: Ocean Dumping of Industrial Wastes, B. H.

Ketchum, D. R. Kester, and P. K. Park (Eds.). Plenum Press, New York, pp. 25–37.

Atlas, E., G. Martinez, and C. S. Giam. 1981. Chemical characterization of ocean-dumped waste materials. *In*: Ocean Dumping of Industrial Wastes, B. H. Ketchum, D. R. Kester, and P. K. Park (Eds.). Plenum Press, New York, pp. 275–293.

Brand, L. E. 1980. Genetic Variability and Differentiation in Niche Components of Marine Phytoplankton Species. Ph.D. Thesis, Woods Hole Oceanographic Institution, Woods Hole, Massachusetts, 227 pp.

Brand, L. E., R. R. L. Guillard, and L. S. Murphy. 1981. A method for rapid and precise determination of acclimated phytoplankton reproduction rates. *Journal of Plankton Research*, **3**, 193–201.

Chandler, A. C., and C. P. Read. 1930. Introduction to Parasitology. Wiley, New York, 237 pp.

Csanady, G. T. 1981. An analysis of dumpsite diffusion experiments. *In*: Ocean Dumping of Industrial Wastes, B. H. Ketchum, D. R. Kester, and P. K. Park (Eds.). Plenum Press, New York, pp. 109–129.

Dunstan, W. M., L. P. Adkinson, and J. Natoli. 1975. Stimulation and inhibition of phytoplankton growth by low molecular weight hydrocarbons. *Marine Biology*, **31**, 305–310.

Fisher, N. S. 1977. On the differential sensitivity of estuarine and open-ocean diatoms to exotic chemical stress. *American Naturalist*, **111**, 871–895.

Guillard, R. R. L., and J. H. Ryther. 1962. Studies on marine planktonic diatoms. I. *Cyclotella nana* Hustedt and *Detonula confervacea* (Cleve.) Gran. *Canadian Journal of Microbiology*, **8**, 229–239.

Hasle, G. R. 1978. Identification problems. *In*: Phytoplankton Manual, A. Sournia (Ed.). United Nations Educational, Scientific and Cultural Organization, Paris, France, pp. 125–128.

Hulburt, E. M., and C. M. Jones. 1977. Phytoplankton in the vicinity of Deepwater Dumpsite 106. *In*: Baseline Report of Environmental Conditions in Deepwater Dumpsite 106. Dumpsite Evaluation Report 77-1, National Oceanic and Atmospheric Administration, National Ocean Survey, Rockville, Maryland, pp. 219–231.

Kohn, B., and G. T. Rowe. 1981. Dispersion of two liquid industrial wastes dumped at 106-mile site. *In*: Assessment Report on the Effects of Waste Dumping in 106-Mile Ocean Waste Disposal Site. Dumpsite Evaluation Report 81-1, Office of Marine Pollution Assessment, National Oceanic and Atmospheric Administration, Boulder, Colorado, pp. 133–155.

Loftus, M. E., and H. H. Seliger. 1975. Some limitations of the *in vivo* fluorescence technique. *Chesapeake Science*, **16**, 79–92.

Menzel, D. W., J. Anderson, and A. Randke. 1970. Marine phytoplankton vary in their response to chlorinated hydrocarbons. *Science*, **167**, 1724–1726.

Mosser, J. L., N. S. Fisher, and C. F. Wurster. 1972. Polychlorinated biphenyls and DDT alter species composition in mixed cultures of algae. *Science*, **176**, 533–535.

Murphy, L. S., and R. A. Belastock. 1980. The effects of environmental origin on the response of marine diatoms to chemical stress. *Limnology and Oceanography*, **25**, 160–165.

Murphy, L. S., P. R. Hoar, and R. A. Belastock. 1981a. The effects of industrial wastes on marine phytoplankton. *In*: Ocean Dumping of Industrial Wastes, B. H. Ketchum, D. R. Kester, and P. K. Park (Eds.). Plenum Press, New York, pp. 399–410.

Murphy, L. S., R. R. L. Guillard, and J. Gavis. 1982. Evolution of resistant phytoplankton strains through exposure to marine pollutants. *In*: Ecological Stress and the New York Bight: Science and Management, G. F. Mayer (Ed.). Estuarine Research Foundation, Columbia, South Carolina, pp. 401–412.

O'Connors, H. B., Jr., C. F. Wurster, C. D. Powers, D. C. Biggs, and R. G. Rowland. 1978. Polychlorinated biphenyls may alter marine trophic pathways by reducing phytoplankton size and production. *Science*, **201**, 737–739.

Ortner, P. B., and L. S. Murphy. 1981. Biological effects of a dump at 106-mile site. *In*: Assessment Report on the Effects of Waste Dumping in 106-Mile Ocean Waste Disposal Site. Dumpsite Evaluation Report 81-1, Office of Marine Pollution Assessment, National Oceanic and Atmospheric Administration, Boulder, Colorado, pp. 245–255.

Parsons, T. R., W. K. W. Li, and R. Walters. 1976. Some preliminary observations on the enhancement of phytoplankton growth by low levels of mineral hydrocarbons. *Hydrobiologia*, **51**, 85–89.

Presley, B. J., J. S. Schofield, and J. Trefry. 1981. Waste material behavior and inorganic geochemistry at the Puerto Rico waste dumpsite. *In*: Ocean Dumping of Industrial Wastes, B. H. Ketchum, D. R. Kester, and P. K. Park (Eds.). Plenum Press, New York, pp. 233–246.

Schwab, C. R., T. C. Sauer, Jr., G. F. Hubbard, H. Abdel-Reheim, and J. M. Brooks. 1981. Chemical and biological aspects of ocean dumping at the Puerto Rico dumpsite. *In*: Ocean Dumping of Industrial Wastes, B. H. Ketchum, D. R. Kester, and P. K. Park (Eds.). Plenum Press, New York, pp. 247–273.

Sharp, J. H., M. J. Perry, E. H. Renger, and R. W. Eppley. 1980. Phytoplankton rate processes in the oligotrophic waters of the central North Pacific Ocean. *Journal of Plankton Research*, **2**, 335–353.

Sheldon, R. W., and W. H. Sutcliffe, Jr. 1978. Generation times of 3 h for Sargasso Sea microplankton determined by ATP analysis. *Limnology and Oceanography*, **23**, 1051–1055.

Spero, H. J., and M. D. Morée. 1981. Phagotrophic feeding and its importance to the life cycle of the holozoic dinoflagellate, *Gymnodinium fungiforme*. *Journal of Phycology*, **17**, 43–51.

Thomas, W. H., and D. L. R. Seibert. 1977. Effects of copper on the dominance and the diversity of algae: Controlled ecosystem pollution experiment. *Bulletin of Marine Science*, **27**, 23–33.

12

COPEPODS AND ICHTHYOPLANKTON: LABORATORY STUDIES OF PHARMACEUTICAL WASTE TOXICITY

Wen Yuh Lee

The University of Texas
Marine Science Institute
Port Aransas Marine Laboratory
Port Aransas, Texas

ABSTRACT

Studies were carried out to determine the toxicity of six ocean-dumped pharmaceutical wastes to marine copepods and redfish (*Sciaenops ocellata*) eggs and larvae. To simulate waste concentrations at the dumpsite, copepods were consecutively immersed in a series of waste dilutions for intervals of 2 min to 2 h: 2 min in 10% waste solution, 5 min in 1%, 1 h in 0.1%, and 2 h in 0.01%. Animals were then transferred to untreated seawater to determine the delayed mortality. Samples from Merck and Pfizer wastes were acutely toxic to marine copepods; mortality was >35% at the end of 3 h exposure. Capri, Squibb, and Upjohn wastes produced low initial mortalities but high delayed mortalities (>60%).

The ichthyoplankton were treated with the Capri, Squibb, and Bristol wastes. During the exposure, observations were made on hatching success of embryos and on morphological deformity, behavioral abnormality, and survival of larvae. Merck and Squibb wastes were acutely toxic to eggs and newly hatched larvae at concentrations of 0.5–1%. The corresponding toxic levels for Bristol waste were 0.07% for eggs and 0.045% for larvae. The higher toxicity of Bristol waste may have been caused by its major component, *N,N*-dimethylaniline, which is more persistent in the marine environment than other components in the wastes.

12.1. INTRODUCTION

Previous research has shown that composite waste materials from Puerto Rico pharmaceutical companies, which are periodically dumped offshore (19°10′ to 19°20′N and 66°35′ to 66°50′W), have detrimental effects on marine organisms, including fish, invertebrates, and microalgae (Van Baalen and Batterton, 1978; Lee and Nicol, 1981; Wohlschlag and Parker, 1981).

The present study was conducted to determine the toxicity of wastes from specific companies to marine zooplankton. The six samples tested were from the following companies: Bristol, Capri, Merck, Pfizer, Squibb, and Upjohn. Each company manufactures different pharmaceutical products; hence, the waste composition varies with the manufacturer. The major reported waste products from each company are summarized in Table 12.1.

Two experiments were designed to test the toxicity of each individual waste. In one experiment, marine copepods were treated with a series of dilutions of six effluents for various periods, and then transferred to untreated seawater. This experiment was carried out to simulate dilution process occurring at the dumpsite and to determine both the potential acute effects associated with early dilution and the short-term effects thereafter. In the other experiment, the toxicities of three

Table 12.1. Major Compounds Reported in Individual Waste Samples Discharged at the Puerto Rico Industrial Dumpsite

		Reported Wastes		
Company	Products	Reference[a]	Reference[b]	Rate of Input (metric tons y^{-1})
Bristol	Ampicillin Penicillin derivatives	Triethylamine, methylisobutyl ketone, phenylglycine, β-naphthalene	Ca^{2+}, K^+, Mg^{2+}, Na^+, dimethylaniline, methylaniline	12,200
Capri	Levamisole Tetramisole	Data not available	Ca^{2+}, K^+, Mg^{2+}, Na^+, (±)-2,3,5,6-tetrahydro-6-phenyl imidazol [2,1-6]-thiazole	–
Merck	L-Methyldopa	Vanillin, ethylamine, dimethylsulfoxide	Ca^{2+}, Fe^{2+}, Mg^{2+}, Na^+, methoxy-4-hydroxyphenyl acetone	84,300
Pfizer	Sinequow Diabinese Bonine	Tetrahydrofuran, xylene	Ca^{2+}, K^+, Mg^{2+}, Na^+, unidentified nitrogen compounds	21,300
Squibb	Penicillin derivatives Cephradine	Methanol, acetone, xylene, methylisobutyl ketone, isopropyl alcohol, triethylamine	Data not available	8,500
Upjohn	Lincomycin Clinomycin	Dimethylformamide, cuminic aldehyde, palmitic acid, pyridine, methanol	Ca^{2+}, K^+, Mg^{2+}, Na^+, fatty acids, fatty acid esters	152,400

[a]T. O'Connor (personal communication).
[b]Atlas et al. (1981) and Presley et al. (1981).

effluents (Bristol, Merck, and Squibb) to redfish (*Sciaenops ocellata*) larvae and eggs were compared. A standard laboratory toxicity test was performed to determine the waste concentration at which 50% of the test population was killed in a 48 h exposure (48-h LC_{50}). Behavioral and morphological aberrations of larval fish observed during the experiments were also described.

12.2. MATERIALS AND METHODS

12.2.1. Experiments with Copepods

Each stock solution supplied by the six pharmaceutical companies was diluted with seawater to the following concentrations: 10%, 1%, 0.1%, and 0.01% (by volume). The three subsamples of each dilution were designated A, B, and C, and a control was also used during the experiments.

The seawater (36‰) used came from offshore Florida, and was prefiltered through a 1-μm glass fiber filter, and aerated for 20 min. Copepods were collected about 27 km east of central Florida off Fort Pierce, using a 0.5-m net of 153-μm mesh. Ten plankton hauls were made at a depth of about 2 m on 10 May 1979 between 1000 and 1400 h. Each tow lasted for 15 min. On board ship, the samples were divided into 21 parts, and each part was transferred to a 12 cm length × 6.8 cm diameter plexiglass cylinder with 93-μm mesh netting attached to one end. The cylinder was then immersed in a 1-liter wide-mouth bottle, containing test solution or the prefiltered seawater (control).

To simulate the changes in waste concentration occurring at the dumpsite immediately after a dumping operation, copepods were consecutively exposed to a series of waste dilutions, from high to low, for various time intervals. The duration of exposure increased with decreasing waste concentration: 2 min in 10% dilution, 5 min in 1%, 1 h in 0.1%, and 2 h in 0.01%. At the end of this exposure, subsample A was stained with neutral red, and then fixed by a method described by Crippen and Perrier (1974). Copepods in the remaining B and C subsamples were transferred to clean seawater for a 24-h period of depuration before staining and fixing. The observed mortality in subsample A indicated the acute toxicity of each pollutant, while the difference between the mortality of subsample A and the mean mortality of subsamples B and C (replicates) represented the delayed toxicity.

The copepods in subsample A were not fed because the exposure lasted only about 3 h. Animals in subsamples B and C were starved for the first 6 h and then fed with a mixture of *Platymonas* sp. and *Dunaliella salina* four times at intervals of 5–6 h.

12.2.2. Experiments with Redfish Eggs and Larvae

The wastes used in this part of the study were from Merck, Squibb, and Bristol. Two experiments were carried out on each test waste to determine their toxicities at various concentrations to redfish eggs and larvae. The concentrations of waste in the initial test solutions, in duplicate, for both experiments were 0% (control), 0.1%, 0.5%, 1%, and 5%. Wastes were diluted with glass-fiber (0.2–10 μm) filtered seawater (26‰), which had been treated with the antibiotics penicillin G and streptomycin sulfate and aerated for 30 min before the experiments.

Redfish larvae were obtained from fertilized eggs of *S. ocellata* from the Marine Fisheries Laboratory, University of Texas, Port Aransas, Texas. The eggs were already in late gastrula and were incubated for an additional 18 h until hatching. Twenty-five larvae were then transferred by large-bore pipette to the waste dilutions. Survival counts were made at 6-h intervals from the time of exposure. A larva was considered dead when it had no detectable heartbeat. Dead larvae were removed from the bowls at the time of the counts. The experiment proceeded until the end of the 48-h yolksac phase; thus, it was unnecessary to feed the larvae. Values of 48-h LC_{50} and their 95% confidence intervals were estimated by the method of Litchfield and Wilcoxon (1949).

The majority of the redfish eggs were in late gastrula when obtained from the Marine Fisheries Laboratory. Batches of 30–50 eggs were transferred to a petri dish for microscopic observation. Eggs that were deformed, unfertilized, or otherwise not in late gastrula were discarded. The remaining eggs were transferred to the test media. Each sample contained 25 eggs.

The first count was made 24 h after exposure to the waste solution. The resultant eggs and larvae were recorded as hatched-living larvae, hatched–dead larvae, unhatched-living embryos, and unhatched-dead embryos. All dead specimens were discarded. The surviving larvae were then transferred from the test solution to clean seawater. Subsequently, observations on the survival of organisms were made at 6 h intervals until the end of the yolksac stage.

12.3. RESULTS

12.3.1. Experiments with Copepods

12.3.1a. Waste Toxicity

The acute and delayed toxicities of a sample were variable (Table 12.2). Survival at the end of 3-h exposure suggested that the Merck waste was the most toxic with mortality about 50% of the test population. For the remaining wastes, the order of decreasing toxicity was Pfizer > Capri > Squibb > Upjohn. Survival of copepods in the Upjohn waste was high; 94.8% compared with 96% in the control.

The last column in Table 12.2 shows the difference between survival of organisms at the end of 3-h exposure to waste and at the end of 24 h after being transferred to clean seawater. Although there were high mortalities in the controls (25%), delayed mortalities in all wastes were still significant. The highest delayed mortality was observed in the Upjohn sample and the lowest was in the Merck sample. The order of toxicity for these two wastes was the reverse of that recorded at the end of 3-h exposure. Nevertheless, the overall toxicity of each test

Table 12.2. Toxicity, Expressed as Percentage Survival, of the Six Pharmaceutical Wastes to Marine Copepods

	Subsample				Delayed Toxicity
Company	A[a]	B[b]	C[b]	$\frac{B+C}{2}$	$A - \left(\frac{B+C}{2}\right)$
Control	97	69	75	72	25
Bristol	—[c]	—	—	—	—
Capri	72	16	6	11	61
Merck	52	5	2	4	48
Pfizer	63	0	16	8	55
Squibb	73	0	1	1	72
Upjohn	95	4	5	5	90

[a]A is percentage survival at the end of 3-h exposure to waste dilutions.
[b]B and C (replicates) are percentage survival at the end of 24-h depuration in clean seawater.
[c]Samples were discarded because the copepods were covered with oily brown materials, which made live–dead determination impossible.

sample was similar; survival of copepods in most wastes was < 10% at the end of the experiment (Table 12.2).

Copepods in the Bristol wastes were covered with oily brown materials. This not only prevented the specimens from being adequately stained by the neutral red but also caused difficulty in copepod identification. All specimens in Bristol samples were therefore discarded.

12.3.1b. Copepod Composition

The majority of zooplankton in the test samples were copepods. Several dominant species and their numerical abundance are listed in Table 12.3. The cyclopoids *Farranula carinata* and *Corycaeus* sp. together with the calanoids *Paracalanus* spp. and copepod nauplii make up about 75% of the total number counted in the 18 samples tested. Other species such as the harpacticoids *Macrosetella gracilis* and *Miracia efferata* were much less abundant and their numbers were always < 30 in a sample of about 250 individuals. Except for the cyclopoid *F. carinata*, the other species varied widely from sample to sample in their order of dominance. For example, copepod nauplii were the most abundant taxon in the Pfizer B sample but one of the least abundant in the Merck C sample.

12.3.2. Experiments with Redfish Larvae

12.3.2a. Squibb Wastes

The mortality of redfish larvae increased with both increasing concentration of the Squibb waste and duration of exposure. The estimated 48-h LC_{50} was 0.42% (by volume) with the 95% confidence intervals between 0.34 and 0.52%.

Table 12.3. Abundance and Occurrence of Common Copepods in the 18 Test Solutions[a]

	Control			Pfizer			Capri			Upjohn			Squibb			Merck				
Species	A	B	C	A	B	C	A	B	C	A	B	C	A	B	C	A	B	C	Total	%
Farranula carinata	175	162	192	150	51	176	203	125	215	187	82	95	111	26	102	107	106	208	2473	47.0
Paracalanus spp.	6	14	3	29	30	3	7	25	2	47	62	68	54	102	26	74	44	25	621	11.8
Copepod nauplii	2	19	14	25	145	1	6	74	16	7	83	8	0	88	2	6	96	0	592	11.3
Corycaeus sp.	19	19	25	25	9	26	2	18	8	7	36	9	26	20	34	23	11	17	334	6.4
Oncaea spp.	21	39	6	21	5	26	3	5	6	17	1	35	7	1	4	36	3	10	246	4.7
Corycaeus amazonicus	9	23	29	29	6	27	15	16	3	23	1	25	6	0	5	2	5	5	229	4.4
Macrosetella gracilis	17	9	11	19	9	15	18	9	16	13	2	11	10	2	22	13	4	12	212	4.0
Corycaeus speciosus	14	0	4	3	1	4	3	2	2	0	1	5	1	0	5	3	0	7	55	1.0
Others	16	23	6	11	38	5	4	18	34	5	31	47	12	37	123	43	30	12	495	9.4

[a]The last column represents the percentage of each taxon in total count (5257).

Concentrations ≥ 0.5% were toxic to the redfish larvae; all the larvae in the 5% solutions and more than half of those in the 1% had expired at the end of 6-h exposure. The dead larvae in the 1% were corkscrew-shaped from the yolksac to the caudal region. Deformed larvae were not observed in the 5% solutions. It is possible that at 5% they died very soon after exposure.

After 12-h exposure, some behavioral abnormalities were noted among the larvae in the 0.5% mixtures. The majority of the larvae drifted about, but displayed little or no escape response when approached with the tip of a pipette. However, death did not occur in this concentration for an additional 18 h.

All larvae in the 0.1% dilution survived throughout the yolksac phase. Even at the end of 62 h, the mortality of the larvae did not differ significantly from that in the controls. No behavioral or morphological abnormalities were observed during the experiment.

12.3.2b. Merck Wastes

Because of the dark color of this waste, it was impossible to recover any larvae from the 5% solutions for observation. All these larvae were assumed dead. Three crippled larvae were found in the 1% solution. Each was bent approximately 90° laterally; they were dead at 12 h. The survivors in the 1% solutions lay motionless on the bottom of the culture bowls at 18 h. For the remaining 30 h the only detectable sign of life for these larvae was the heartbeat.

Less severe effects were observed for the larvae in the concentrations ≤ 0.5%. The mortality of larvae in the 0.1% was low; only 8 of 50 were lost at 48 h. The estimated 48-h LC_{50} was 0.49%, with the 95% confidence intervals from 0.39 to 0.61%.

12.3.2c. Bristol Wastes

A preliminary test on the Bristol waste showed it to be much more toxic than the other wastes. Therefore, concentrations lower than the levels tested for the Squibb and Merck wastes were used for the Bristol sample. For the latter, concentrations ranged from 0.01 to 0.5%. Significant mortalities were recorded at concentrations ≥ 0.02%. At concentrations ≥ 0.1%, complete mortality occurred within 12 h. The estimated 48-h LC_{50} was 0.045% (confidence intervals, 0.039–0.050%).

Morphological aberrations were noted in the 0.1% solution at 6 h. All the larvae were twisted and knotted from the yolksac to the caudal region. No muscle contraction was noted, except for the heartbeats, which were faint and irregular.

Behavioral abnormalities were observed in the 0.02% group. At the end of 6-h exposure, the length of escape swimming path was about 0.5–1 cm compared with 5–10 cm in the controls. The larvae became rigid and no movement other than heartbeat was detected from 18 h to the end of the experiment.

Table 12.4 shows the toxic effects of the Squibb, Merck, and Bristol wastes on the redfish larvae, and includes the following categories: values of 48-h LC_{50}, morphological aberrations, and behavioral abnormalities.

Table 12.4. **Summary of the Toxic Effects of the Three Pharmaceutical Wastes on Redfish Larvae**

Pharmaceutical Waste	48-h LC_{50} and 95% Confidence Intervals	Morphological Aberrations	Behavioral Abnormalities
Squibb	0.42% (0.34–0.52)	Larvae exposed to the 1% concentration had a corkscrew configuration (6 h).	Most larvae in the 0.5% concentration drifted about and displayed little or no escape response (12 h).
Merck	0.49% (0.39–0.61)	Several larvae in the 1% concentration were bent 90° laterally (6 h).	Larvae exposed to the 0.5% concentration had an escape swimming length < 3 cm. An occasional quivering was also observed (18 h).
Bristol	0.045% (0.039–0.050)	All larvae in the 0.1% concentration were twisted and knotted (6 h). Larvae in the 0.02% concentration became rigid 18 h after exposure.	Larvae in the 0.02% concentration showed no inducible movement; they drifted or lay on the bottom of culture bowls (12 h).

12.3.3. Experiments with Redfish Eggs

12.3.3a. Squibb Wastes

Egg and larval mortality increased with increasing concentrations of waste, as indicated in Table 12.5. More than 85% of the eggs in the 0.1% solutions were hatched, and survival of these larvae after transfer to clean seawater was high and also comparable to that of the larvae from eggs exposed to 0.1% Merck waste. At higher concentrations (0.5 and 1%), most larvae were hatched but apparently died shortly after hatching. The two survivors in the 0.5% solutions were motionless on the bottom of the culture bowl and could not be induced to move. At 5%, none of the eggs hatched.

12.3.3b. Merck Wastes

As in the Squibb waste, no hatching occurred in the 5% solution (Table 12.5). At the end of 24-h exposure, all the embryos in this concentration were dead and none showed signs of development after the initial exposure.

Extensive morphological and behavioral aberrations were observed in the 1% solution. Of the 42% that hatched, about 5% of the larvae were deformed. In each case, the yolksac was highly distended, while the posterior half of the body was atrophied. Movement was limited to an occasional twitching of the body. Many

Table 12.5. Hatching Success of Redfish Eggs after 24-h Exposure to Pharmaceutical Wastes[a]

	Hatched		Unhatched	
Waste Concentration (%)	Live Larvae	Dead Larvae	Live Embryos	Dead Embryos
Squibb				
0.1	22	0	0	3
0.5	1	21	0	3
1.0	0	15	0	10
5.0	0	0	0	25
Merck				
0.1	23	1	0	1
0.5	24	0	0	1
1.0	10	0	0	15
5.0	0	0	0	25
Bristol				
0.02	24	1	0	0
0.04	21	1	1	2
0.06	21	3	0	1
0.08	0	22	0	3

[a]Results are the mean of two observations.

of the noncrippled larvae had abnormal swimming patterns. When touched with a pipette, the larvae would swim in small circles rather than dart away from the pipette as in the controls.

After the transfer of hatched larvae in the 1% solution to clean seawater, the progress of the behavioral aberrations was arrested. The larvae gradually became more active and the abnormal circular swimming patterns were no longer observed after 18 h in fresh seawater. However, the exposed larvae never became as active as those in the controls. Their distance of travel in the escape response was 1-2 cm compared with 8-12 cm in the controls.

No behavioral aberrations were evident in the 0.5 and 0.1% waste solutions. Survival of all larvae that hatched from exposed eggs was high in clean seawater, as was also observed in the experiment on the Squibb sample.

12.3.3c. Bristol Wastes

Egg hatching was > 85% in all test concentrations. At the highest concentration tested (0.08%), an average of 90% of the exposed eggs hatched. However, all the larvae in the 0.08% solution died soon after hatching, while there was > 80% survival in the groups exposed to 0 (control), 0.02, 0.04, and 0.06% concentrations (Table 12.5).

Morphological and behavioral aberrations occurred in larvae at all concentrations tested. In the 0.02% solution, all larvae lay motionless on their sides, with movement limited to an occasional quivering along the spine. About 8% of the larvae in this concentration were severely malformed, having the corkscrew-shaped configuration from the yolksac to the caudal region. The remaining 92% had a distinct S-shaped spine compared with the relatively straight spine observed in the specimens of the controls.

In the 0.04 and 0.06% waste solutions, the hatched larvae exhibited more severe damage than was noted in the 0.02% solution. A quarter of the larvae were corkscrew-shaped and the remaining 75% had a pronounced S-shaped spine. In addition, all larvae in these two concentrations had faint and irregular heartbeats.

Mortality of redfish larvae after transfer to clean seawater was low for the first 36 h and then began increasing. After 48 h, only $1/3$ of the hatched larvae at 0.06% and $2/3$ at 0.04% concentrations were still alive. For the entire 48-h posthatching period, there was no evidence that the larvae recovered from the morphological or behavioral aberrations that were present at the time of hatching.

Table 12.6 summarizes the toxicity of the Squibb, Merck, and Bristol wastes to the redfish eggs and the resultant larvae in terms of percentage of eggs hatched at the 0.5% solution and status of the hatched larvae.

12.4. DISCUSSION

Of the five pharmaceutical wastes tested, samples from Merck and Pfizer were acutely toxic to marine copepods. Survival in these two samples was < 65% at the

Table 12.6. Hatching (%), Survival (%), and Some Behavioral and Morphological Characteristics of Redfish Larvae in Three Pharmaceutical Wastes

Sample	Concentration (v/v) at Which No Eggs Hatched	% of Eggs Hatched at 0.5% Concentration	% Hatched as Live Larvae at 0.5% Concentration	% Survival of Larvae at 0.5% Concentration[a] (at 48 h)	Behavioral or Morphological Abnormalities	Recovery from Damage	Delayed Mortality
Squibb	≥ 5	88	4	2.0 (0.5)[b]	Larvae from eggs exposed to 0.5% waste were motionless on the bottom of the culture bowls.	No	Yes
Merck	≥ 5	98	98	82.0 (20.5)	Several larvae from eggs exposed to 1% waste were deformed; the yolksac was highly distended, while the posterior half of the body was atrophied. When touched, many of the noncrippled larvae swam in small circles.	Yes	No
Bristol	≥ 0.2	0	0	0	All larvae from eggs exposed to 0.02% waste were motionless and deformed. About 92% of the larvae had a distinct S-shaped spine, and the remaining 8% had a corkscrew configuration.	No	Yes

[a] The percentage survival of larvae was estimated relative to the initial number of exposed eggs.
[b] Indicates the average number of surviving larvae 48 h after they were transferred to clean seawater.

end of the 3-h exposure. Animals in the other three wastes had low initial mortalities but high delayed mortalities. Twenty-four hours after copepods were transferred to clean seawater, the highest percentage of survival (10.8%) was observed in the group exposed to the Capri sample. These data suggest that wastes from the different companies varied in both acute and delayed toxicity.

Initial survival of the controls was high, 97% at 3h. At the end of 27 h, the percentage of survival dropped to about 72%, which was a 25% loss during a 24-h period. This loss could be caused either by inadequate food or by the changing temperatures (25° to 29.5°C) encountered during the experiment. Aboard the research vessel, the study was conducted under ambient conditions where the temperature changed according to the position of the sun. The high delayed mortalities observed in all samples were probably caused by the combined stress of pollutants and varying temperatures.

The toxicities of the three wastes to larval fish are in the order Bristol>Squibb>Merck. The values of 48-h LC_{50} for Squibb and Merck were close (0.42 and 0.49%, respectively), while the 48-h LC_{50} for Bristol was 0.045%, approximately 10 times lower than either of the other two samples. Our previous study (Lee et al., 1978) found the composite waste samples to be toxic to anemones at a concentration of 0.49% (96-h LC_{50}), to juvenile amphipods at 0.25% (96-h LC_{50}), to isopods at 1.4% (96-h LC_{50}), to jellyfish at 1% (96-h LC_{50}), and to mixed coastal zooplankton at 1% (24-h LC_{50}). These values cannot be directly compared with those of the present study because the two experiments differed in two factors, namely, test animals and test pollutants (individual versus composite waste material).

Although the chemical composition of the three waste samples varied considerably (see Table 12.1), the symptoms of their toxic effects on the eggs and larvae were similar. The severity of the symptoms and frequency of occurrence depend on many factors, including the concentration of pollutant, its inherent toxicity, and duration of exposure. Generally, a short-term exposure of redfish larvae or eggs to a concentration $\geq$ 0.5% of any company's waste would result in reduced swimming activity, deformity, irregular heartbeat, and also some deaths. Those larvae that hatched from eggs exposed to the Merck sample appeared to partially revive after several hours in clean seawater, but those exposed to samples of Squibb and Bristol wastes suffered irreversible damage.

Biological consequences of waste dumping at the dumpsite depend on many factors, such as the toxicity of dumped materials, the rate of dilution, the structure of the *in situ* biological community, and the oceanographic conditions. It is impossible to select one setting to describe all the possible waste plume behaviors or the structure of the plankton community. However, field observations on the plume growth and the chemical analysis of *N,N*-dimethylaniline in seawater suggested that the minimum dilution factors for the first 3 h might range from 5×10^3 to about 2×10^4 (Simpson et al., 1981), and the maximum concentration of waste at one day to one week after a dump could be near 10 μg liter^{-1} (O'Connor, 1979). Compared with these predictions, the concentrations employed in the copepod bioassay for the first hour of exposure were too high, while the concentrations for the next 2 h were reasonable. Hence, the results obtained from the copepod study may not represent the outcome at the large dumping area, but at the region near the center of the waste plume, where at least half of the copepod populations could be killed at the end of 24 h.

In terms of acute effects, the estimated maximum concentration of waste (about 10 μg liter^{-1}) at the station was also below the levels that cause mortality and morphological changes in the redfish larvae. Yet, the higher toxicity of the Bristol sample compared with the samples from the other companies is of concern and interest. Atlas et al. (1981) reported that *N,N*-dimethylaniline (DMA) was the major component of the Bristol sample, and Brooks et al. (Chapter 9) report the presence of DMA at the Puerto Rico dumpsite. It is possible that the higher toxicity of Bristol waste is attributable to the compound DMA. Furthermore, DMA was found to be relatively persistent in the marine environment (Hatcher and Harvey, 1978; Schwab et al., 1981) (Chapter 9). The acute toxicity of this compound and it sublethal effects on marine organisms should be studied in future programs.

12.5. CONCLUSION

Pharmaceutical wastes from the six manufacturers, Bristol, Capri, Merck, Pfizer, Squibb, and Upjohn, are acutely toxic to marine copepods. Immersion of copepods in a series of increasing waste dilutions and exposure intervals induced $>$ 85% mortality.

Hatching, survival, morphology, and activity of redfish larvae in pharmaceutical wastes depend on the pollutant concentration, exposure duration, and source of wastes. A short-term exposure of redfish larvae or eggs to a waste concentration $\geq$ 0.5% generally resulted in low hatching success, deformity, and reduced swimming activity. The estimated 48-h LC_{50} is 0.42% for Squibb waste, 0.49% for Merck waste, and 0.045% for Bristol waste.

The compound *N,N*-dimethylaniline was reported to be the major component of the Bristol waste, and is relatively persistent at the dumpsite. Its chronic toxic effects, especially on feeding and reproduction of marine zooplankton, warrant further investigation.

ACKNOWLEDGMENTS

I am indebted to Dr. C. Arnold, who always provided redfish eggs generously. I thank Dr. R. S. Jones, the Director of the Johnson Science Laboratory, Harbor Branch Foundation in Fort Pierce, Florida, for providing the shiptime and much assistance. I am also grateful to many of his colleagues, especially Drs. M. Youngbluth, J. Miller, and J. Prentice and Misses L. Eyster and R. Neville, who helped me in various aspects either during my stay in the Johnson Science Laboratory or on board the *R/V Johnson*. I wish to thank especially my colleagues, A. Morris and S. Anderson, who assisted with laboratory and field work, respectively. Without their help, this zooplankton project would have been impossible to complete. Finally, I thank Drs. J. A. C. Nicol, T. W. Duke, and C. Kitting for their critical reading of the manuscript. This study was supported by the National Oceanic and Atmospheric Administration, Ocean Dumping and Monitoring Program, grant No. 04-8-M01-54. It is the University of Texas Marine Science Institute Contribution number 464.

REFERENCES

Atlas, E., G. Martinez, and C. S. Giam. 1981. Chemical characterization of ocean-dumped waste materials. *In*: Ocean Dumping of Industrial Wastes, B. H. Ketchum, D. R. Kester, and P. K. Park (Eds.). Plenum Press, New York, pp. 275–293.

Crippen, R. W., and J. L. Perrier. 1974. The use of neutral red and Evans blue for live-dead determinations of marine plankton. *Stain Technology*, **49**, 97–104.

Hatcher, P. G., and G. R. Harvey. 1978. Trace Organic Studies at the Puerto Rico Pharmaceutical Waste Dumpsite. Unpublished Report, Atlantic Oceanographic and Meteorological Laboratories, National Oceanic and Atmospheric Administration, Miami, Florida. Submitted to Ocean Dumping Program, Office of Marine Pollution Assessment, National Oceanic and Atmospheric Administration, Rockville, Maryland, 17 pp.

Lee, W. Y., and J. A. C. Nicol. 1981. Toxicity of biosludge and pharmaceutical wastes to marine invertebrates. *In*: Ocean Dumping of Industrial Wastes, B. H. Ketchum, D. R. Kester, and P. K. Park (Eds.). Plenum Press, New York, pp. 439–454.

Lee, W. Y., N. Hannebaum, and J. A. C. Nicol. 1978. Toxicity of Puerto Rican Organic Waste Materials to Marine Invertebrates. Unpublished Report, Marine Science Institute, University of Texas, Port Aransas. Submitted to Ocean Dumping Program, Office of Marine Pollution Assessment, National Oceanic and Atmospheric Administration, Rockville, Maryland, 37 pp.

Litchfield, J. T., Jr. and F. Wilcoxon. 1949. A simplified method of evaluating dose-effect experiments. *Journal of Pharmacology and Experimental Therapeutics*, **96**, 99–113.

O'Connor, T. P. 1979. Ocean Dumping Research and Monitoring at Puerto Rico Dumpsite. Unpublished Report, Ocean Dumping Program, National Oceanic and Atmospheric Administration, Rockville, Maryland, 35 pp.

Presley, B. J., J. S. Schofield, and J. Trefry. 1981. Waste material behavior and inorganic geochemistry at the Puerto Rico waste dumpsite. *In*: Ocean Dumping of Industrial Wastes, B. H. Ketchum, D. R. Kester, and P. K. Park (Eds.). Plenum Press, New York, pp. 233–246.

Schwab, C. R., T. C. Sauer, Jr., G. F. Hubbard, H. Abdel-Reheim, and J. M. Brooks. 1981. Chemical and biological aspects of ocean dumping at the Puerto Rico dumpsite. *In*: Ocean Dumping of Industrial Wastes, B. H. Ketchum, D. R. Kester, and P. K. Park, (Eds.). Plenum Press, New York, pp. 247–273.

Simpson, D. C., T. P. O'Connor, and P. K. Park. 1981. Deep-ocean dumping of industrial wastes. *In*: Marine Environmental Pollution, 2, Dumping and Mining, R. A. Geyer (Ed.). Elsevier, New York, pp. 379–400.

Van Baalen, C., and J. C. Batterton. 1978. Effects of Pharmaceutical Wastes on Growth of Microalgae. Unpublished Report, Marine Science Institute, University of Texas, Port Aransas. Submitted to Ocean Dumping Program, Office of Marine Pollution Assessment, National Oceanic and Atmospheric Administration, Rockville, Maryland, 9 pp.

Wohlschlag, D. E., and F. R. Parker, Jr. 1981. Metabolic sensitivity of fish to ocean dumping of industrial wastes. *In*: Ocean Dumping of Industrial Wastes, B. H. Ketchum, D. R. Kester, and P. K. Park (Eds.). Plenum Press, New York, pp. 455–471.

13

FISH: RESPONSE TO OCEAN-DUMPED PHARMACEUTICAL WASTES

Donald E. Wohlschlag and Faust R. Parker, Jr.

The Universtiy of Texas
Marine Science Institute
Port Aransas, Texas

ABSTRACT

Metabolic levels and swimming performances of *Lutjanus campechanus* (red snapper) at 20°C and a salinity of 35‰ in dilutions of a composite pharmaceutical waste indicated fatal toxicity for fish required to swim in 24 h at a concentration of 0.25% (v/v) in seawater. For fish held two days at 0.125 and 0.0625%, both swimming performance and active metabolism dropped as waste concentration increased. Standard (maintenance) metabolism was fairly constant. Metabolic scope (the difference between active and standard metabolic rates) declined in proportion to the sublethal waste concentration for two-day exposures.

Exposure-recovery experiments at 28°C and a salinity of 35‰ with *Cynoscion nebulosus* (spotted seatrout) were conducted with composite and six individual industry wastes. The first visible signs of stress appeared in 2 h or less at concentrations between 0.5 and 0.00625%, depending on the initial toxicity of each waste source. When the fish were transferred to clean seawater the metabolic levels were highly variable. The standard metabolism always increased compared to controls, and swimming performance declined. The active metabolic levels did not decrease uniformly because of the stimulatory effects of the Upjohn waste and possibly of the Merck, Bristol, and Pfizer wastes. The metabolic scope, except in the Upjohn experiments, remained depressed at the end of the recovery period. The unusually great depression of scope in the composite experiment may have been caused by negative interactions among waste components. Signs of morbidity after a 2- to 4-d recovery indicated that initial exposure concentrations were too high and too long for complete recovery. The extent of cumulative or delayed lethality in ocean dumping needs to be investigated in terms of population suppression effects on fishes that may be exposed in short-term pulses of subacute toxic levels of wastes every few days.

13.1. INTRODUCTION

The purpose of this study was to ascertain the degree of toxicity to fish of various dilutions of ocean-dumped wastes from Puerto Rican pharmaceutical industries. The rationale of the study presumed that fish in relatively few days would show measurable biological reactions to sublethal chemical stresses, however obscure the chemical etiology of the relations might be. This rationale is generally assumed if fish are to be used for monitoring wastes (Hocutt and Stauffer, 1980; Wohlschlag and Parker, 1981).

There is considerable information on the chemical nature of the Puerto Rican pharmaceutical wastes (Chapter 9) and on the degree and rapidity of dilution after dumping. Atlas et al. (1981) indicate that there is a high degree of chemical complexity in each of the wastes from the individual industries as well as in the composite mixtures involved in ocean dumping and that many of the major organic components are known. Schwab et al. (1981) and Brooks et al. (Chapter 9) note that the major nonaqueous fractions of the wastes are volatile, which implies difficulties in assessing waste component concentrations to which organisms would be exposed. Presley et al. (1981) point out that the degree of dilution of the inorganic components is rapid, with little evidence of short-term effects on zooplankton. However, there is a general lack of direct evidence of the long-term quantitative and qualitative effects that these various components, or interactions among them, might have on marine fish. The complexities of the chemicals in the wastes, their changes following dumping, and their ultimate effects on long-living organisms such as fish suggest the utility of biological assessment or biomonitoring.

The advantages of using fish for the assessment of low-level stresses, especially when the chemical nature of the toxic materials and their interactions are not known completely, are important and frequently overlooked. Fish are relatively long-lived organisms and their populations respond integratively over long periods of time, so that even insignificant short-term effects of stresses can be cumulatively devastating.

One of the more sensitive ways to evaluate quantitatively the stress effects of toxic materials on organisms is to measure the reduction in potential aerobic metabolic performance under stress, especially at sublethal levels. The scope for activity, that is, the difference between the active metabolic rate at maximum sustained swimming and the standard (maintenance) metabolic rate at complete rest, tends to decline under stress, because either the active rate decreases, or the standard rate increases, or both happen simultaneously. Metabolism is usually expressed in terms of oxygen consumption rates adjusted for body weight. The physiological principle of metabolic scope reduction under stress is described in Fry (1947, 1971); an application to scope reduction in marine fish exposed to a low-level pollutant is described in Wohlschlag and Parker (1981).

Many kinds of stresses reduce metabolic scope, with accompanying reductions in functions such as food conversion efficiency, growth rate, swimming performance, and other metabolic manifestations (Smith et al., 1967; Wohlschlag and Cameron, 1967; Krueger et al., 1968; Mount and Stephan, 1969; Carlson, 1971; Warren, 1971; Pickering and Gast, 1972; Grant and Mehrle, 1973; Webb and Brett, 1972). However, there is a limited literature to explain sublethal stress effects on larger marine fish either in the first case in which fish are exposed to low-level chronic stresses of known or unknown toxic wastes or in the second case in which fish are exposed but briefly to subacute toxic conditions and then allowed to recover. The first case occurs where persistent ocean dumping results eventually in a widespread sublethal concentration of toxic materials. The first case may also occur temporarily surrounding the immediate vicinity of dumping before extensive dilution. The second case occurs at the time and place of dumping with exposure of fish before there is extensive dilution and with their recovery after rapid dilution to nontoxic levels. These two cases provide the objectives for assessing the effects of pharmaceutical wastes on fish.

The first objective was to determine what levels of composite pharmaceutical waste concentrations would be sublethal but detectable by the attenuation of the metabolic scope of the red snapper (*Lutjanus campechanus*). The second objective was to determine the change in the metabolic scope of the spotted seatrout (*Cynoscion nebulosus*) when exposed briefly to pharmaceutical wastes at subacute toxic levels and then allowed to recover in unpolluted water. The resultant displacement of the active and standard metabolic levels and the change in magnitude of the scope compared to controls were studied for the purpose of determining recovery times or the magnitude of delayed (residual) effects.

13.2. METHODS

13.2.1. Sublethal Exposure Experiments with Red Snapper (*Lutjanus campechanus*)

All fish were from Gulf of Mexico waters immediately off Port Aransas, Texas, where ambient salinities were about 35‰ and temperatures near 20°C. Fish were then held for at least 24 h at these conditions before experimentation, followed by another 24-48 h at 35‰ and 20°C after exposure to pollutant treatment. Experimentation was in two parts: determination of metabolic rates at standard (complete resting) conditions, and determination of metabolic rates at maximum sustained swimming speeds.

Standard rates were determined by use of four acrylic plastic tubular flow-through chambers (13.8 cm diameter × 61 cm long). The chambers were immersed in a large temperature-controlled aquarium. Excitement of fish was reduced by use of black plastic shields around each tube and by black curtains around the entire aquarium. Measurements of flow rates of water and oxygen content at the inlet and outlet of each chamber over 1-2 d permitted the calculation of minimal resting rates and estimation of standard rates. Throughout this and the following experiments, oxygen content of the well-oxygenated waters was measured by polarographic oxygen electrode probes, appropriately calibrated and temperature compensated.

Active metabolic rates were determined with a specially constructed 207-liter Blazka-type chamber (Blazka et al., 1960; Fry, 1971), which has been described by Wohlschlag and Wakeman (1978) and Wakeman and Wohlschlag (1978). The chamber is essentially an enclosed 19-cm tube within a 29-cm tube with linearized recirculating flow maintained by a 10 horsepower constant-torque variable-speed motor. The Blazka chamber itself is suspended in a large tank, which is a part of an interconnecting system that also includes a large circular conditioning tank and appropriate filtration and temperature control units. These interconnected units were designed to permit acclimatization of fish in the same water in which experiments are conducted.

After the acclimation period following capture, and after fasting if necessary, small groups of fish were put into the circular tank and conditioned to swim for 24-48 h at about 1 body length (L) per second. This procedure was used both for controls and for various dilutions of the wastes prior to metabolic and swimming rate determinations in the Blazka chamber. The "training" for swimming was required for precision and replicability (Webb, 1975). When individual fish were transferred to the Blazka chamber, they were first run at a series of increasing speeds for about 1 h each until a maximum sustained speed could be held for 2 h or longer. During the course of determinations at the maximum sustained swimming (active) rate, samples of water were withdrawn for oxygen analysis at frequent intervals so that the oxygen consumption rate could be calculated in units of mg 0_2 h^{-1}. After the determination of an active rate, the fish were usually run at an intermediate speed to detect any erratic swimming that would result from fatigue and the residual anaerobic effects of excessively prolonged activity. Throughout the acclimatization and experimental procedures the tanks and Blazka chamber were well aerated; within the Blazka chamber the oxygen level was not allowed to drop much below 75% saturation.

For both resting and active metabolic rate measurements, red snappers were exposed to dilutions of the composite waste samples, which were shipped in plastic bottles from Puerto Rico. At a concentration of 0.5% (v/v), all fish died within a few hours; at 0.25%, resting fish survived, but swimming fish died. After 48 h there was heavy mortality at concentrations of 0.125%, but not at 0.0625%. A separate comparison at 0.0625% with waste shipped in a glass container was also made. Control runs were made with both the flow-through chambers and the Blazka chamber. At the end of all runs, the weights (g) and the lengths (cm) of the fish were recorded to accompany the oxygen rate measurements (mg O_2 h^{-1}), and where applicable, the swimming rates (L s^{-1} or $L^{0.5}$ s^{-1}).

Red snapper swimming rates were transformed to a square-root basis ($L^{0.5}$ s^{-1}) due to the fairly great length range and the pronounced tendency for the arithmetic swimming rate (L s^{-1}) to be inversely and geometrically correlated with fish length (L). The hydrodynamic explanation of this transformation is in Webb (1975) and in Wakeman and Wohlschlag (1978).

13.2.2. Exposure-Recovery Experiments with Spotted Seatrout (*Cynoscion nebulosus*)

The second set of experiments involved the spotted seatrout, a coastal and estuarine species, in a brief exposure to subacute levels of both the composite and the individual industrial wastes. The fish were then put into clean water. By measurement of active and standard metabolic rates for several days thereafter, the course of metabolic scope fluctuation could be compared to the controls. This species was held at 28°C and 35‰ salinity throughout, and was fasted 48 h before experimentation. Time of exposure for each waste was initially based on time to first visible signs of stress at a concentration of 0.5%, but the Bristol waste was so toxic that its concentration was reduced to 0.00625%. The time of onset of stress was visually evident when the first of a group of fish showed difficulty with equilibrium or respiration at a stage considerably before irreversible supination occurred. The fish were immediately transferred to the clean waters of the Blazka exercise conditioning tank. After several hours, respiration rates were determined at active, intermediate, and resting rates as described above. The process was repeated during the second and third days in clean water.

13.2.3. Multiple Regressions

Because respiratory metabolic rates of fish in terms of oxygen consumption rates and swimming propensities are related to body size, comparative data were made available by use of multiple regressions that relate the observed oxygen consumption rate ($\hat{Y}$) of a fish to its body weight X_w and its swimming rate X_s or X_v as follows:

$$\hat{Y} = a + b_w X_w + b_s X_s$$

$$\hat{Y} = a + b_w X_w + b_v X_v$$

where

$\hat{Y}$ = expected (calculated) O_2 consumption rate in log mg O_2 h^{-1},
a = constant,
X_w = log weight of fish (g),
b_w = partial regression coefficient of X_w to $\hat{Y}$,
X_s = swimming rate ($L^{0.5}$ s^{-1}, where L is length in cm), and
b_s = partial regression coefficient of X_s to $\hat{Y}$;

or where

X_v = swimming rate (L s^{-1}), and

b_v = partial regression coefficient X_v to $\hat{Y}$.

The statistical calculation procedures are given in most standard texts and are incorporated in many pretested computer programs. When X_s or X_v was zero for resting fish or for the Brett (1964) method of estimating standard metabolic rates, only the X_w and b_w terms were utilized. The swimming rates were used in X_s form for the red snapper experiments (but the X_v form for the spotted seatrout experiments) in order to compensate for the effect of disproportionate change in maximum sustained swimming velocity with size of the red snappers (Webb, 1975; Wakeman and Wohlschlag, 1978).

13.3. RESULTS

13.3.1. Sublethal Exposure Experiments with Red Snapper

The flow-through experiments to estimate resting and standard metabolic rates involved the concentrations of the composite wastes and the weights of the fish as shown in the upper half of Table 13.1. For the active metabolic rates in the Blazka chamber, the concentrations, averages and ranges of the weights and active swimming rates, and the number of fish observed at each dilution or control are shown in the lower part of Table 13.1.

The regression equations for the red snapper results for both the flow-through and the Blazka chamber experiments are summarized in Table 13.1. The equations for the standard and active rates of the respective experiments were utilized to calculate oxygen consumption rates at the mean weights of the fish and, where pertinent, at their mean maximum sustained swimming rate, after which conventional rates per unit weight (usually kg) can be utilized. This procedure avoids the biases in making comparisons of oxygen rates per unit weight from different experiments in which the mean fish weights vary considerably and permits the comparison among the log oxygen consumption rates and log weights of the fish from different experiments by the simple expedient of comparing the b_w regression values. Similarly differences among the b_s values may be compared. The biases result both from the fact that the oxygen consumption rates per unit weight are higher for smaller fish and from the more complex statistical properties of ratios of oxygen consumption rates per unit body weights.

The metabolic levels for the control and various concentrations of composite wastes are given in Fig. 13.1 for all levels calculated on the basis of the mean weights and plotted conventionally in mg O_2 kg^{-1} h^{-1}. The arrows in Fig. 13.1 denote the metabolic scope, the difference between active and standard metabolic levels; the dashed line indicates the estimate of the highest individual maximum performance level.

The standard metabolic rate remains relatively constant while the mean maximum sustained active rate decreases, as does the scope, with increasing concentrations of the waste. The highest individual active metabolic rates as in Fig. 13.1 would accentuate even more the relative decrease in scope. Note that there was no active metabolism (or scope) possible for fish at 0.25% concentrations of the composite waste.

13.3.2. Exposure–Recovery Experiments with Spotted Seatrout

Spotted seatrout were exposed to various concentrations of composite and individual pharmaceutical wastes for various times until first obvious signs of stress were subjectively observed. The fish were immediately removed to pure seawater for the evaluation of their recovery.

The concentrations and exposure times of the composite and various individual pharmaceutical wastes, the mean weights and weight ranges of the fish, and the swimming velocity means and ranges for all the data and for the maximum sustained swimming rates only are given in Table 13.2. Table 13.3 contains the regressions from the Blazka chamber respiratory data for control, composite waste, and individual waste experiments. The active rates are from the same type of regression calculations explained above for the red snapper experiments. The standard rates are from the Brett (1964) method of calculation of the line through the lowest measured oxygen consumption rates at observed weights and adjustment to zero swimming rate. From the equations in Table 13.3, the active and standard metabolic rates were calculated for a mean weight of 172 g for both control and experimental groups, converted to conventional kilogram units, and then plotted on the histograms of Fig. 13.2 as the solid horizontal lines. The arrows in Fig. 13.2 represent the scope for activity; the lower horizontal dashed line is the calculated resting metabolic rate at 172 g and a zero swimming rate; the upper horizontal dashed line is an estimated upper limit calculated at 172 g and at the highest observed swimming rate for each group of data. The individual upper limit rates in the composite and Upjohn waste experiments were relatively much higher than their corresponding maximum sustained rates compared to the other experiments. In both cases the highest rate of oxygen consumption was by fairly sluggish fish. This was interpreted to mean that the effect of the composite and Upjohn wastes was hyperstimulatory to respiration but not swimming performance.

Table 13.1. Weights and Swimming Rates with Regression Equations for Resting, Standard, and Active Respiratory Metabolic Rates of Red Snapper (*Lutjanus campechanus*) exposed to Pharmaceutical Waste Composites at 20°C and 35‰ Salinity

Waste Concentration % (v/v)	No. of Fish	Weight (g) Mean	Weight (g) Range	Swimming Rate[a] ($L^{0.5}\ sec^{-1}$) Mean	Swimming Rate[a] ($L^{0.5}\ sec^{-1}$) Range	Swimming Rate[a] ($L\ sec^{-1}$) Range	Equations[b,c]
				Flow-Through Chamber Experiments			
Control	40	386	47–746	—	—	—	Rest $\hat{Y} = -0.752 + 0.820X_w$
							Std. $\hat{Y} = -0.847 + 0.820X_w$
0.25[d]	5	329	188–540	—	—	—	Rest $\hat{Y} = -1.107 + 1.010X_w$
							Std. $\hat{Y} = -1.166 + 1.010X_w$
0.125	18	399	188–747	—	—	—	Rest $\hat{Y} = -0.911 + 0.918X_w$
							Std. $\hat{Y} = -1.063 + 0.918X_w$
0.0625[e]	21	147	64–668	—	—	—	Rest $\hat{Y} = -0.898 + 0.873X_w$
							Std. $\hat{Y} = -0.953 + 0.873X_w$
0.0625	21	137	72–440	—	—	—	Rest $\hat{Y} = -0.659 + 0.783X_w$
							Std. $\hat{Y} = -0.786 + 0.783X_w$

				Blazka Chamber Experiments			
Control	73	282	128–860	13.57	0–20.40	0–4.5	Act. $\hat{Y} = -0.517 + 0.779X_w + 0.037X_s$
							High $\hat{Y} = +1.563 + 0.779X_w + 0.037X_s$
0.125	34	424	222–734	14.76	0–16.28	0–3.1	Act. $\hat{Y} = +0.159 + 0.603X_w + 0.017X_s$
							High $\hat{Y} = +1.906 + 0.603X_w + 0.017X_s$
0.625	50	158	68–649	14.84	0–18.42	0–4.3	Act. $\hat{Y} = +0.448 + 0.488X_w + 0.016X_s$
							High $\hat{Y} = +1.664 + 0.488X_w + 0.016X_s$
0.0625[e]	46	140	72–447	13.64	0–16.84	0–4.0	Act. $\hat{Y} = +0.082 + 0.716X_w + 0.018X_s$
							High $\hat{Y} = +1.577 + 0.716X_w + 0.018X_s$

[a]Relative to length of fish (L).

[b]From flow-through chamber data, average resting oxygen consumption rates, Rest $\hat{Y}$, which were calculated from observations, are too high for estimates of standard oxygen consumption rates, Std. $\hat{Y}$. Standard rates (without elevated respiration due to spontaneous, nonlocomotory activity) are estimated by method of Brett (1964) as adapted by Wohlschlag and Parker (1981), which reduces regression to pass through lowest oxygen consumption rates.

[c]From Blazka chamber data, active oxygen consumption rates, Act. $\hat{Y}$, were calculated directly over range of swimming rates. The upper rate, High $\hat{Y}$, was calculated through observations with the highest swimming and oxygen consumption rates in order to estimate a maximum physiological performance capability (Wohlschlag and Parker, 1981).

[d]Heavy mortality.

[e]Separate sample shipped in glass container.

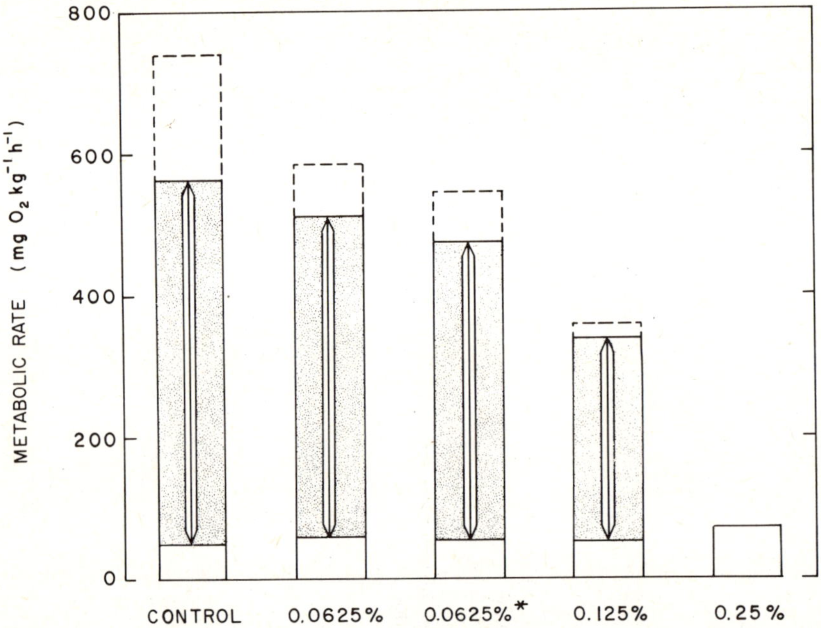

Figure 13.1. Histograms of red snapper (*Lutjanus campechanus*) metabolic rates at 20°C and 35 ‰ salinity, and at mean weights of each control and composite waste experimental group. Upper dashed line is calculated highest rate, upper solid line is metabolic rate at average maximum sustained activity, and lower solid line is estimate of standard rate. Arrow represents scope for activity. Asterisk denotes separate experiment with waste transported in glass instead of plastic container.

To permit comparison of each composite or individual industrial waste effect with the control on the basis of equal weights of the fish, the data were recalculated at the average weights for each experimental group as given in Table 13.2. The pertinent metabolic levels and the scopes for each experimental–control group are plotted in Fig. 13.3, which may be used for comparison with Fig. 13.2, in which all levels are based on a single average weight. Because of the possible excessive stimulatory effect on respiration but not activity by some wastes on some individual fish, the upper limit metabolic rates were not plotted.

13.4. DISCUSSION

13.4.1. Composite Waste Effects on the Red Snapper

These experiments agree well with those reported by Wohlschlag and Parker (1981) for this species. The fact that the maximum sustained swimming metabolic level and scope for activity decreased uniformly with the concentration of the composite pharmaceutical waste (Fig. 13.1) is evidence that the technique is highly sensitive. It is noteworthy that the estimated highest individual rates (Fig. 13.1) declined

Table 13.2. Means and Ranges of Variables Used in Regression Equations for Exposure–Recovery Respiration Rate Experiments on Spotted Seatrout (*Cynoscion nebulosus*) under Control Conditions and Various Concentrations of Pharmaceutical Wastes at 28°C and 35‰ Salinity

Description	Concentration, v/v (%)	Number of Fish	Weight (g)		Velocity ($L\ s^{-1}$)		Max. Sustained Velocity ($L\ s^{-1}$)		Equation Number[a]
			Range	Mean	Range	Mean	Range	Mean	
Control	0% (0)[b]	34	72–174	114	0–3.6	1.7	2.9–3.6	3.3	0.0
Composite	0.5% (60 min)	36	61–377	175	0–2.7	1.3	1.0–2.7	2.0	2.0
Upjohn	0.5% (120 min)	14	321–674	461	0–2.5	1.1	1.6–2.5	2.0	3.1
Merck	0.5% (25 min)	25	99–371	163	0–3.1	1.5	2.2–3.1	2.8	3.2
Bristol	0.00625% (120 min)	13	67–152	109	0–3.4	1.7	2.6–3.4	3.1	3.3
Capri	0.5% (120 min)	19	86–410	172	0–3.1	1.5	2.0–3.1	2.6	3.4
Squibb	0.5% (110 min)	15	87–258	137	0–3.3	1.6	2.6–3.3	2.9	3.5
Pfizer	0.5% (120 min)	23	75–332	144	0–2.9	1.5	2.2–2.9	2.6	3.6

[a]See Table 13.3.

[b]Exposure time in parentheses.

Table 13.3. Regression Equations for Spotted Seatrout (*Cynoscion nebulosus*) Respiratory Metabolic Responses to Control and Short-Term Exposure to Pharmaceutical Waste Samples at Various Active Swimming (Act.) and Standard (Std.) Rates and at 28°C and 35‰ Salinity[a]

Description	Concentration, v/v (%)	Number of Fish	Equations	Equation Number
Control	0[b]	34	Act. $\hat{Y} = +0.070 + 0.712X_w + 0.112X_v$ Std. $\hat{Y} = -0.037 + 0.712X_w$	0.0
Exposure–recovery (Composite)	0.5 (60 min)	36	Act. $\hat{Y} = +0.287 + 0.679X_w + 0.083X_v$ Std. $\hat{Y} = +0.187 + 0.679X_w$	2.0
Exposure–recovery (Upjohn)	0.5 (120 min)	14	Act. $\hat{Y} = -0.057 + 0.740X_w + 0.178X_v$ Std. $\hat{Y} = -0.044 + 0.740X_w$	3.1
Exposure–recovery (Merck)	0.5 (25 min)	25	Act. $\hat{Y} = +0.397 + 0.647X_w + 0.080X_v$ Std. $\hat{Y} = +0.328 + 0.647X_w +$	3.2
Exposure–recovery (Bristol)	0.00625 (120 min)	13	Act. $\hat{Y} = +0.910 + 0.392X_w + 0.059X_v$ Std. $\hat{Y} = +0.799 + 0.392X_w$	3.3
Exposure–recovery (Capri)	0.5 (120 min)	19	Act. $\hat{Y} = +0.311 + 0.643X_w + 0.099X_v$ Std. $\hat{Y} = +0.161 + 0.643X_w$	3.4
Exposure–recovery (Squibb)	0.5 (110 min)	15	Act. $\hat{Y} = +0.389 + 0.634X_w + 0.056X_v$ Std. $\hat{Y} = +0.292 + 0.634X_w$	3.5
Exposure–recovery (Pfizer)	0.5 (120 min)	23	Act. $\hat{Y} = +0.579 + 0.550X_w + 0.082X_v$ Std. $\hat{Y} = +0.465 + 0.550X_w$	3.6

[a] Values rounded to third decimal place.
[b] Exposure time in parentheses.

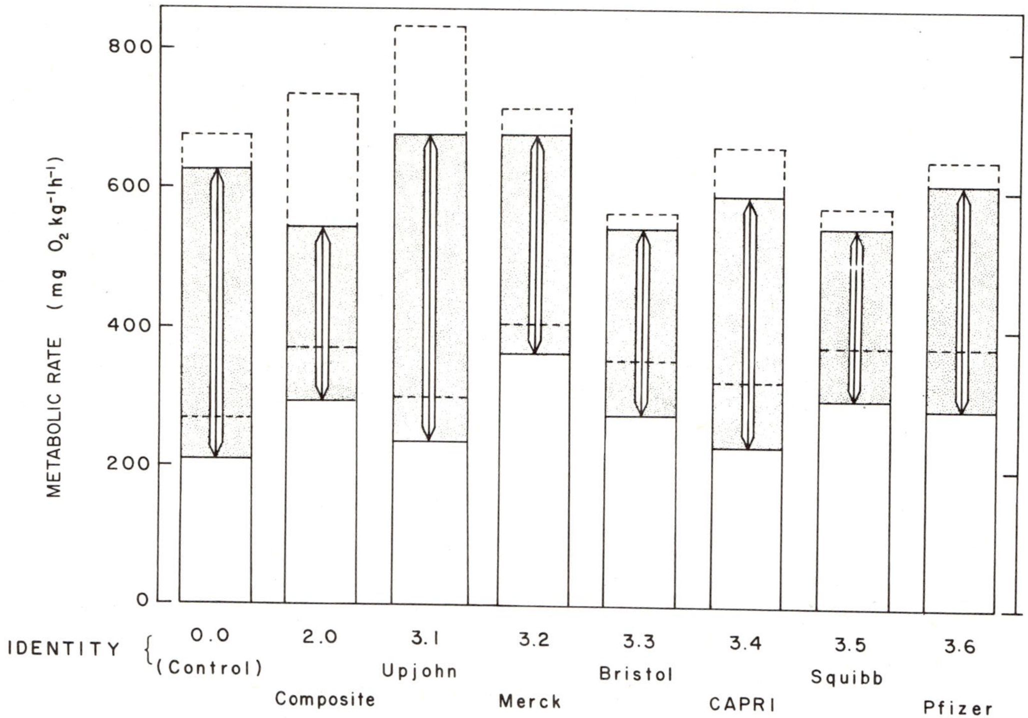

Figure 13.2. Plots of calculated metabolic rates for the spotted seatrout (*Cynoscion nebulosus*) at 28°C and 35 ‰ salinity and at mean weight of 172.04 g for all experimental control data. Upper dashed horizontal line is highest individual sustained active rate. Upper solid line is average maximum sustained rate; lower solid line is standard rate; arrow represents scope for activity. The dashed line at the intermediate level represents a resting rate (swimming rate = 0) calculated from active fish data.

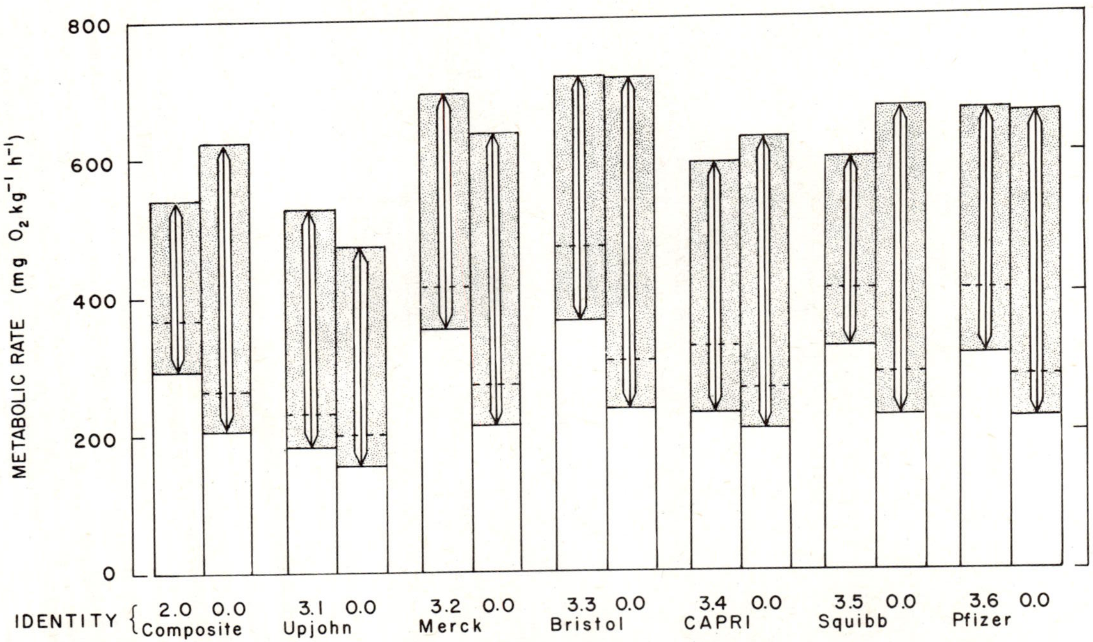

Figure 13.3. Plots of spotted seatrout (*Cynoscion nebulosus*) metabolic rates calculated at 28°C and 35 ‰ salinity and at the mean weight of fish in each industrial waste experiment and paired with calculated rates at that weight for control (equation 0.0) fish. Histogram components same as in Fig. 13.2.

even more than did the scopes with increasing waste concentrations. Exposing the fish for longer than 48 h to dilutions greater than 0.0625% would allow for even greater sensitivity, but 48 h is near the maximum time for fasted fish to be retained without considering a feeding regime and also the waste probably loses its toxicity at great dilutions after 48 h or longer. Both these exceptions could be accommodated in future experiments by adding wastes to flowing seawater to maintain a given equilibrium dilution and by alternately feeding and fasting the fish.

Statistics for the regressions and the individual variables b_w and b_s are generally highly significant ($P < 0.001$) with multiple correlations of $r = 0.84$ to 0.98. The statistical variability of the regression equations for the fish in waste transported in glass was slightly higher than that of the corresponding equations for wastes transported in plastic containers. The variability of the corresponding b_w was also slightly higher, which also indicated a slightly increased toxicity and reduced scope (Fig. 13.1). Dodson and Mayfield (1979) noted a similar increase in variability for kinetic components when rainbow trout initially experience low concentrations of a toxic substance. At all dilutions beyond 0.25%, the standard metabolic level remained essentially constant (Fig. 13.1), which indicates that the red snapper apparently did not adjust to sublethal materials with an increase in maintenance costs.

13.4.2. Exposure-Recovery Experiments with Spotted Seatrout

The control regressions of Table 13.3 agree with those for this species at 35‰ and 28°C published by Wohlschlag and Wakeman (1978). The regression equations for the composite and individual industrial wastes (equations 2.0 and 3.1 to 3.6 in Table 13.3) should be considered as arbitrary averages of conditions over the 2–4 d of recovery in clean seawater. No statistically useful trends were evident in metabolism–time plots over the short recovery interval. Although the numbers of fish in each of the separate experiments were small, the lowest multiple correlation was $r =$ 0.64 ($P \simeq 0.05$) for the 13 fish in the highly diluted Bristol waste (equation 3.3); the unusually small b_w for this experiment was not statistically significant. Other multiple correlation R values were much higher, and all the remaining b_w and b_v coefficients were statistically significant, usually at $P < 0.01$. The erratic metabolism–weight response is associated with the expectation that the larger fish tend to be more depressed under stress than the smaller ones (Wohlschlag and Cameron, 1967; Kloth and Wohlschlag, 1972). Such size-related responses to stress are probably the cause of the generally low b_v in Table 13.3. The b_v coefficients lower than about 0.1 result from some individual fish swimming very slowly compared to their potential in clean water; high b_v for the Upjohn waste (equation 3.1) experiment indicates that the fish were larger than average and swam in a "labored" manner with high energy expenditures for slow swimming rates.

Preliminary experiments with the concentrations and exposure times required to produce somewhat equivalent visible stress effects with the separate wastes were selected as summarized in Table 13.2. All the fish in a single trial did not behave alike. There were inexplicably different reactions to the different individual diluted wastes as well. The reaction ranged upon initial exposure from nearly no observable effects to slightly obvious erratic swimming, minor ventilatory rate increases, to appearance of partial anaesthesia. By the end of 2 h (or 1 h for the composite waste, 25 min for Merck, and 110 min for Squibb), the various reactions were quite pronounced supination, erratic swimming, and other manifestations of disequilibrium; also increasing with exposure time were pronounced coughing, head shaking, and increased ventilatory rate. In no cases were any fish allowed to die at the conditions in Table 13.2 before transfer to clean waters. The initial metabolic rates tended to be high and then decrease, while the swimming rates tended to be depressed initially and then over 2–4 d of experiments sometimes improved, although generalities from one toxic waste to the next were not possible from plots of the metabolic and swimming activity levels against elapsed time in clean water.

The results for the spotted seatrout are about the "worst possible" situation consistent with recovery. The fish at the end of the exposure–recovery experiments were approaching morbidity with high incidences of fin and tail rot, which is a common occurrence after prolonged stress. Too few metabolic measurements were possible at the end of the experiments for a clear-cut assessment of the metabolic and swimming rates at that time alone.

At the exposure–recovery conditions plotted in Fig. 13.2 with all data adjusted to the same weight, the composite waste recovery levels indicate a great reduction in scope both by a drop in the active level and by an increase in the standard level. Possibly the relatively small scope of the composite waste experiments, compared with the single waste experiments, is due to interactions among all the different wastes. The standard rates in the earlier experiment with the composite waste on the red snappers at 20°C, however, did not increase. The differences between the two experiments suggest differences between the toxicities of the wastes in 2 y. All other standard metabolic levels in Fig. 13.2 are higher than the control level. Why the active levels of the Upjohn (equation 3.1) and Merck (equation 3.2) wastes are higher than the controls, even though the subject fish swam more slowly on the average than did the controls (Table 13.3) is not altogether clear, but it is possible that the wastes did contain stimulatory substances. Stimulation by low-level toxicants has been known for some time. Webb (1978) has summarized several instances from the literature to show such stimulation effects. However, in the experiments with the spotted seatrout, concentrations of toxic components were all sufficiently high to make ultimate recovery marginal, if the amount of fin rot is taken as evidence for morbidity at the termination of the recovery periods. Thus, there is reason to believe that the high metabolic scope for the Upjohn (equation 3.1) experiment and the high active metabolic rates (with poor swimming performance) for both Upjohn and Merck (equation 3.2) were simply transitory manifestations or "burnouts" that would eventually be followed by depressed active rates and scopes, had it been possible to continue beyond 4 d. The unusually high standard rate of the Merck experiment represents a stress level that would be difficult to maintain for the spotted seatrout under normal 28°C conditions (Wohlschlag and Wakeman, 1978).

When the controls and the experimental exposure–recovery results are compared on an equivalent weight basis as in Fig. 13.3, the active rates appear to be stimulated by the Upjohn wastes (as in Fig. 13.2) and by the very low concentration of the Bristol (equation 3.3) and the Pfizer (equation 3.6) wastes. In both Fig. 13.2 and Fig. 13.3, however, the scope for all but the Upjohn (equation 3.1) experiment are reduced.

The relationship of weight to metabolism is very important when comparisons are made between unstressed and stressed fishes. In multiple regression analyses of the types shown in Table 13.3, for example, about 80% of the variability of $\hat{Y}$ (log oxygen consumption rate) would be associated with X_W (log weight) while the remaining variability would be associated with swimming behavior, the kinds of stress, and uncontrolled or random variability. For this reason, it is essential that metabolic comparisons be made on an adjusted weight basis, either as in Fig. 13.2 (all at the same weight) or as in Fig. 13.3 (each experiment–control comparison at its characteristic weight). Each alternative (Fig. 13.2 or Fig. 13.3) is important in its own right.

13.4.3. Relationships of Biological Effects to Chemistry of Wastes

Biological effects tend to be noticed before the chemical nature of sublethal toxic materials is identified and defined, except in cases where materials of known composition and previously established toxicity are released into the environment. In this study, the premise was that the chemical composition of the composite or individual wastes was not known, but that the sublethal effects could be detected both by metabolic scope diminution and by impaired active swimming performance. Actually considerable information is available on chemical methodology (Atlas et al., 1981) and analyses (Presley et al., 1981; Schwab et al., 1981; Brooks et al, Chapter 9) of these wastes. Among the organic fractions, it is noteworthy that most would probably be acutely detrimental to fish only in much higher concentrations than the usual dilution several hours after dumping at sea. Literature-based speculation is possible on which of the more persistent and noxious identified compounds (for example, dimethylaniline) at the dumpsite might be sublethal. However, there is little direct experimental evidence available to answer such questions as whether (1) suspected or identified compounds actually are sublethal to marine fish, (2) the compounds maintain their chemical identities after dumping, (3) there are interactions among compounds at very low levels that would have a great effect on fish, and (4) there are cumulative or delayed effects on organisms by any of the compounds. In lieu of chemical experiments to gain answers to such questions, Hocutt and Stauffer (1980) suggest biological approaches for detection of sublethal stresses.

Because it is difficult to relate separate chemical components of complex organic wastes to specific physiological effects in an oceanic environment, Wohlschlag and Parker (1981) suggested that a simple, modified Blazka respirometer system be utilized to study marine fish at a site where seawater and a continuous supply of test wastes could be added or diverted from the respirometer at will. In all the above experiments, metabolic scope and the maximum sustained swimming rates are closely correlated, which suggests that swimming performance alone could be used directly to monitor stress without the oxygen consumption rate determinations.

Biological monitoring as an early warning system, as has long been advocated by Prof. John Cairns, Jr., and his associates [for example, Cairns and Gruber (1979)], would permit predictions of long-term population effects.

13.4.4. Relationships of Stresses to Fish Populations

For open ocean dumping of such toxic materials as were used in these studies, two kinds of population effects are important if dumping occurs at intervals of about one week at a given locality. The first is when toxicities at somewhat lower than the levels indicated in the red snapper experiments have a depressing effect on such related instantaneous rate functions as growth, assuming no cumulative effects leading to direct lethality. This effect can be persistent, depending on dispersion over the area where the sublethal dilution level persists. In the second effect, all the fish subjected briefly to waste concentrations of the orders illustrated in the spotted seatrout experiments, or higher, shortly after the time of dumping would experience a deleterious effect for up to 4 d even after all trace of waste has disappeared. Should such dumping continue with regularity, it is conceivable that entire fish populations could eventually be decimated over a wide area. Future application of metabolic stress data, as shown in this study, will be useful for the various models of population dynamics.

Wohlschlag and Parker (1981) have suggested that metabolic and swimming rates have important general relationships to population biology in the following areas: evaluations of environmental optima, assessment of stress effects, explanations of niche theory, assessment of growth–foraging energetics, clarification of various fishery dynamics models, etc. Similarly responses to all environmental variations, as pointed out by Fry (1971), are ultimately expressed at the total metabolic level. Fry (1971) relates response categories of fish to the following environmental factors: lethal, controlling, limiting, masking, and directive. No doubt all these factors and their physiological responses were involved in the study of toxic waste effects on fishes reported here. The practical challenge in ecological and physiological terms remains to separate the effects by these factors among these response categories.

ACKNOWLEDGMENTS

We are grateful for the support received from the National Oceanic and Atmospheric Administration (NOAA grant no. 04-8-M01-54) and for aid given by NOAA's Ocean Dumping Program personnel who shipped the pharmaceutical wastes from Puerto Rico and provided advice and encouragement for this work.

We extend special thanks to research assistants M. Dobbs and E. Findley, graduate student assistants M. Gunter and R. Ilg, boat captains D. Gibson (*R/V Long-*

horn) and E. Wingfield (*R/V Lorene*), the *R/V* crew members, and typist H. Garrett. This study is the University of Texas Marine Science Institute Contribution No. 471.

REFERENCES

Atlas, E., G. Martinez, and C. S. Giam. 1981. Chemical characterization of ocean-dumped waste materials. *In*: Ocean Dumping of Industrial Wastes, B. H. Ketchum, D. R. Kester, and P. K. Park (Eds.). Plenum Press, New York, pp. 275–293.

Blazka, P., M. Volf, and M. Cepela. 1960. A new type of respirometer for the determination of metabolism of fish in the active state. *Physiologia Bohemoslovenica* **9**, 533–558.

Brett, J. R. 1964. The respiratory metabolism and swimming performance of young sockeye salmon. *Journal of the Fisheries Research Board of Canada*, **21**, 1183–1226.

Cairns, J., and D. Gruber. 1979. The coupling of mini- and micro-computers to biological early warning systems. *BioScience*, **29**, 665–669.

Carlson, A. R. 1972. Effects of long-term exposure to carbaryl (sevin) on survival, growth and reproduction of the fathead minnow (*Pimephales promelas*). *Journal of the Fisheries Research Board of Canada*, **29**, 583–587.

Dodson, J. J., and C. I. Mayfield. 1979. The dynamics and behavioral toxicology of Aqua-Kleen® (2,4-D butoxyethanol ester) as revealed by the modification of rheotropism in rainbow trout. *Transactions of the American Fisheries Society*, **108**, 632–640.

Fry, F. E. J. 1947. Effects of the environment on animal activity. University of Toronto Studies of Biology, *Publications of the Ontario Fisheries Research Laboratory*, **68**, 1–62.

Fry, F. E. J. 1971. The effect of environmental factors on the physiology of fish. *In*: Fish Physiology, Vol. 6, W. S. Hoar and D. J. Randall (Eds.). Academic Press, New York, pp. 1–98.

Grant, B. F., and P. M. Mehrle. 1973. Endrin toxicosis in rainbow trout (*Salmo gairdneri*). *Journal of the Fisheries Research Board of Canada*, **30**, 31–40.

Hocutt, C. H., and J. R. Stauffer, Jr. (Eds.) 1980. Biological Monitoring of Fish. Lexington Books, D. C. Heath, Lexington, Massachusetts, 416 pp.

Kloth, T. C., and D. E. Wohlschlag. 1972. Size-related metabolic responses of the pinfish, *Lagodon rhomboides*, to salinity variations and sublethal pollution. *Contributions in Marine Science*, **16**, 125–137.

Krueger, H. M., J. B. Sadler, G. A. Chapman, I. J. Tinsley, and R. R. Lowry. 1968. Bioenergetics, exercise and fatty acids of fish. *American Zoologist*, **8**, 119–129.

Mount, D. I., and C. E. Stephan. 1969. Chronic toxicity of copper to the fathead minnow (*Pimephales promelas*) in soft water. *Journal of the Fisheries Research Board of Canada*, **26**, 2449–2457.

Pickering, Q. H., and M. H. Gast. 1972. Acute toxicity of cadmium to the fathead minnow (*Pimephales promelas*). *Journal of the Fisheries Research Board of Canada*, **29**, 1099–1106.

Presley, B. J., J. S. Schofield, and J. Trefry. 1981. Waste material behavior and inorganic geochemistry at the Puerto Rico waste dumpsite. *In*: Ocean Dumping of Industrial Wastes, B. H. Ketchum, D. R. Kester, and P. K. Park (Eds.). Plenum Press, New York, pp. 233–246.

Schwab, C. R., T. C. Sauer, Jr., G. F. Hubbard, H. Abdel-Reheim, and J. M. Brooks. 1981. Cehmical and biological aspects of ocean dumping at the Puerto Rico dumpsite. *In*: Ocean Dumping of Industrial Wastes, B. H. Ketchum, D. R. Kester, and P. K. Park (Eds.). Plenum Press, New York, pp. 247–273.

Smith, L. L., R. H. Kramer, and D. M. Oseid. 1966. Long-term effects of conifer-groundwood paper fiber on walleyes. *Transactions of the American Fisheries Society*, **95**, 60–70.

Wakeman, J. M., and D. E. Wohlschlag. 1978. Salinity stress and swimming performance of spotted seatrout. *Proceedings of the Annual Conference of the Southeastern Association of Fish and Wildlife Agencies*, **31**, 357–361.

Warren, C. E. 1971. Biology and Water Pollution Control. Saunders, Philadelphia, 434 pp.

Webb, P. W. 1975. Hydrodynamics and energetics of fish propulsion. *Bulletin of the Fisheries Research Board of Canada*, **190**, x + 158 pp.

Webb, P. W. 1978. Partitioning of energy into metabolism and growth. *In*: Ecology of Freshwater Fish Production, S. D. Gerking (Ed.). Wiley, Halsted Press, New York, pp. 184–214.

Webb, P. W., and J. R. Brett. 1972. Effects of sublethal concentrations of sodium pentachlorophenate on growth rate, food conversion efficiency, and swimming performance in underyearling sockeye solmon (*Oncorhynchus nerka*). *Journal of the Fisheries Research Board of Canada*, **30**, 499–507.

Wohlschlag, D. E., and J. N. Cameron. 1967. Assessment of a low level stress on the respiratory metabolism of the pinfish (*Lagodon rhomboides*). *Contributions in Marine Science*, **12**, 160–171.

Wohlschlag, D. E., and F. R. Parker. 1981. Metabolic sensitivity of fish to ocean dumping of industrial wastes. *In*: Ocean Dumping of Industrial Wastes, B. H. Ketchum, D. R. Kester, and P. K. Park (Eds.). Plenum Press, New York, pp. 455–472.

Wohlschlag, D. E., and J. M. Wakeman. 1978. Salinity stresses, metabolic responses and distribution of the coastal spotted seatrout, *Cynoscion nebulosus*. *Contributions in Marine Science*, **22**, 171–185.

PART V: THE MID-ATLANTIC AND NEW YORK BIGHT DUMPSITES

14

HISTORY OF OCEAN DISPOSAL IN THE MID-ATLANTIC BIGHT

William C. Muir

Environmental Impacts Branch
U.S. Environmental Protection Agency (Region III)
Philadelphia, Pennsylvania

ABSTRACT

During the 1960s and 1970s a variety of industrial and municipal wastes were dumped in the ocean on the mid-continental shelf bordering Maryland and Delaware. U.S. Environmental Protection Agency (U.S. EPA) Region III managed four disposal sites in this area between 1972 and 1980.

With the passage of the Marine Protection, Research, and Sanctuaries Act of 1972 and the subsequent regulations, the U.S. EPA's Region III began a systematic review of each dumper for the need to dump, the feasible alternatives to ocean disposal, and the impacts of the dumping on the marine environment. As a result of the Region's efforts and the eventual assistance of the dumpers themselves, all four disposal sites have been successfully phased out. Further, in all instances environmentally acceptable land-based alternatives were developed. In several cases these alternatives were less costly than ocean disposal, once implemented.

14.1. INTRODUCTION

About 10% of the pollutants entering the world's oceans are introduced in some form by ocean dumping (U.S. Environmental Protection Agency, 1972a). Until 1972 the legal framework governing ocean disposal consisted of a complex maze of jurisdictional overlaps of international, federal, state, and local systems. Smith and Brown (1971) summarized the legal aspects of waste disposal at sea through 1971; for all intents and purposes, no environmental safeguards existed. The U.S. Army Corps of Engineers, which had the most authority, was concerned only with impediment and obstruction to navigation. States had little if any say in ocean disposal.

The revelations of the late 1960s and 1970s concerning the dumping of nerve gas and petroleum off the Florida coast, the dumping of enormous quantities of sewage sludge each day from New York and Philadelphia, and a number of similar instances, spawned the Marine Protection, Research, and Sanctuaries Act of 1972 (MPRSA). Relying heavily on the Council of Environmental Quality's recommendations (U.S. Council of Environmental Quality, 1970), Congress declared in Section 2(b) of the MPRSA (U.S. Congress, 1972) "it is the policy of the United States *to regulate* and *to prevent* or *strictly limit* the dumping into ocean waters of any materials which would adversely affect human health, welfare, or amenities, or the marine environment, ecological systems, or economic potentialities" (emphasis added).

Under the provisions of the MPRSA the U.S. Environmental Protection Agency's (U.S. EPA) Region III was given jurisdiction over disposal activities in the Middle Atlantic Bight (U.S. Environmental Protection Agency, 1973a). Region III covers Pennsylvania, Maryland, Delaware, Virginia, West Virginia, and the District of Columbia.

At the onset of the ocean disposal program in April 1973, EPA's Region III administered four disposal sites shown in Fig. 14.1 (U.S. Environmental Protection Agency, 1973b). The old sewage sludge site was authorized by the U.S. Army Corps of Engineers in 1961 for the disposal of Philadelphia's municipal sewage sludge. In 1966, Camden, New Jersey, began using the same site. The old sewage sludge site was closed on 8 May 1973, and both Philadelphia and Camden sludges were barged to a site called the interim sewage sludge dumpsite. The interim dumpsite closed on 31 December, 1980. E. I. du Pont de Nemours & Co.'s Edge Moor, Delaware plant, hereafter referred to as DuPont–Edge Moor, used the acid-iron waste disposal site from 1969 to 1977, at which time its dumping was moved off the Continental shelf to Region II's industrial wastes dumpsite (DWD-106) (see Fig. 14.1.). The fourth site, the arsenic waste dumpsite, was used intermittently for

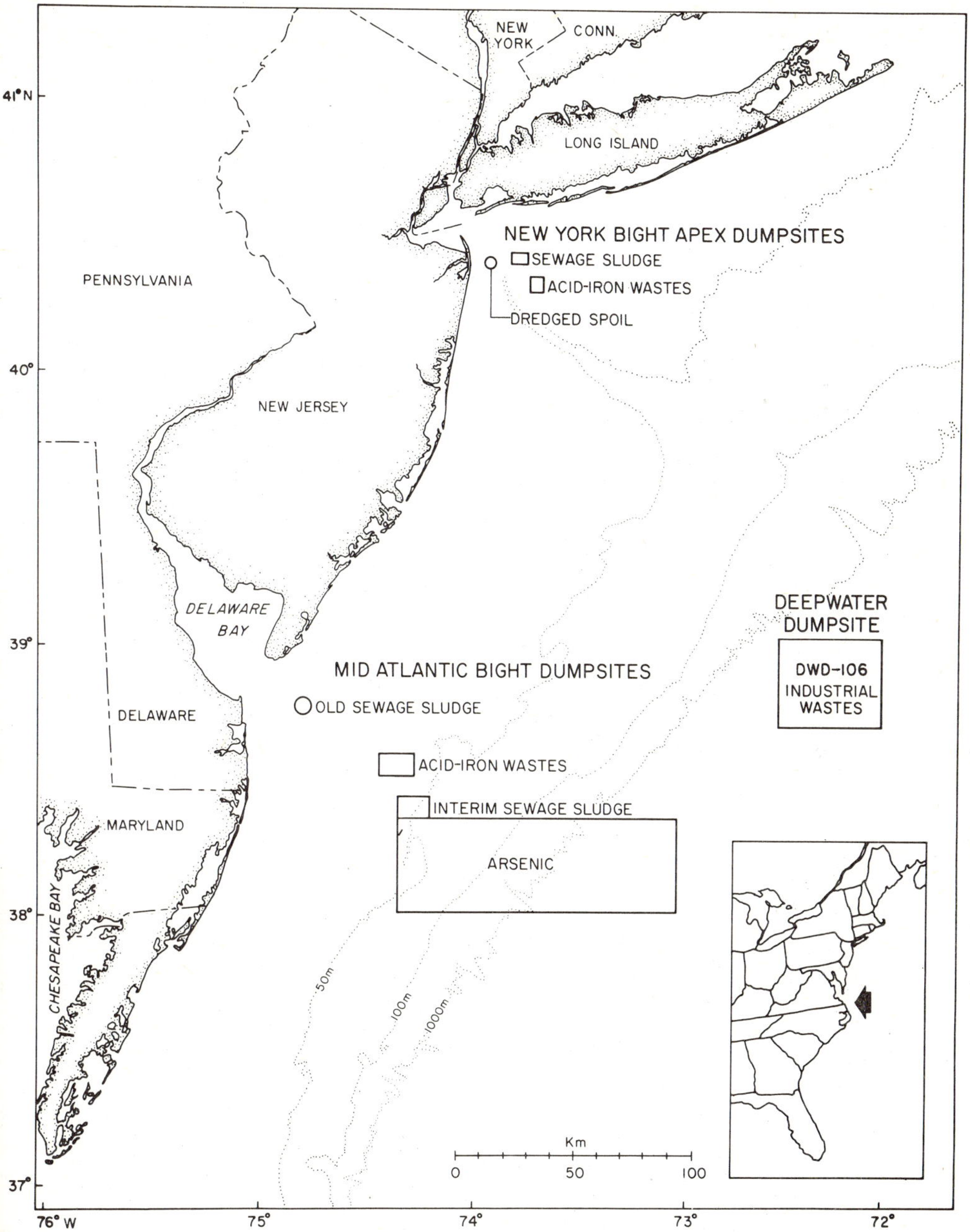

Figure 14.1. Map of the Mid-Atlantic Bight study area showing the dumpsites, EPA Region III controls.

Table 14.1. Summary of U.S. EPA Region III Dumpsite–Total Wastes Dumped

Dumpsite	Waste Type	Coordinates (Latitude, Longitude)	Quantity (thousand metric tons)
Old sewage sludge site (1961–1973)	Sewage sludge	38°45′N, 74°47′W	3750
Interim sewage sludge site (1973–1980)	Sewage sludge	38°20′–38°25′N, 74°10′–74°20′W	4110
Acid-iron waste site (1969–1977)	Acid	38°30′–38°35′N, 74°15′–74°20′W	5060
Arsenic waste site (1966–1974)	Caustic	38°00′–38°20′N, 73°00′–74°20′W	110

a number of years by various industries, but no records were kept until 1966, when Sun Oil Company's Marcus Hook, Pennsylvania, plant began dumping caustic and white-water oil emulsions. The arsenic dumpsite was closed in 1974. A summary of the total site usage is shown in Table 14.1.

U.S. EPA's Region III has rigorously enforced the mandate set forth by Congress. Table 14.2 shows all permit applications within Region III. Each application re-

Table 14.2. U.S. EPA Region III Ocean Dumping Permit Activities[a]

State	Applicant	Issued/Denied	Phase Out
Delaware	DuPont–Edge Moor	Issued	1983
	Rollins Environmental Services, Wilmington	Issued	1975
	Barcroft Company, Wilmington	Denied	
District of Columbia	U.S. Army Corps of Engineers	Denied	
New Jersey	City of Camden	Issued	1978
	New Jersey Zinc Company, Gloucester City	Denied	
Ohio	U.S. Air Force, Patterson	Denied	
Pennsylvania	City of Philadelphia	Issued	1980
	Sun Oil Company, Marcus Hook	Issued	1975
	Modern Transportation Company–Sewage, Kearney	Issued	1975
	Modern Transportation Company–Industrial	Issued	1975
	ABM Disposal Services, Lester	Denied	
	Integrity Shipping Company, Philadelphia	Denied	
	Whitmoyer Laboratories	Denied	
	McAdoo & Allen, Incorporated	Denied	
	Crompton & Knowles, Incorporated	Issued	1979
Virginia	Hampton Roads Sanitation District, Virginia Beach	Denied	
West Virginia	DuPont, Belle	Issued	1976
Wisconsin	Ansul Company	Denied	

[a]U.S. EPA Region III ocean dumping files, Philadelphia, Pennsylvania.

ceived a critical review to determine if there was a real need to dump, what the potential environmental effects of dumping would be, and how the effects could be minimized if dumping was necessary. All permits that were issued contained requirements that the dumper develop alternatives to dumping as soon as possible.

There are now no active dumpsites in Region III. Except for DuPont-Edge Moor, which now dumps its wastes at DWD-106 (Fig. 14.1), all other dumping that originated within Region III has been phased out.

14.2. REGIONAL ACTIVITIES

14.2.1. Old Sewage Sludge Dumpsite

The old sewage sludge dumpsite (Fig. 14.1) was first designated in the *Federal Register* as an interim site (U.S. Environmental Protection Agency, 1973b). The site covered approximately 12.5 km^2 and was located 19.3 km east of the Delaware Bay in 10-15 m of water.

The site was chosen in 1961 by Philadelphia to dispose of sewage sludge generated at its three sewage treatment plants. The Northeast sewage plant employed primary and intermediate wastewater treatment followed by mesophilic anaerobic digestion of the residual solids. The Southeast and Southwest sewage treatment plants used primary wastewater treatment with mesophilic anaerobic digestion of the residual solids (Wisdom, 1973).

Philadelphia decided to dump its sludge at sea after a study of alternatives by the consulting firm of Greeley and Hansen, Chicago, Illinois. Together with the city's Water Department, they analyzed 14 alternative methods of sludge disposal and concluded that ocean dumping was the most practical and economical (Baxter, 1959).

The location of the dumpsite at the mouth of Delaware Bay (Fig. 14.1) was made after considering favorable ocean currents, depth of water, and the marginal value of shellfishing in the area (Guarino, 1976). The dumpsite was just beyond Delaware and New Jersey's 5-km and the United States' 20-km jurisdictions. The U.S. Army Corps of Engineers authorized the site as it was not considered a hazard to navigation.

In 1966, Camden, New Jersey, began using the same disposal dumpsite. Camden barged sewage sludge from its primary sewage treatment plant. Its sludge, however, was undigested. Ocean disposal was chosen when its municipal sewage sludge incineration system became inoperative and ocean disposal was found to be the most economical alternative (Porter and Ripa, 1974).

Because of increased public concern for the environment in the late 1960s, the ecological soundness of sewage sludge dumping came under review and criticism. As a result, Philadelphia contracted with The Franklin Institute Research Laboratory and Jefferson Medical College, both of Philadelphia, Pennsylvania, to study the dumpsite and determine if any harmful effects had occurred (Davey, 1972). The study consisted of one year of oceanographic cruises to the old sewage sludge dumpsite and the laboratory analysis of 126 and 63 samples of suspended and bottom sediment samples, respectively. The conclusion from the study was that the sewage sludge had a negligible effect on the ocean in the discharge area. No evidence was found to indicate an accumulation of sewage sludge on the ocean bottom, and there was no appreciable percentage of man-contributed pollutants in the bottom sediments.

The environmental effects of barge disposal at the old sewage sludge dumpsite were also studied by Buelow et al. (1968) and Verber et al. (1969). Because of the preliminary nature of their field investigations, neither Buelow et al. nor Verber et al. were able to approve the dumpsite nor could they recommend its relocation. However, Buelow et al. did conclude that the old sewage sludge dumpsite posed a potential threat to existing commercial surf clam beds and shellfish beds located south and west of the site.

The only follow-up monitoring of the dumpsite was done by Oostdam et al. (1975) (see Chapter 16). They could find no significant differences in heavy metal contents of sediments or of clams inside and outside the dumpsite. They concluded that it was not possible to tell whether the area had been impacted and then recovered or had never been impacted.

A summary of Philadelphia and Camden's dumping at the old sewage sludge dumpsite is illustrated in Fig. 14.2. and a summary of the total loadings of contaminants is given in Table 14.3. It is evident that Philadelphia provided the majority of pollutants to the dumpsite. However, both exceeded U.S. EPA's interim disposal criteria (U.S. Environmental Protection Agency, 1973b), for mercury and cadmium and were therefore treated equally.

The EPA's Region III took jurisdiction of the dumpsite in April 1973 and issued

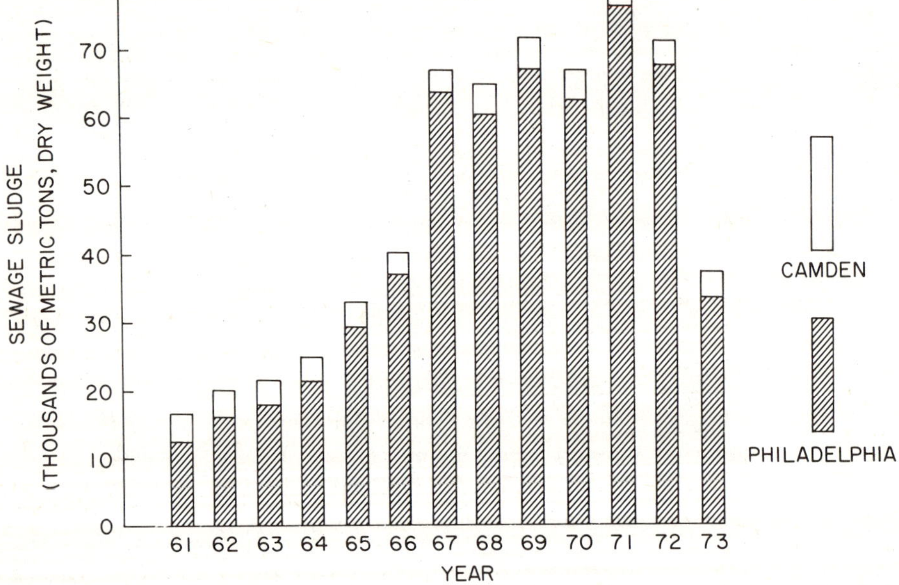

Figure 14.2. Dumping at old sewage sludge dumpsite.

Table 14.3. Old Sewage Sludge Dumpsite–Total Loading (metric tons) of Waste Constituents[a]

Constituent	Philadelphia[b]	Camden[c]
Arsenic	12	1
Beryllium	12	1
Cadmium	23	1
Chromium	1050	8
Copper	600	8
Lead	1640	12
Mercury	47	1
Nickel	169	1
Selenium	207	1
Vanadium	15	1
Zinc	3250	39

[a]EPA Region III ocean dumping files, Philadelphia, Pennsylvania.
[b]Philadelphia: average two trips per month, 7570 m^3 barge; discharge rate: 150 m^3 min^{-1} while moving at 11 km h^{-1}.
[c]Camden: average one trip per month, 6620 m^3 barge; discharge rate: 150 m^3 min^{-1} while moving at 11 km h^{-1}.

only one permit at the site to Philadelphia on 23 April 1973. The permit required Philadelphia to cease dumping at the site by June 1973 and move to a new interim dumpsite 50 km southeast of the old sewage sludge dumpsite (Fig. 14.1). Camden never received a permit at the old dumpsite because of delays in processing its application and was moved along with Philadelphia in June 1973.

14.2.2. New Interim Sewage Sludge Dumpsite

The new sewage sludge dumpsite (Fig. 14.1) was designated in the *Federal Register* as an interim dumpsite (U.S. Environmental Protection Agency, 1973b). The dumpsite covers approximately 65 km^2 and is located 55 km due east of the Maryland/Delaware border in 45-65 m of water.

The site was first used by DuPont-Edge Moor as an interim disposal site between September 1968 and July 1969 (Falk et al., 1972), while base line studies were conducted at the new acid waste disposal site. DuPont-Edge Moor dumped primarily the acid waste residue from the production of titanium dioxide. A summary of the amount dumped and waste characteristics is shown in Table 14.4 and is compared with other users of the site in Fig. 14.3. Although the site had been used for only a short period and also on an interim basis, it was still considered an active disposal site.

The EPA made the decision in April 1973 to move sludge dumping from the old sewage sludge dumpsite. The decision was based partly on the studies of Buelow (1968) and Verber et al. (1969) and also at the insistence of the beach communities, and especially Senator William V. Roth of Delaware (U.S. Environmental Protection Agency, 1972b), one of the sponsors of the MPRSA of 1972.

Table 14.4. Interim Sewage Sludge Dumpsite–Total Loadings (metric tons) of Waste Constituents[a]

Constituent	DuPont–Edge Moor[b]	Philadelphia[c]	Camden[d]	Modern Transportation[e]	Total
Arsenic	1	1	1	1	1
Beryllium	1	1	1	1	1
Cadmium	1	30	1	1	31
Chromium	30	703	6	1	739
Copper	5	696	7	1	708
Iron	18,262	8,780	–	–	27,040
Lead	6	1,300	10	1	1,330
Manganese	476	857	–		1,330
Mercury	1	15	1	1	15
Nickel	4	155	1	1	160
Selenium	1	7	1	1	7
Titanium	675	–	–	–	675
Vanadium	57	17	1	1	75
Zinc	16	2,811	30	1	2,857

[a]EPA Region III ocean dumping files, Philadelphia, Pennsylvania; dash indicates not analyzed.
[b]DuPont: average 20 trips per month, 3875 m^3 barge; discharge rate: 64 m^3 min^{-1} while moving at 11 km h^{-1}.
[c]Philadelphia: average 8 trips per month, 7570 m^3 barge; discharge rate: 25 m^3 min^{-1} while moving at 11 km h^{-1}.
[d]Camden: average one trip per month, 7570 m^3 barge; discharge rate: 25 m^3 min^{-1} while moving at 11 km h^{-1}.
[e]Modern Transportation: average one trip per month, 5677 m^3 barge; discharge rate: 387 m^3 min^{-1} while moving at 11 km h^{-1}.

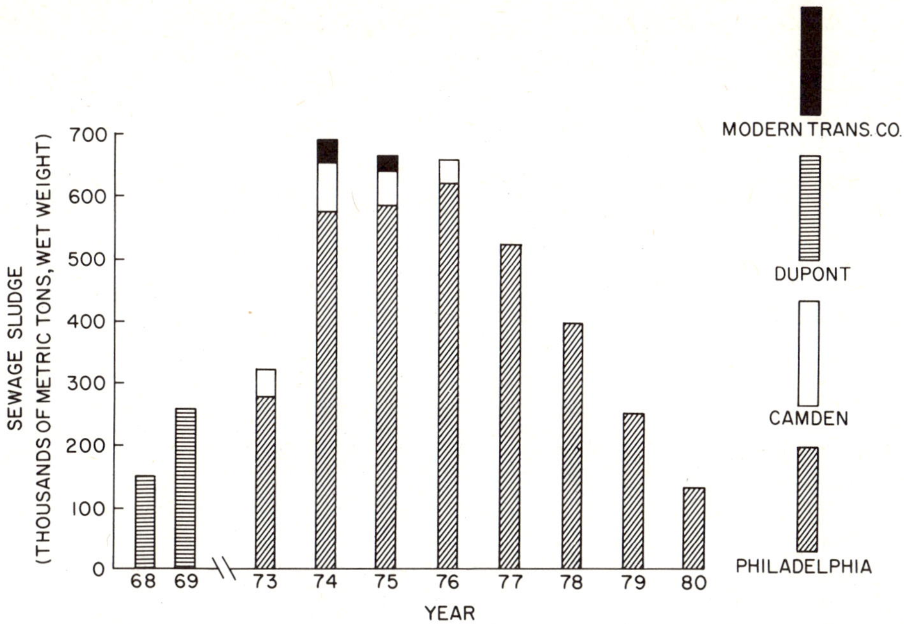

Figure 14.3. Dumping at the interim sewage sludge dumpsite.

The decision on selection of the dumpsite as a replacement for the old sewage sludge dumpsite came at a meeting on 12 April 1973 at U.S. EPA's National Marine Water Quality Laboratory (Narragansett, Rhode Island). A number of problems arose in the selection of a new site. The U.S. EPA's policy was to use only existing dumpsites until a final criteria could be published on new site selection. Therefore, the only permissible dumpsites within a reasonable area were the following: (1) the industrial wastes Deepwater Dumpsite-106 (DWD-106); this was eliminated because at the time there was a lack of information on the effects of deep water on sewage sludge (Gross, 1969); (2) the acid waste dumpsite; this was eliminated because of possible synergistic effects and the difficulty in monitoring dissimilar wastes; (3) the arsenic dumpsite; this was eliminated because of its excessive size; and, (4) the new interim sewage sludge dumpsite.

Some of the favorable reasons for the selection of the interim sewage dumpsite were the ease in monitoring the effects as it was close to the acid-iron waste site (Fig. 14.1); the existence of baseline data collected by DuPont for the acid-iron waste dumpsite; the lack of commercially harvested shellfish in the area; and the ability of the U.S. Coast Guard to conduct surveillance of the dumpers to assure compliance with permit conditions.

From 5 May to 12 May 1973, the EPA's Region III conducted a brief baseline survey of the interim sewage dumpsite (Palmer and Lear, 1973), and in June 1973 Camden and Philadelphia began dumping sewage sludge at the interim sewage sludge dumpsite. For a short period in 1974 and 1975 the Modern Transportation Company (Philadelphia, Pennsylvania) dumped sewage sludge collected from northern New Jersey communities in the interim dumpsite. This practice was quickly phased out.

Philadelphia dumped at the interim sewage sludge dumpsite from June 1973 to December 1980, when, under a Consent Decree (U.S. Federal District Court, 1979), they were forced to end all ocean dumping. Camden dumped from June 1973 to November 1976 and was forced by Federal Court Order (U.S. Federal District Court, 1976) to move to DWD-106 (Fig. 14.1). Camden ceased all ocean dumping in 1978.

Philadelphia began developing alternatives to dumping in 1974 after publication of EPA's final regulations on ocean dumping (U.S. Environmental Protection Agency, 1973c), and also instituted a program to attempt to meet the criteria. Section 227.6 of the criteria required in part that the waste not contain more than 0.75 mg kg^{-1} of mercury and 0.6 mg kg^{-1} of cadmium in the solid phase. As shown in Table 14.5, Philadelphia exceeded those criteria, having 85 mg kg^{-1} and 41 mg kg^{-1} of mercury and cadmium, respectively. After conducting an extensive investigation of the sources of metal contamination, Philadelphia established pretreatment requirements for industries discharging into the city's sewer system. The project was moderately successful in reducing the concentrations of metal in the sewage sludge, especially for mercury, which decreased from 85 mg kg^{-1} to 2 mg kg^{-1}, but pretreatment had no overall impact on cadmium. Since Philadelphia was unable to meet the criteria, EPA's Region III required the end of all dumping by January 1981. The end date was contested by Philadelphia, but they ultimately

Table 14.5. New Sewage Sludge Dumpsite–Comparison of Waste Characteristic between Philadelphia and Camden

	Philadelphia (mg kg^{-1})[a]				Camden (mg kg^{-1})[b]
Constituent	1973	1975	1977	1979	1973–1976
Arsenic	22	1	–	–	1
Beryllium	22	2	–	–	1
Cadmium	41	70	43	54	72
Chromium	1,880	930	1,140	840	58
Copper	1,080	980	1,790	1,020	63
Iron	19,300	11,300	16,500	16,400	–
Lead	2,940	1,950	2,370	1,720	90
Manganese	1,800	895	2,080	1,400	–
Mercury	85	4.4	2.4	2	3
Nickel	300	230	250	380	12
Selenium	371	12	–	–	0
Vanadium	27	40	–	–	5
Zinc	5,840	4,040	5,380	4,510	270

[a]Dry weight; U.S. EPA Region III ocean dumping records, Philadelphia, Pennsylvania; based on analysis of every fifth barge. Dash indicates not analyzed.

[b]Dry weight; U.S. EPA Region III ocean dumping records, Philadelphia, Pennsylvania; based on analysis of every fifth barge.

developed and implemented a complex plan which focused on the nutrient value of sewage sludge as a soil enricher (Townsend and Guarino, 1975).

In a similar manner, Camden in 1975 was placed on an implementation schedule to cease dumping by 1979. In December, 1975 high numbers of coliform bacteria were found in sediments around the sewage sludge dumpsite. Although Camden dumped one-tenth the volume of Philadelphia (Fig. 14.3), Camden was considered a major source of the bacteria because the sludge was undigested (Table 14.5). The sludge digestion process kills a significant portion of the bacteria and viruses, and also reduces the nutrient content, thereby reducing the viability of the bacteria in the ocean (Berg, 1976; Verber, 1976). Because of this, Camden was moved to DWD-106 in December 1976. Philadelphia, however, was allowed to remain at the new interim sewage sludge dumpsite because its sludge was digested, and it was not feasible to transport all of Philadelphia's sewage sludge to the DWD-106.

On 1 January 1981, the interim sewage sludge dumpsite was closed. Philadelphia had made its last trip on 23 November 1980, over a month ahead of schedule. A summary of the annual waste dumping for the site is illustrated in Fig. 14.4 and the total loadings of contaminants is shown in Table 14.4.

14.2.3. Acid-Iron Waste Dumpsite

The acid-iron dumpsite (Fig. 14.1) was designated in the *Federal Register* as an interim site (U.S. Environmental Protection Agency, 1973b). This dumpsite covers approximately 65 km^2 and is located 50 km east of Indian River Inlet, Delaware, in 40–60 m of water.

The DuPont–Edge Moor titanium dioxide plant was the only dumper to use the acid-iron waste dumpsite. DuPont–Edge Moor began dumping acid-iron sulfate waste at sea in September 1968 as part of a commitment to Delaware and to the Delaware River Basin Commission to reduce acid discharges to the Delaware River. From September 1968 to July 1969, the previously mentioned interim sewage sludge site was used while DuPont collected baseline data on the acid-iron waste disposal site. The acid-iron waste dumpsite was used from July 1969 to March 1977 (Fig. 14.4), when DuPont–Edge Moor dumping was again moved, this time to the DWD-106 dumpsite where it is scheduled to be phased out in 1983.

The original selection of the acid waste dumpsite (Fig. 14.1) was made by the U.S. Bureau of Commercial Fisheries, which is now part of the U.S. National Marine Fisheries Service and the U.S. Fish and Wildlife Service (Falk et al., 1972). The dumpsite was recommended for a number of reasons. First, the dumpsite was considered to have a very low potential for harvesting of the surf clam, *Spisula solidissima*, as the water was deeper than the optimal for that species. Further, while the ocean quahog, *Arctica islandica*, was present, it was not a commercially harvested clam at that time (Champ, 1973). The selection was also based on an earlier study conducted by Ketchum (1953), who had evaluated the potential for ocean disposal of industrial wastes off Delaware Bay.

The DuPont–Edge Moor wastes were the by-product of the manufacture of titanium dioxide, a white pigment used in the manufacture of paints, paper, print-

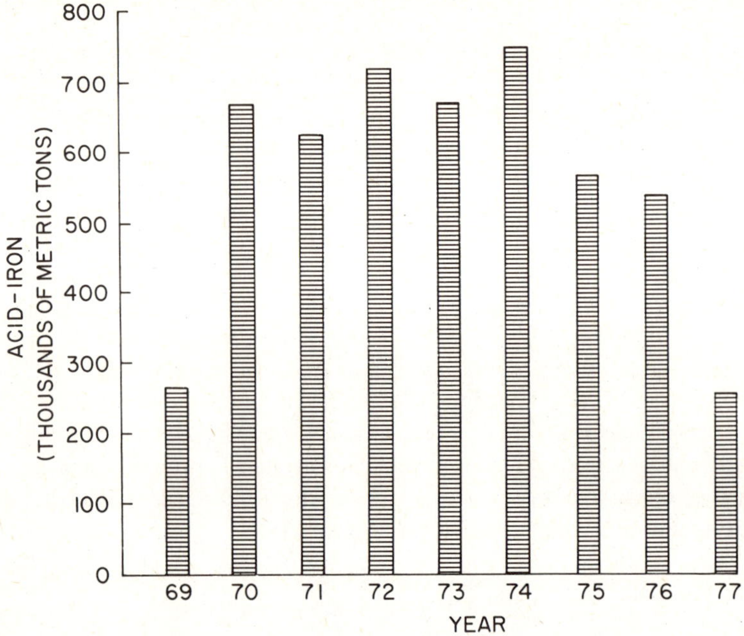

Figure 14.4. Dumping at acid-iron waste dumpsite.

ing inks, plastics, textile fibers, food additives, and better grades of cosmetics. Titanium dioxide is a replacement for the now infrequently used and highly toxic lead-based white pigment. Two separate processes for producing titanium dioxide were employed at DuPont's Edge Moor plant. One was a batch type called the sulfate process which was phased out between 1968 and 1974. The other was the continuous chloride process. Briefly, the processes are as follows (U.S. Environmental Protection Agency, 1975): In the sulfate process, titanium-bearing ores containing large quantities of iron oxide are mixed with and dissolved in fuming sulfuric acid (oleum) to produce a titanyl sulfate intermediate. The acidic solution is clarified and purified. The purified titanyl sulfate is hydrolyzed to form a white nonpigmentary hydrate. The hydrate is calcined to form the pigmentary crystal, after which it is surface treated, milled, and packaged for sale. The wastes from this process include the iron removed during purification of the titanyl sulfate and the mother liquor from the hydrolysis step. In the chloride process, titanium-bearing ores are mixed and treated with chlorine to produce a titanium tetrachloride intermediate, which is purified and then combined with gaseous oxygen to form a pigmentary grade of titanium dioxide. This material is dried, then wet-treated, milled, and packaged for sale. The residual waste liquid from the titanium tetrachloride purification contains metal chlorides.

The EPA issued its first permit to DuPont-Edge Moor on 23 April 1973. Consistent with the ocean dumping regulations, an implementation schedule was developed for the eventual cessation of ocean dumping. In accordance with its compliance schedule, DuPont-Edge Moor has completed the following steps (U.S.

Environmental Protection Agency, 1980): In 1974, installation of facilities for the recovery and upgrading of iron chloride from the waste occurred; also initiated was the development of a market for the sale of iron chloride for use in water and wastewater treatment. In 1975, conversion to the chloride process occurred, thus significantly reducing the amount of waste produced per unit of titanium dioxide. In 1976, there was the installation and start-up of wastewater sludge dewatering facilities, thus ceasing the ocean dumping of the solid components of the waste and an expansion of iron chloride sales facility. And finally, in 1977 there was the installation of the hydrochloric acid recycling plant.

Review of the above schedule and comparison with Table 14.6 indicates the following:

1. Upon upgrading from the sulfate to the chloride process, there was a significant reduction in the concentrations of arsenic, aluminum, copper, antimony, tin, and titanium.
2. The elimination in 1976 of the solid components of the waste drastically reduced the metal loadings. The residue was taken to a lined landfill.

In 1974 when DuPont-Edge Moor was first studying alternatives to ocean dumping, U.S. EPA's Region III assisted in developing a long-range solution. In the conversion from the sulfate to the chloride process, ferrous chloride became a major component of the waste. It was noted that if the ferrous chloride could be converted to ferric chloride, a solution was available. Ferric chloride was being used as a flocculating agent in advanced sewage treatment plants; it was especially useful in the removal of phosphates. By 1977, DuPont-Edge Moor was selling a portion of its waste as ferric chloride to sewage treatment plants. It is paradoxical that DuPont-Edge Moor was partially dependent on EPA's enforcement of sewage treatment plant upgradings throughout the country to provide a sufficient market of users for the ferric chloride.

In March 1977, at DuPont-Edge Moor's request, dumping operations were moved to the DWD-106 dumpsite. The facility is scheduled to end all dumping in 1983.

14.2.4. Arsenic Waste Dumpsite

The arsenic waste dumpsite (Fig. 14.1) was designated in the *Federal Register* as an interim site (U.S. Environmental Protection Agency, 1973b). The dumpsite covers approximately 1400 km^2 and is located 100 km east of Assateague Island in 60 to 600 m of water. The dumpsite was first identified by Smith and Brown (1971). It is not known how the dumpsite got its name, as there are no records of arsenic being dumped there. From 1966 to 1974 the arsenic dumpsite was used to dispose of caustic and white-water oil emulsions from the Sun Oil Company of Pennsylvania, which is an oil refinery located at Marcus Hook, Pennsylvania. The wastes generated came from two processes described as follows (U.S. Environmental Protection Agency, 1973d):

White water is the term used to describe stable oil-on-water emulsions in the

Table 14.6. Acid Waste Dumpsite—Characteristics of Waste and Total Loadings[a]

	DuPont–Concentrations (mg kg^{-1})[b]				DuPont–Total Loading (metric tons)
Constituent	1969–1972	1973–1974	1975	1976	
Arsenic	1	1	1	1	1
Beryllium	1	1	1	1	1
Cadmium	1	1	1	1	3
Chromium	37	87	410	200	545
Copper	4	14	11	8	36
Iron	58,000	52,000	133,000	50,000	285,100
Lead	13	17	100	52	138
Manganese	1,120	1,400	4,900	1,700	7,790
Mercury	1	1	1	1	1
Nickel	10	14	50	16	74
Silver	1	1	1	1	2
Titanium	2,030	1,900	10,000	3,000	13,900
Vanadium	171	165	1,000	220	1,250
Zinc	37	47	265	110	354
pH[c]	0.06	0.1	0.01	0.01	
Specific gravity[d]	1.18	1.22	1.35	1.5	

[a]DuPont–Edge Moor, Delaware; EPA Region III ocean dumping records, Philadelphia, Pennsylvania; based on analysis of every fifth barge; average 20 barges per month; 3875 m^3 barge; discharge rate: 64 m^3 min^{-1} at 11 km h^{-1}.

[b]1969–1972: 90% sulfate waste, 10% chloride waste (rutile); 1973–1974: 25% sulfate wastes, 75% chloride waste (rutile); 1975: 100% chloride (ilmenite); 1976: 100% chloride (beneficiated rutile). Ilmenite contains a high degree of impurities and low titanium content; rutile is high-grade titanium ore.

[c]pH, calculated value; pH = – log(H^+).

[d]Specific gravity (g cm^{-3}).

refinery. This creamy grayish-colored light oil/water emulsion consists of approximately 85% water, 15% oil, and contains small amounts of sodium antioxidants, organic acids, alcohol, and heavy metals. The pH is usually between 8 and 13. The largest volume of white water is generated from caustic washing operations that follow the acid treatment of specialty oils at the plant. Smaller quantities result from two other sources; the bottom layer accumulations in the slop tank at the blending and packaging area, and spent emulsions from chemical cleaning activities in refinery process areas. Spent caustic is the term used for process wastewater containing approximately 0.1 to 7.5% (by weight) free sodium hydroxide with possible trace quantities of sulfides, phenols, and heavy metals. While generally clear and colorless, it can vary from a cloudy translucent to a clear, grayish blue, light green, or light orange. The pH averages from 9.5 to 14. The spent caustic soda results from caustic scrubbing for the removal of sulfur compounds.

The U.S. EPA's Region III took jurisdiction of the arsenic waste dumpsite in April 1973 and issued the first ocean dumping permit to Sun Oil Company on 7 August 1973. A major condition of that permit was that the company develop alternatives to replace ocean dumping. With the assistance of U.S. EPA engineers, Sun Oil Company developed an abatement program by November 1973. The refinery was able to reduce drastically the volume of caustic generated by increasing the efficiency of the plant. This was done without major plant modifications or additional expenditures. Oil separation equipment was used to treat the whitewater solution and the oil was fed back into the plant as feedstock. The neutral-

Table 14.7. Arsenic Waste Dumpsite–Total Loading (kg) of Waste Constituents[a]

Constituent	Sun Oil Company[b]	Allied Chemical[c]
Aluminum	5480	1
Arsenic	3	1
Barium	2192	–[d]
Beryllium	110	1
Cadmium	1	1
Chromium	6	16
Copper	16	22
Lead	6	2
Mercury	1	1
Nickel	6	3
Selenium	11	1
Vanadium	110	1
Zinc	–[d]	10

[a]EPA Region III ocean dumping files, Philadelphia, Pennsylvania.
[b]Sun Oil Company: average three trips per month, 1335 m^3 barge; discharge rate: 3 m^3 min^{-1} while moving at 28 km h^{-1}.
[c]Allied Chemical Corporation: average two trips per month, 1730 m^3 barge; discharge rate: 25 m^3 km^{-1}.
[d]Not analyzed.

ization of spent caustic was accomplished using the waste stream from Sun Oil Company's naphthenic acid waste stream.

The U.S. EPA issued a second ocean dumping permit to Sun Oil Company on 31 July 1974 to allow sufficient time to complete plant modifications to treat all of its waste. By December 1974 the Sun Oil Company had ceased ocean dumping at a cost saving to the industry.

In April 1975, Allied Chemical Corporation, Elizabeth, New Jersey, made a single dump of mixed chemical wastes (Table 14.7) at the arsenic dumpsite. The company normally uses the DWD-106 dumpsite controlled by the U.S. EPA's Region II, New York City. However, the National Oceanic and Atmospheric Administration was conducting monitoring studies and this one dump was therefore allowed.

A summary of the loadings of contaminants from the site are given on Table 14.7, and the total dump is illustrated in Fig. 14.5. It should be noted that, although the site borders the new sewage sludge site (Fig. 14.1), the wastes were dissimilar enough not to interfere with the sludge monitoring. Also, as a permit condition on the dumping of Sun Oil Company's wastes, only the portion of the site east of 73°30′W longitude was used for dumping.

14.3. CONCLUSIONS

In October 1972 when the MPRSA was enacted, there were four ocean dumpsites in U.S. EPA's Region III (Fig. 14.1). On 1 January 1981 that number had been re-

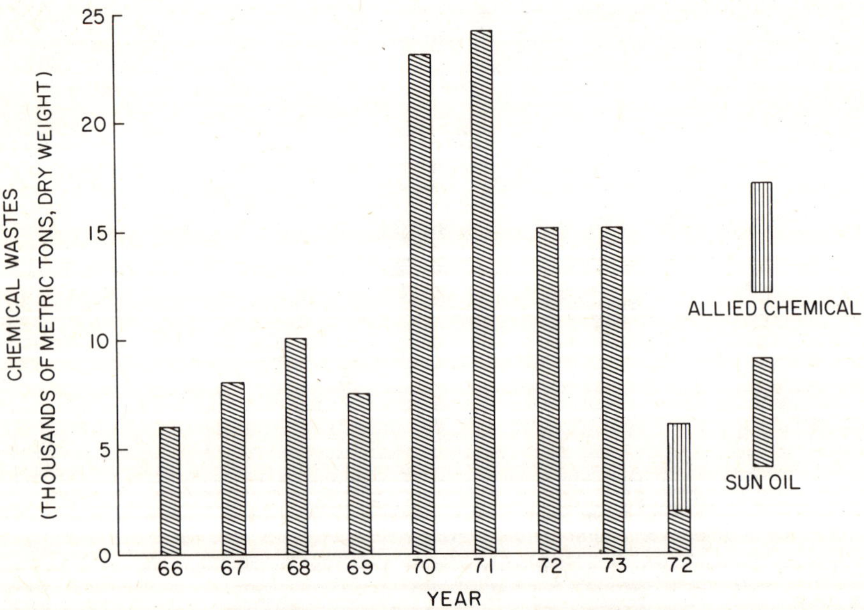

Figure 14.5. Dumping at arsenic dumpsite.

duced to zero. During that period over 13,000 metric tons of wastes were dumped at sea from Region III's cities and industries. Although it will be difficult to assess all of the impacts due to dumping, the quantitive loadings for the major pollutants at the dumpsites are available.

The major emphasis of Region III's permit program was in the development of alternatives. Each dumper received a thorough evaluation of its entire facility in the determination of the need for dumping. Thus, Sun Oil Company had been dumping since 1966; within six months of its first permit in 1973, Sun Oil Company had developed an abatement plan involving the modification of existing equipment. The technology was already available to recover oil and recycle the spent caustic at an eventual cost savings to the company. DuPont-Edge Moor presently sells a large portion of the ferric chloride previously dumped as waste and recycles most of the hydrochloric acid which was also a waste. Philadelphia has reclaimed over 20 km^2 (5000 acres) of barren strip mine land in middle Pennsylvania with sewage sludge which would have gone to sea. Each dumper represented a unique situation.

It has been shown that U.S. EPA's environmental regulations are sufficient to control and limit ocean dumping. Further, industries and municipalities do have environmentally and economically sound alternatives to ocean dumping.

ACKNOWLEDGMENTS

I would like to thank D. Lear and M. O'Malley of the Annapolis Field Office for their continuing support in making the program a success in Region III. I would also like to thank G. Horwitz who played an integral part in the development of alternatives to ocean dumping. Last, I would like to acknowledge the invaluable assistance of L. Alessi for typing the manuscript.

REFERENCES

Baxter, S. S. 1959. Waste Disposal in Philadelphia. *Journal of the Sanitary Engineering Division, Preceedings of the American Society of Civil Engineers*, **4**, 127–134.

Berg, G. 1976. Removal of viruses from sewage, effluents and waters. Present and future trends. *Bulletin of World Health Organization*, **49**, 461–465.

Buelow, R. W. 1968. Ocean Disposal of Waste Material. *In*: Transactions: National Symposium on Ocean Sciences and Engineering of the Atlantic Shelf, Philadelphia, March 19–20, 1968. Marine Technology Society, Washington, D.C., pp. 311–337.

Buelow, R. W., B. H. Pringle, and J. L. Verber. 1968. Preliminary Investigation of Sewage Sludge Dumping off Delaware Bay. Unpublished data, Narragansett, Rhode Island, Northeast Marine Health Sciences Laboratory, January 1968, 20 pp.

Champ, M. A. 1973. A Survey of Three Atlantic Ocean Disposal Sites: The City of

Philadelphia Sewage Sludge Dump Site, the DuPont Acid Waste Dump Site, and the New York (DWD-106) Toxic Industrial Waste Site. Program Publication No. 1, Center for Earth Resources and Environmental Studies, American University, Washington, D.C. 169 pp.

Davey, C. T. 1972. Assessment of the Effects of Digested Sewage Sludge Disposal off the Mouth of Delaware Bay. Technical Report F C 2970, The Franklin Institute, Philadelphia, Pennsylvania, 95 pp.

Falk, L. L., T. D. Meyers, and R. V. Thomann. 1972. Waste Dispersion Characteristics and Effects in an Oceanic Environment. Technical Report EAW-12020, U.S. Environmental Protection Agency, Washington, D.C., 393 pp.

Gross, M. G. 1969. New York City–A Major Source of Marine Sediments. Technical Report No. 2, Marine Sciences Research Center, State University of New York, Stony Brook, New York, 24 pp.

Guarino, C. F. 1976. Statement before the U.S. House of Representatives Subcommittee on Oceanography and the Subcommittee on Fisheries and Wildlife Conservation, Jan. 22, 1976. Washington, D.C., 15 pp.

Ketchum, B. H. 1953. Preliminary Evaluation of the Coastal Water off Delaware Bay for Disposal of Industrial Wastes. Reference No. 53–51, unpublished manuscript, Woods Hole Oceanographic Institution Woods Hole, Massachusetts, 26 pp.

Oostdam, B. L., F. K. Szucs, and R. N. Swift. 1975. Heavy metals in sediments and surf clams (*Spisula solidissima*) in the former Philadelphia treated sewage sludge dump area off Delaware Bay. Report published by Marine Science Consortium, Millersville College, Pennsylvania, 181 pp.

Palmer, H. D., and D. W. Lear (Eds.). 1973. Environmental Survey of an Interim Ocean Dump Site, Middle Atlantic Bight. EPA 903/9-73-010A, U.S. Environmental Protection Agency, Philadelphia, Pennsylvania, 134 pp.

Porter, D., and R. R. Ripa. 1974. Regional Wastewater Management Plan for Camden County, Alternatives Assessment Report. Camden County Municipal Utilities Authority, Camden, New Jersey, 350 pp.

Smith, D. D., and R. P. Brown. 1971. Ocean Disposal of Barge-delivered Liquid and Solid Wastes. Solid Waste Management Series No. SW-192 series. U.S. Environmental Protection Agency, Washington, D.C., 119 pp.

Townsend, S. J., and C. F. Guarino. 1975. Ten Point Master Plan for the City of Philadelphia–Sludge Alternatives. Submitted Nov. 10, 1975 to U.S. Environmental Protection Agency, Philadelphia, Pennsylvania, 16 pp.

U.S. Congress. 1972. Marine Protection, Research, and Sanctuaries Act of 1972. Public Law 92-532, 86 STAT. 1052.

U.S. Council of Environmental Quality. 1970. Ocean Dumping: A National Policy. Report 0-404-547, U.S. Government Printing Office, Washington, D.C., 45 pp.

U.S. Environmental Protection Agency. 1972a. Report of the United States Delegation to the Intergovernmental Conference on the Conventions of Ocean Dumping of Wastes at Sea, London, England, Oct. 30–Nov. 13, 1972. Washington, D.C., 10 pp.

U.S. Environmental Protection Agency. 1972b. Appendix 2. *In*: Ocean Disposal Practices and Effects–A Report to the Administrator of the Environmental Protection Agency of a recent meeting held by the President's Water Pollution

Control Advisory Board, Sept. 26–29, 1972. U.S. Environmental Protection Agency, Washington, D.C., 20 pp. + references.

U.S. Environmental Protection Agency. 1973a. Ocean Dumping—Interim Regulations Governing Transportation for Dumping, and Dumping of Material into Ocean Water. *Federal Register*, April 5, 1973, Part III, **33**, 8726–8730.

U.S. Environmental Protection Agency. 1973b. Ocean Dumping Criteria. *Federal Register*, May 16, 1973, Part II, **38**, 12872–12877.

U.S. Environmental Protection Agency. 1973c. Ocean Dumping—Final Regulations and Criteria. *Federal Register*, October 15, 1973, Part II, **38**, 28610–28621.

U.S. Environmental Protection Agency. 1973d. Application for an Ocean Dumping Permit Submitted by the Sun Oil Company at Pennsylvania, April 20, 1973, Philadelphia, Pennsylvania.

U.S. Environmental Protection Agency. 1975. Application for an Ocean Dumping Permit Submitted by E. I. du Pont de Nemours & Company, Edge Moor, Delaware, Nov. 15, 1975, Philadelphia, Pennsylvania.

U.S. Environmental Protection Agency. 1980. Application for a Special Permit for Ocean Dumping at DWD-106—Submitted by E. I. du Pont de Nemours & Company, Edge Moor, Delaware, March 1, 1980, New York, New York.

U.S. Federal District Court. 1976. *U.S. Environmental Protection Agency v. Camden, New Jersey*, Fifth District Court, Nov., 1976, Camden, New Jersey.

U.S. Federal District Court. 1979. *U.S. Environmental Protection Agency v. Philadelphia, Pennsylvania*, Third District Court, April, 1979, Philadelphia, Pennsylvania.

Verber, J. L. 1976. Philadelphia Sludge Data—Ocean Disposal Site Study. Technical Memorandum of March 11, 1976, U.S. Public Health Service, Davisville, Rhode Island, 10 pp.

Verber, J. L., R. W. Buelow, and B. H. Pringle. 1969. Ocean disposal of waste material—Off Delaware Bay and New York Bight. *In*: Proceedings: Second Mid-Atlantic Industrial Waste Conference, 18–20 November 1969. Drexel Institute of Technology, Philadelphia, Pennsylvania, pp. 127–153.

Wisdom, D. A. 1973. Waste Disposal for the City of Philadelphia and the U.S. Army Corps of Engineers. Publication No. 39, Pennsylvania Department of Environmental Resources, Harrisburg, Pennsylvania, 68 pp.

15

EFFECTS OF SEWAGE SLUDGE DUMPING ON CONTINENTAL SHELF BENTHOS

Donald W. Lear and Marria L. O'Malley

Environmental Protection Agency and National Oceanic
and Atmospheric Administration Oceanographic Unit
Annapolis, Maryland

ABSTRACT

The fate and effects of the ocean dumping of sewage sludge at a mid-continental shelf dumpsite were studied by comparison of the defined area of deposition with a reference area away from prevailing flows. Concentrations of lead, copper, and organic carbon, and the percentage fine sediment fraction (silts plus clays) were significantly greater in the dumpsite grid compared with the reference grid. Swales in the dumpsite grid were areas of highest concentrations of contaminants, and the pollution-sensitive ampeliscid amphipods were atypically absent from these sites. The pollution-tolerant polychaete *Capitella capitata* was found only in contaminated swales.

This study identified at least one locus of pollution due to ocean dumping, and the biological response characteristic of such contamination.

15.1. INTRODUCTION

The disposal of wastes into the marine environment has been shown to have measurable and sometimes deleterious effects on marine benthic organisms in several geographic areas (Pearce et al., 1976; Smith and Greene, 1976; Pearson and Rosenberg, 1978; Botton, 1979; Word, 1979). In 1973 a field program was initiated by the U.S. Environmental Protection Agency, Region III, to develop methods to assess the effects of ocean dumping on the temperate mid-continental shelf 72 km off the coasts of Maryland and Delaware (Fig. 15.1) where two dumpsites approximately 20 km apart were being used. The northern site was the release area for industrial acid-iron wastes with residual metals from a titanium ore extraction process. The site was active from 1968 until 1977. The southern site, referred to as the interim sewage sludge dumpsite (Chapter 14), was the release zone for municipal sewage sludge, primarily from the City of Philadelphia. Dumping began at this site in June 1973 and ended in November 1980. The major constituents of the dumped

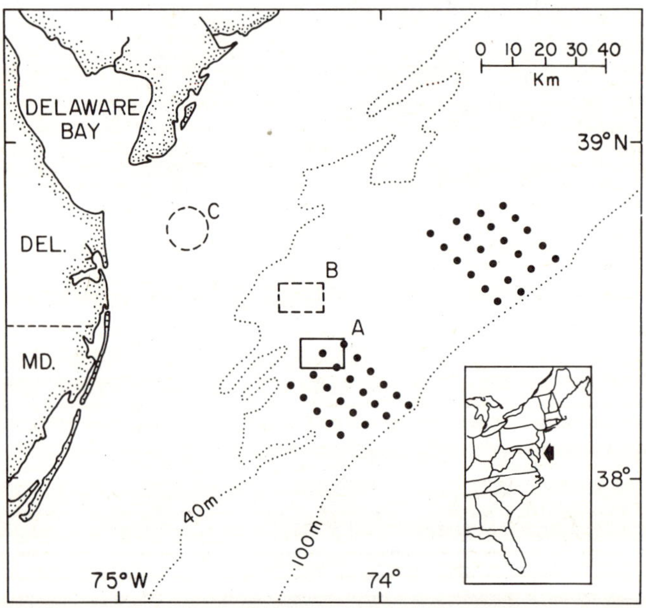

Figure 15.1. Location of interim sewage sludge dumpsite (A, solid line rectangle), the acid-iron wastes dumpsite (B, dashed line rectangle), and old sewage sludge dumpsite (C, dashed line circle) and dumpsite and reference grid sampling stations.

wastes are summarized by Muir (Chapter 14). A base line assessment and previous studies were reported by Palmer and Lear (1973), Lear (1974), Lear and Pesch (1975), and Lear et al. (1974, 1977, 1981).

The information developed in this monitoring program indicated that the sediments in the south portion of the sewage sludge dumpsite, and immediately adjacent southward, had significantly elevated concentrations of contaminants. On a cruise in August 1977, a grid of sampling stations was established centered on latitude 38°15′N, longitude 74°10′W (grid A, Fig. 15.1). A reference or control grid of similar design, dimension, and habitat was established 60 km northeast centered on latitude 38°40′N, longitude 73°30′W. This reference area was known from the hydrodynamics (Klemas et al., 1977) and previous observations (Lear et al., 1977) to be outside the influence of dumping activity.

15.2. EXPERIMENTAL

15.2.1. Description of Study Area

The hydrographic structure of the area is vertically isothermal in winter and stratified under summer conditions with a well-developed thermocline at 16–20 m (Meyers, 1974; Lear and O'Malley, 1979; Welch and Ruzecki, 1979). Salinities and temperatures below the thermocline at the sewage sludge site are characteristic of the "cold pool" waters described by Beardsley et al. (1976). The presence of the pycnocline presumably retards deposition of wastes to the bottom and allows for greater dispersal with current regimes (E.I. du Pont de Nemours Co. et al., 1972).

The circulation in the area is characterized by ellipsoidal tidal currents with velocities generally 12 cm s^{-1} (Palmer and Lear, 1973). These tidal currents are of sufficient magnitude to transport the surficial sediments of the area (Palmer et al., 1976). The net transport of water masses appears from bottom drifter studies (Bumpus, 1973; Lear and O'Malley, 1979) to be to the south or southwest. Earlier works also indicated a net southwesterly drift through this area (Ketchum, 1953; Norcross and Stanley, 1967).

A mathematical model predicted accumulations of sewage sludge in the south portion of the sludge release site and southward for about 9 km (Demenkow and Wiekramartane, 1976). The model did not consider bottom topography or storm events.

The sediments in the area are sands with the silt plus clay fractions generally less than one percent. They are reworked Holocene materials with little or no modern inputs (Swift et al., 1972). The topography is dominated by subtle northeast-southwest trending ridges and swales (Duane et al., 1972) that may account for variability in sediment grain sizes (Johnson and Wood, 1977).

15.2.2. Sampling and Analysis

Two grids of 4 X 6 stations (Fig. 15.1), at intervals of 8.8 km (928 km^2 total area), coincided with LORAN C lines on the U.S. Coast and Geodetic Survey chart 12200.

At each station one Smith–McIntyre grab (0.1 m^2) was taken and the undisturbed surficial sediment subsampled for metals, organic carbon, and organohalogens. Three subsamples for metal analysis were taken in separate 250-ml polyethylene cups, capped and stored for later analysis. Three subsamples for organic carbon were taken in prewashed glass vials and immediately frozen on dry ice. A sample for organohalogen analysis was placed in a hexane-rinsed, baked pint Mason jar with Teflon liner, and refrigerated at 4°C until analysis.

The metals lead, copper, zinc, chromium, nickel, and silver were analyzed by mild acid extraction of the sediment sample and atomic absorption spectrophotometry (Lear and Pesch, 1975). Organic carbon was determined in an Oceanography International organic carbon analyzer (American Public Health Association, 1976). Organohalogens (PCBs) were analyzed by extraction of sediment in 2:1 hexane-acetone and then using a gas chromatograph (Palmer and Lear, 1973). PCBs reported are the Arochlor® 1254 fraction.

Triplicate grab samples for benthic infauna were taken using a 0.04 m^2 Shipek bottom grab. A small aliquot for sediment-size analysis was removed from each grab and the remainder of the sample preserved in 10% buffered formalin. In the laboratory the samples were washed through a 0.5-mm screen and the retained organisms stained with rose bengal and preserved in 70% ethanol. Organisms were identified and counted, to species where possible (Marine Research, Inc., 1979). The grain size of sediment samples was determined by sieve analysis using a Ro-Tap mechanical shaker (Ingram, 1969).

Criteria were established for topographic characterization of each station for data analysis purposes, using fathometer readings, LORAN C locations on the National Oceanic and Atmospheric Administration special bathymetric chart (0807N-56, 1976), sediment parameters, and visual observations from field sheets. Swales were identified as topographic depressions, with more than 50% of the sand fraction greater than 3 phi (very find sand), the percentage fine fraction (greater than 4 phi) more than 3%, and organic carbon content greater than 1000 mg kg^{-1} dry wt. Erosional swales showed essentially the same characteristics as swales, but with a larger percentage of coarse sand and with shell fragments. Ridges were identified by positioning and bathymetry as above, but with coarse to medium sands (−1.0 to 1.0 phi), percent fines less than 1%, and organic carbon less than 900 mg kg^{-1} dry wt. These criteria are similar to those used by Boesch (1979). Flank habitats were intermediate between ridges and swales, with sands predominantly 1.0–2.0 phi, percentage fines less than 3%, and organic carbon content less than 1000 mg kg^{-1} dry wt. This is mesoscale topography, with ridges and swales in the order of tens to hundreds of meters in length, and amplitudes of 10–15 m (Folger et al., 1979).

15.3. RESULTS AND DISCUSSION

15.3.1. Comparisons between Grids

Direct comparisons of the dumpsite and reference grids established the similarities of the two areas and detected differences for further examination. The physical and chemical parameters are shown in Table 15.1 together with the results of the comparative statistical testing.

The mean water depth in the dumpsite grid was 61.5 m and the mean depth in the reference grid was 65.6 m. Two shallow stations from the dumpsite grid and two deep stations from the reference grid were dropped for data analysis purposes, to make the grids statistically comparable in depth.

The median grain sizes of sediments in the two grids were not significantly different; median grain size in the dumpsite grid averaged 1.25 phi, while median grain size in the reference area averaged 1.53 phi. Generally a medium sand substrate predominated, but coarse and fine sands were found in both grids as indicated by the ranges of median phi size. The percentage fine (silt plus clay) fraction of sediments, greater than 4.0 phi, was significantly greater in the dumpsite grid (mean = 2.7%) than in the reference area (mean = 0.7%).

The means of lead, copper, chromium, and zinc concentrations were all greater in the dumpsite grid than in the reference grid. Lead and copper concentrations were significantly greater as indicated by the *t*-test of the means of all stations in each grid. Nickel concentrations were comparable in both grids. Silver concentrations were below the detection limit of the analytical procedure at all stations.

The organic carbon content of sediments in the dumpsite grid was significantly higher than in the reference grid. Mean organic carbon concentration in the dumpsite area was 1330 mg kg^{-1} while organic carbon in the reference area averaged 830 mg kg^{-1}.

The concentration of PCB was higher in the dumpsite grid (mean = 18.9 ng g^{-1}) than in the reference grid (mean = 11.8 ng g^{-1}), but the means were not statistically different. PCB analysis was carried out on only one subsample taken at each station. The highest concentration was found close to the dumpsite.

A topographic classification for each station was determined based on the criteria discussed above (Section 15.2.3). Stations were identified as ridge, flank, or swale. Analysis of the two grids showed similar distributions of these topographic features (Table 15.2).

Overall comparisons of the benthic biological communities were made using conventional diversity indices (Marine Research, Inc., 1976). Comparisons of Margalef, Simpson, and Shannon-Weaver indices, as well as number of taxa per station and the average density of organisms, indicated that no major differences in community structure were evident when all stations in each grid were considered together (Table 15.3).

These results show that the two grids were essentially similar habitats as indicated by depth, topographic features, sediment fractions, and the biological community indices.

The differences between the two grids in lead, copper, and organic carbon concentrations, and in the fine fraction of surficial sediments were further examined through the interrelationships of these parameters. Correlation coefficients for these parameters for each study area are shown in Table 15.4. Lead and copper concentrations were each correlated with organic carbon in both the reference grid and the dumpsite grid. Organic carbon concentration was also correlated with the sediment fine fraction in both areas, but to a higher degree in the dumpsite grid.

Table 15.1. Comparison of Physical and Chemical Parameters in Dumpsite and Reference Grids

Parameters	Dumpsite Grid			Reference Grid			t-Statistic	Degrees of Freedom
	Mean	Standard Deviation	Range	Mean	Standard Deviation	Range		
Depth (m)	61.5	11.0	44.2–85.4	65.6	10.0	48.8–83.9	1.57	41
Median grain size (phi)	1.25	1.07	-1.60–2.51	1.53	0.35	0.85–2.12	1.06	31
Percentage fines	2.66	3.31	0.16–12.1	0.71	0.87	0.13–3.71	2.45[a]	31
Lead (mg kg^{-1})	3.28	1.59	0.74–6.70	1.82	0.92	0.40–4.80	3.66[a]	41
Copper (mg kg^{-1})	1.52	0.56	0.93–2.90	1.07	0.21	0.87–1.67	3.45[a]	41
Chromium (mg kg^{-1})	3.57	1.01	2.20–5.87	3.47	0.82	2.23–5.17	0.34	41
Zinc (mg kg^{-1})	6.69	2.84	2.87–13.4	6.43	1.86	3.93–11.3	0.35	41
Nickel (mg kg^{-1})	1.21	0.86	0.23–3.47	1.24	0.61	0.31–2.50	0.13	41
Silver (mg kg^{-1})	<0.1			<0.1				
Organic carbon (mg kg^{-1})	1330	1000	270–4480	830	440	380–2160	2.10[a]	41
PCB (ng g^{-1})	18.9	9.2	7.5–34.0	11.8	4.9	7.1–18.4	1.68	10

[a]Significant at $P>0.05$.

Table 15.2. Occurrence of Topographic Features in the Dumpsite Grid and Reference Grid

Grid	Ridge	Flank	Swale	Chi-square
Dumpsite	4	8	9	2.46
Reference	1	13	6	

Lead and copper concentrations were each correlated with the fine fraction in the dumpsite area, but not in the reference area.

These data indicate that dumping activity was contributing not only lead, copper, and organic carbon to the dumpsite area but also materials analyzed as the percentage fine fraction, normally considered as silts and clays. The composition of the fine fraction is not known, but the organic materials and grit in the sludge, and the insoluble metal precipitates formed by the addition of sludge to seawater, may augment the fine fraction of the sediments. A similar observation was made by Harris (1976) in the New York Bight.

The distribution of lead and copper in sediments as a function of topography was examined. Table 15.5 shows the mean concentrations of these materials on ridges, flanks, and swales in each grid. An analysis of variance (Table 15.6) shows that both lead and copper were found in significantly higher concentrations in swales in the dumpsite study area, but no significant differences were found in the reference area.

The significantly greater concentrations of lead and copper in the dumpsite area,

Table 15.3. Comparison of Biological Indices in Two Study Areas

Index	Dumpsite Grid		Reference Grid		t-Statistic	Degrees of Freedom
	Mean	Range	Mean	Range		
Number of taxa per station	43	29–62	37	22–58	1.82	37
Mean density per station	94	38–179	75	25–133	1.75	37
Margalef index, d^a	7.48	4.94–10.05	6.81	4.12–9.88	1.48	37
Simpson index, D'^b	0.09	0.04–0.14	0.11	0.04–0.30	0.98	37
Shannon–Weaver index, H'^c	2.96	2.20–3.43	2.83	1.92–3.38	1.14	37

[a]Margalef index, $d = (s-1)/\ln(N)$; s = number of species, N = total individuals in population.

[b]Simpson index, $D' = 1 - \sum_{i=1}^{s} [n_i(n_i-1)]/[N(N-1)]$; n_i = abundance of ith species.

[c]Shannon–Weaver index $H' = \sum_{i=1}^{s} p_i \ln(p_i)$; p_i = percentage of population represented by ith species (Marine Research, Inc., 1976).

Table 15.4. Correlation Coefficients of Lead, Copper, Organic Carbon, and Sediment Fine Fraction in the Dumpsite and Reference Grids

	Dumpsite Grid			Reference Grid		
	Correlation Coefficient r	t-Statistic	Degrees of Freedom	Correlation Coefficient r	t-Statistic	Degrees of Freedom
Lead: organic carbon	0.57[b]	3.15	20	0.66[a]	3.91	19
Copper: organic carbon	0.62[b]	3.52	20	0.72[a]	4.52	19
Organic carbon: percentage fines	0.76[b]	4.08	12	0.56[c]	2.84	17
Lead: percentage fines	0.77[b]	4.25	12	0.14	0.59	17
Copper: percentage fines	0.87[a]	6.40	12	0.35	1.58	17

[a]Significant at $P > 0.001$.
[b]Significant at $P > 0.01$.
[c]Significant at $P > 0.05$.

Table 15.5. Distribution of Lead and Copper as a Function of Topography in the Dumpsite and Reference Areas

	Dumpsite Grid			Reference Grid		
	Ridge	Flank	Swale	Ridge	Flank	Swale
Lead (mg kg^{-1}, dry wt)	2.26	2.54	4.46	1.90	1.82	1.86
Copper (mg kg^{-1}, dry wt)	1.33	1.28	1.89	0.93	1.07	1.13

the significant correlations of lead and copper with the percentage fine fraction in the dumpsite area, and their significantly higher concentration in dumpsite swales, but not in swales in the reference area, are evidence of their anthropogenic origin. Both metals are known inputs to this environment (Muir, Chapter 14), and the increased concentrations can be attributed to ocean dumping activity. The persistence of lead and copper in surficial sediments, but not zinc, chromium, or nickel, can be partially attributed to their lesser solubility in seawater (Krauskopf, 1956; Goldberg, 1963; Morel et al., 1975).

Concentrations of metals in surficial sediments vary considerably in different geographic areas. The levels reported here appear to be considerably lower than in nearshore marine habitats studied in England (Taylor, 1974), Southern California (Bruland et al., 1974; Hershelman et al., 1977), Greece (Chester and Voutsinou, 1981), and in the New York Bight apex (Carmody et al., 1973). These differences in concentrations of metals may be due to local geochemical background, sediment composition and texture, local inputs, and analytical methods.

15.3.2. Comparisons within Grids

The comparisons between grids having shown statistical differences in concentrations of lead, copper, percentage fines, and organic carbon concentrations, the loci of lead, copper, and organic carbon within grids were discriminated by the Duncan multiple range test. The stations circled in Fig. 15.2 had significantly higher concentrations than all other stations in both grids. These figures show the areas of contamination to be in swales immediately south of the dumpsite (Fig. 15.3).

Previous observations have indicated high concentrations of the dumped materials in the southern section of the sewage sludge release area and immediately southward. Higher concentrations of metals and organic carbon were associated with topographic depressions in this area (Lear et al., 1977). A recent summary of studies of microbial contamination of these surficial sediments by bacteria of sewage sludge origin (O'Malley et al., 1982) showed the exclusively anthropogenic contamination of this mid-shelf site, and indicated contamination of sediments extending to 37 km northeast and southwest of the release site. The incidence of microbial contamination of sediments was inversely related to distance from the release site, while the occurrences of coliform and fecal coliform bacteria were statistically greater in topographic lows or swales. This provided information on the extent of contamination by a nonconservative parameter, and agrees with the results reported here.

Table 15.6. Analysis of Variance

	Dumpsite Grid				Reference Grid			
	Degrees of Freedom	Sum of Squares	Mean Square	*F* Ratio	Degrees of Freedom	Sum of Squares	Mean Square	*F* Ratio
Lead								
Between	2	21.02	10.51	5.95[a]	2	0.01	0.006	0.006
Within	18	31.78	1.76		18	17.40	0.97	
Total	20	52.80			20	17.41		
Copper								
Between	2	1.83	0.92	3.73[a]	2	0.06	0.03	0.65
Within	18	4.41	0.26		18	0.87	0.05	
Total	20	6.24			20	0.93		

[a]Significant at $P>0.05$.

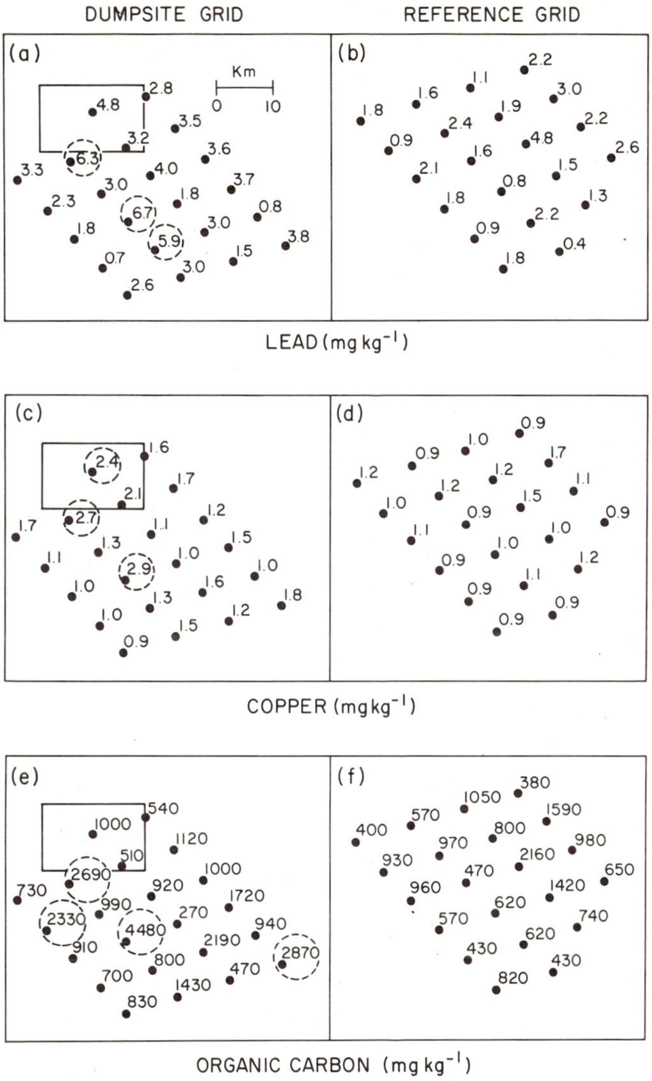

Figure 15.2. Distribution of lead, copper, and organic carbon in surficial sediments of the dumpsite and reference grids. All values are the means of three replicate samples. Values circled are significantly higher concentrations as discriminated by the Duncan multiple range test.

Stations having sediments with a median grain size greater than 2.0 phi, fine fraction (>4.0 phi) generally greater than 3%, more than 50% fine sands (>3.0 phi), and a high organic carbon content (>1000 mg kg^{-1} dry wt) were located in swales. The distributions of these parameters are shown in Figs. 15.2e and f, and 15.3c-f for both grids. Comparable swale areas were found in both grids (Figs. 15.3a and b). However, highest values for both total organic carbon and percent fines were found at the swale stations in the dumpsite area.

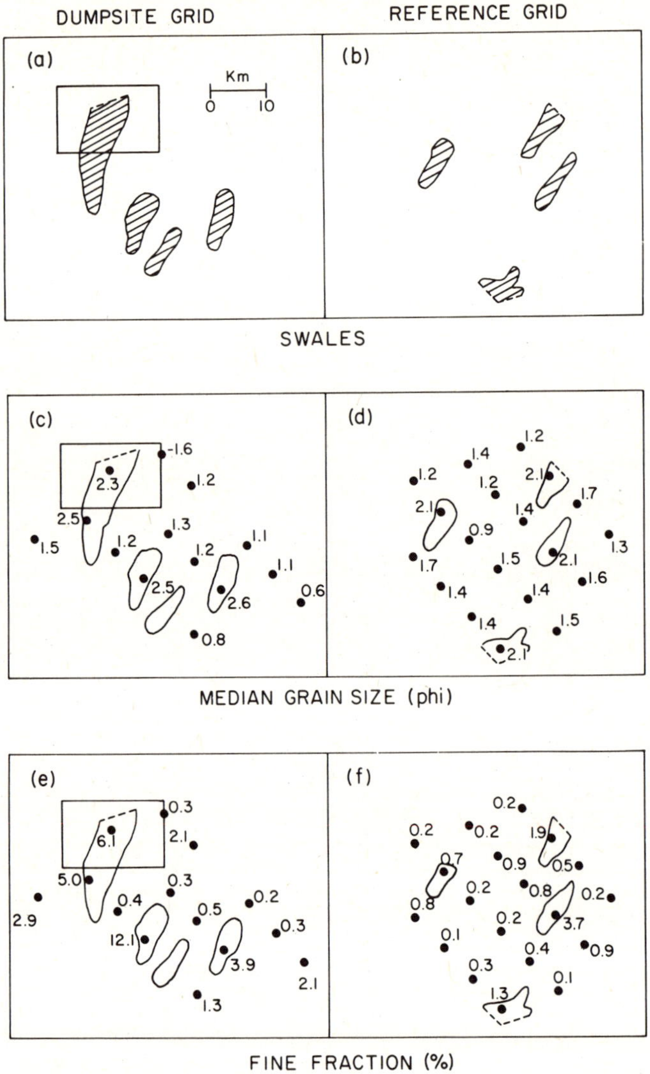

Figure 15.3. Locations of swales and distribution of median grain size (phi) of surficial sediments and of the percent fine fraction of surficial sediments in the dumpsite and reference grids.

The distributions of certain benthic infaunal species, in contrast to distributions of communities, were examined in both the dumpsite and the reference areas. Boesch (1979) has shown that the distributions of certain benthic species are influenced by the mesoscale topography, with some species characteristic of ridge habitats, and others characteristic of the finer, more stable substrates found in swales, and gradations between.

The polychaetes *Goniadella gracilis* and *Lumbrinereis acuta*, both interstitial macrofaunal feeders, are considered highly diagnostic of the dynamic coarse sands of ridges (Boesch, 1979). The distributions of these organisms in both study areas are shown in Figs. 15.4a–d. Within each grid, *G. gracilis* and *L. acuta* occurred at stations with coarse or medium sands (median phi <2.0), low percent fines (<3%), and low organic carbon content (<1000 mg kg^{-1} dry wt). The biological, chemical, and bathymetric parameters clearly identified the ridge systems in both grids. Other species similarly distributed were the syllid polychaetes *Parapionosyllis longicirrata* and *Sphaerosyllis erinaceus*. These organisms have also been shown to be preferentially distributed on ridges with medium and coarse sands (Boesch, 1979). It has been hypothesized that ridge habitats are swept clean of ocean-dumped materials by bottom currents (Lear et al., 1981); thus, these habitats should show little or no effects of ocean disposal of sewage sludge on the benthic infauna. Neither *G. gracilis* nor *L. acuta* was absent from ridge systems in either the dumpsite or the reference areas.

The chemical results indicated that swales, especially those in and immediately

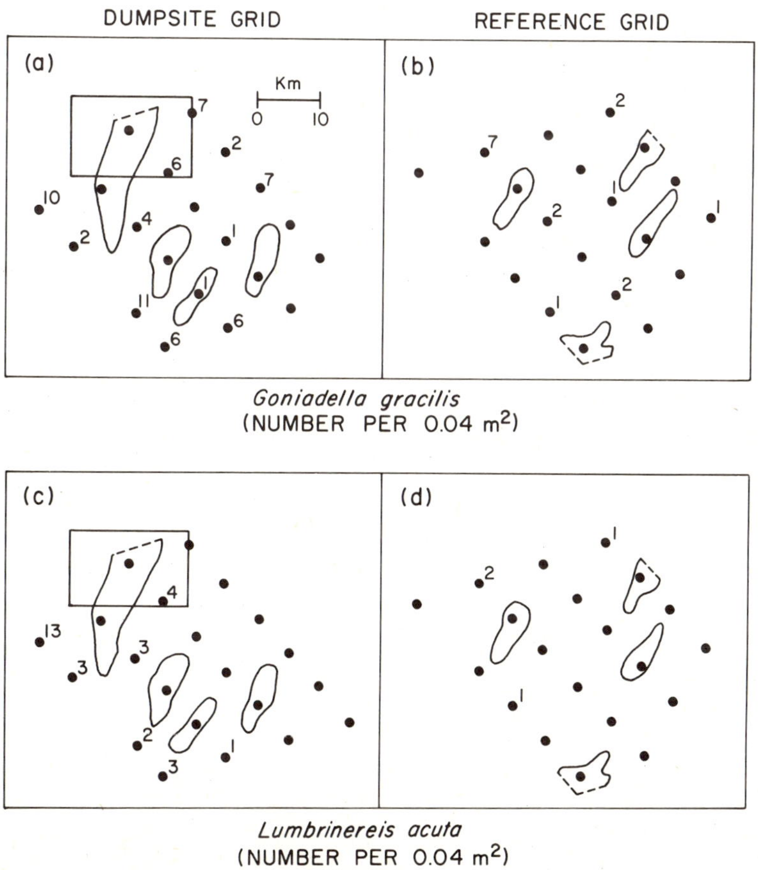

Figure 15.4. Distribution of the characteristic ridge organisms *Goniadella gracilis* and *Lumbrinereis acuta* in the dumpsite and reference grids. All numbers are the means of three replicate samples. Contours refer to the position of the swales identified in Figs. 15.3a and 15.3b.

south of the dumpsite, were areas of accumulation of sludge-derived materials; consequently, distributions of organisms characteristic of swales were examined in the dumpsite grid and the reference grid.

The distributions of the amphipod *Ampelisca agassizi* (Figs. 15.5a and b) and the polychaete *Lumbrinereis impatiens* (Figs. 15.5c and d) paralleled the distributions of finer sediments (median phi >2.0) with a fine fraction (>3%) and high organic carbon concentrations (>1000 mg kg^{-1} dry wt). The finer, more stable substrate and the co-occurrences of *A. agassizi* and *L. impatiens* identified the swale habitats in both areas (Boesch, 1979). These swale habitats were further indicated by similar distributions of the molluscs *Astarte undata, Cyclocardia borealis, Nucula delphinodonta,* and *Arctica islandica*, the amphipods *Ampelisca vadorum* and *Unciola irrorata*, and the polychaete *Euchone incolor*.

An important exception to the expected occurrence of *A. agassizi* was the absence of this organism at two stations in the swale that passes through the dumpsite

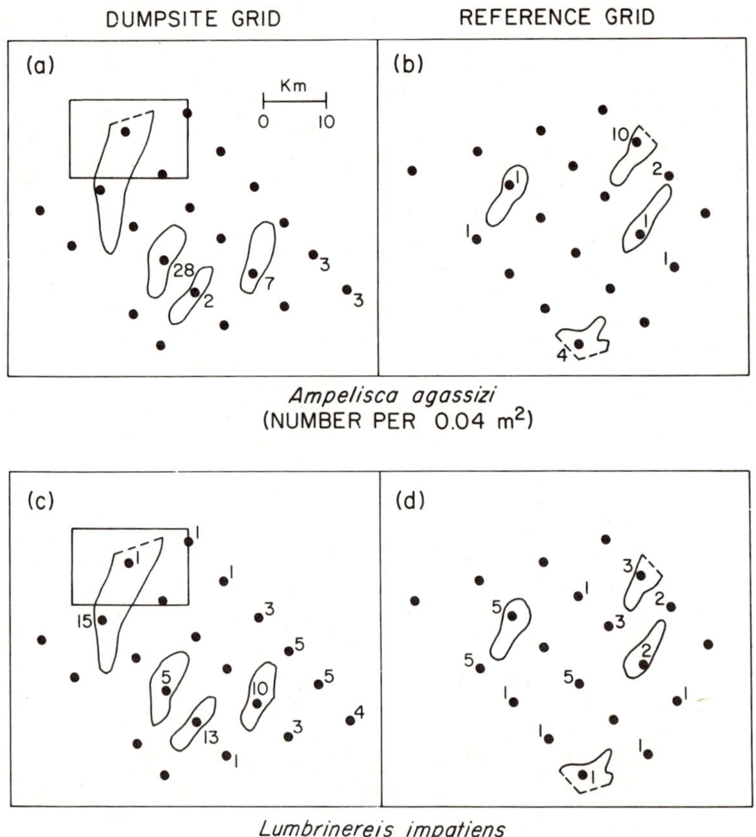

Figure 15.5. Distribution of the characteristic swale organisms *Ampelisca agassizi* and *Lumbrinereis impatiens* in the dumpsite and reference grids. All numbers are the means of three replicate samples. Contours refer to the position of the swales identified in Figs. 15.3a and 15.3b.

in a southwesterly direction (Fig. 15.5a). This swale had sediments comprised of fine sands (50% $>$3.0 phi), median phi greater than 2.0, with a high percent fine fraction (5.5%), and organic carbon concentrations greater than 1500 mg kg^{-1} dry wt. Lead and copper concentrations were also shown to be higher in this swale (Figs. 15.2a and c). *A. agassizi* occurred in all other swales in both grids. The distribution pattern of another ampeliscid amphipod, *Ampelisca vadorum*, more ubiquitous than *A. agassizi*, showed this species also to be absent only from this dumpsite swale. These ampeliscid amphipods have been shown to be sensitive to the presence of sewage sludge in the New York Bight (Botton, 1979) and were absent in samples from the highly organic sediments of the New York Bight apex (Pearce et al., 1976). Stirn et al. (1975) has reported that organic effluents, for example, sewage sludge, have resulted in the loss of entire taxonomic groups that are intolerant of organic pollution.

Of the 48 stations sampled in triplicate in both grids, the polychaete *Capitella capitata* was found only at one, the contaminated swale station in the dumpsite. This organism is one of the most characteristic of marine pollution-tolerant species (Word et al., 1977). The occurrence of this organism and the unexpected absences of *A. agassizi* and *A. vadorum* in the same location suggest that changes to benthic conditions characteristic of marine pollution have occurred at this site.

Earlier work at the sludge dumpsite in 1975 (Lear et al., 1977) reported deviations from the normal of selected benthic infauna at loci in the same swale in the immediate vicinity of dumping. Decreased abundances of the ampeliscids were apparent when compared to a similar swale to the east. It appears that the loci of impaction have not varied considerably in the two years from 1975 to 1977.

15.4. CONCLUSIONS

These studies show by direct comparisons that dumping of sewage sludge over a period of two to four years resulted in demonstrable contamination by lead, copper, organic materials, and particulates of sewage sludge origin in surficial sediments in a hydrodynamically active mid-continental shelf site. There was discernible biological response in the absence of the pollution-sensitive ampeliscid amphipods and the occurrence of the pollution-tolerant polychaete *Capitella capitata* at these locations.

These studies show that the rate of loading of sewage sludge at this deep and hydrodynamically active site produced characteristic indications of marine pollution near the dumpsite in a short period.

ACKNOWLEDGMENTS

This program was initiated under the aegis of the U.S. Environmental Protection Agency, Region III, Philadelphia, Pennsylvania, and continues cooperatively with the National Oceanic and Atmospheric Administration, Office of Marine Pollution

Assessment, Rockville, Maryland. We are indebted to the officers and men of the U.S. Coast Guard Cutter *Alert*; to Dr. T. K. Sawyer who gave assistance in sediment size analyses: to Ms. Kris Tholke, Marine Research, Inc., Falmouth, Massachusetts, who did the tedious but vital classification and enumeration of infauna; and to the U.S. Environmental Protection Agency, Annapolis Field Office, Annapolis, Maryland, whose personnel did the inorganic and organic analyses. Many other persons and institutions were and are cooperatively involved, and contributed materially to this program.

REFERENCES

American Public Health Association. 1976. Standard Methods for the Examination of Water and Wastewater, 14th ed. American Public Health Association, Washington, D.C., 1193 pp.

Beardsley, R. C., W. C. Boicourt, and D. V. Hansen. 1976. Physical oceanography of the Middle Atlantic Bight. *In*: Middle Atlantic Continental Shelf and New York Bight, Special Symposia, Vol. 2, M. G. Gross (Ed.). American Society of Limnology and Oceanography, Allen Press, Lawrence, Kansas, pp. 20–34.

Boesch, D. F. 1979. Benthic ecological studies: macrobenthos. *In*: Chemical and Biological Benchmark Studies, Vol. II B. Virginia Institute of Marine Science, Gloucester Point, Virginia, pp. 1–264.

Botton, M. L. 1979. Effects of sewage sludge on the benthic invertebrate community of the inshore New York Bight. *Estuarine and Coastal Marine Science*, **8**, 169–180.

Bruland, K. W., K. Bertine, M. Koide, and E. D. Goldberg. 1974. History of heavy metal pollution in southern California coastal zone. *Environmental Science and Technology*, **8**, 425–432.

Bumpus, D. F. 1973. A description of the circulation on the continental shelf of the East Coast of the United States. *In*: Progress in Oceanography, Vol. 6, Mary Sears (Ed.). Pergamon Press, New York, pp. 111–157.

Carmody, D. J., J. B. Pearce, and W. E. Yasso. 1973. Trace metals in sediments of the New York Bight. *Marine Pollution Bulletin*, **4**, 132–135.

Chester, R., and F. G. Voutsinou. 1981. The initial assessment of trace metal pollution in coastal sediments. *Marine Pollution Bulletin*, **12**, 84–91.

Demenkow, J. W., and P. Wiekramartane. 1976. Far-Field Sewage Release Simulations. Raytheon Corporation, Portsmouth, Rhode Island, 158 pp.

Duane, D. B., M. E. Field, E. P. Meisburger, D. J. P. Swift, and S. J. Williams. 1972. Linear shoals on the Atlantic inner continental shelf, Florida to Long Island. *In*: Shelf Sediment Transport: Process and Pattern, D. J. P. Swift, D. B. Duane, and O. H. Pilkey (Eds.). Dowden, Hutchinson and Ross, Stroudsburg, Pennsylvania, pp. 447–498.

E. I. du Pont de Nemours Co., University of Delaware, and Hydroscience, Inc. 1972. Waste Dispersion Characteristics in an Oceanic Environment. Draft report to the Water Quality Program, U.S. Environmental Protection Agency, Washington, D.C., 393 pp.

Folger, D. W., H. D. Palmer, and R. A. Slater. 1979. Two waste disposal sites on the continental shelf off the Middle Atlantic States: Observations made from submersibles. *In*: Ocean Dumping and Marine Pollution, H. D. Palmer and M. G. Gross (Eds.). Dowden, Hutchinson and Ross, Stroudsburg, Pennsylvania, pp. 163–184.

Goldberg, E. D. 1963. The oceans as a chemical system. *In*: The Sea, Vol. 2, M. N. Hill (Ed.). Wiley, New York, pp. 3–5.

Harris, W. H. 1976. Spatial and temporal variation in sedimentary grain-size facies and sediment heavy-metal ratios in the New York Bight apex. *In*: Middle Atlantic Continental Shelf and New York Bight, Special Symposia, Vol. 2, M. G. Gross (Ed.). American Society of Limnology and Oceanography, Allen Press, Lawrence, Kansas, pp. 102–123.

Hershelman, G. P., T. K. Jan, and H. A. Schafer. 1977. Pollutants in the sediments off Palos Verdes. *In*: Annual Report 1977, Southern California Coastal Water Research Project, El Segundo, California, pp. 63–68.

Ingram, R. L. 1969. Sieve analysis. *In*: Procedures in Sedimentary Petrology, R. E. Carver (Ed.). Wiley-Interscience, New York, pp. 49–67.

Johnson, P. P., and S. A. Wood. 1977. Seasonal variability of sediment texture in the Middle Atlantic Region. *In*: Geologic Studies of the Middle Atlantic Outer Continental Shelf, Vol. III, Geological Studies, D. W. Folger (Ed.). U.S. Geological Survey, Woods Hole, Massachusetts, pp. 1–18.

Ketchum, B. H. 1953. Preliminary Evaluation of the Coastal Water off Delaware Bay for Disposal of Industrial Wastes. Woods Hole Oceanographic Reference No. 53-51, Unpublished manuscript, Woods Hole, Massachusetts, 53 pp.

Klemas, V., G. R. Davis, and R. D. Henry. 1977. Satellite and current drogue studies of ocean-disposed waste drift. *Journal Water Pollution Control Federation*, **49**, 757–763.

Krauskopf, K. B. 1956. Measured concentrations of metals at saturation in sea water. *In*: Handbook of Marine Science, F. G. W. Smith and F. A. Kalber (Eds.). Chemical Rubber Company Press, Cleveland, Ohio, pp. 46.

Lear, D. W. (Ed.). 1974. Environmental Survey of Two Interim Dumpsites, Middle Atlantic Bight–Supplemental Report. EPA903/9-74-010b, U.S. Environmental Protection Agency, Region III, Philadelphia, Pennsylvania, 122 pp.

Lear, D. W., and M. L. O'Malley. 1979. Hydrography and Water Quality: Continental Shelf off Delaware and Maryland, 1978. EPA 903/10-79-032, U.S. Environmental Protection Agency, Region III, Philadelphia, Pennsylvania, 26 pp.

Lear, D. W., and G. B. Pesch (Eds.). 1975. Effects of Ocean Disposal Activities on the Mid-continental Shelf Environment off Delaware and Maryland. EPA 903/9-75-015, U.S. Environmental Protection Agency, Region III, Philadelphia, Pennsylvania, 214 pp.

Lear, D. W., S. K. Smith, and M. L. O'Malley (Eds.). 1974. Environmental Survey of Two Interim Dumpsites, Middle Atlantic Bight. EPA 903/9-74-010a, U.S. Environmental Protection Agency, Region III, Philadelphia, Pennsylvania, 158 pp.

Lear, D. W., M. L. O'Malley, and S. K. Smith. 1977. Effects of Ocean Dumping Activity, Mid-Atlantic Bight 1976, Interim Report. EPA 903/9-77-029, U.S.

Environmental Protection Agency, Region III, Philadelphia, Pennsylvania, 168 pp.

Lear, D. W., M. L. O'Malley, and S. K. Smith. 1981. Effects of ocean dumping on a temperate midshelf environment. *In*: Ocean Dumping of Industrial Wastes, B. H. Ketchum, D. R. Kester, and P. K. Park (Eds.). Plenum Press, New York, pp. 485–502.

Marine Research, Inc. 1976. Analysis of Operation "Midwatch" Benthic Invertebrates. Marine Research, Inc., Falmouth, Massachusetts, 186 pp.

Marine Research, Inc. 1979. Sample Analysis Report, Operations "Hotspot" and "Homerun." Marine Research, Inc., Falmouth, Massachusetts, 197 pp.

Meyers, T. D. 1974. An observation of rapid thermocline formation in the Middle-Atlantic Bight. *Estuarine and Coastal Marine Science*, **2**, 75–82.

Morel, F. M. M., J. C. Westall, C. R. O'Melia, and J. J. Morgan. 1975. Fate of trace metals in Los Angeles county wastewater discharge. *Environmental Science and Technology*, **9**, 756–761.

Norcross, J. J., and E. M. Stanley. 1967. Inferred surface and bottom drift, June 1963 through October 1964. *In*: Circulation of Shelf Waters off the Chesapeake Bight, W. Harrison, J. J. Norcross, N. A. Pore, and E. M. Stanley (Eds.). Professional Paper 3, Environmental Science Services Administration, Washington, D.C., pp. 11–42.

O'Malley, M. L., D. W. Lear, W. N. Adams, J. Gaines, T. K. Sawyer, and E. J. Lewis. 1982. Microbial contamination of continental shelf sediments by ocean disposal of sewage sludge. *Journal Water Pollution Control Federation*, In Press.

Palmer, H. D., and D. W. Lear (Eds.). 1973. Environmental Survey of an Interim Ocean Dumpsite, Middle Atlantic Bight. EPA 903/9-73-010A, U.S. Environmental Protection Agency, Region III, Philadelphia, Pennsylvania, 134 pp.

Palmer, H. D., J. R. Guala, and J. L. Nolder. 1976. Current Meter Reduction, with Comments on Bedload Sediment Transport: Middle Atlantic Bight. WD-6-99-0669-B, Westinghouse Ocean Research Laboratory, Annapolis, Maryland, 17 pp.

Pearce, J. B., J. V. Caracciolo, M. B. Halsey, and L. H. Rodgers. 1976. Temporal and spatial distributions of benthic macroinvertebrates in the New York Bight. *In*: Middle Atlantic Continental Shelf and New York Bight, Special Symposia, Vol. 2, M. G. Gross (Ed.). American Society of Limnology and Oceanography, Allen Press, Lawrence, Kansas, pp. 394–403.

Pearson, T. H., and R. Rosenberg. 1978. Macrobenthic succession in relation to organic enrichment and pollution of the marine environment. *Annual Review of Oceanography and Marine Biology*, **16**, 229–311.

Smith, R. W., and C. S. Greene. 1976. Biological communities near submarine outfall. *Journal Water Pollution Control Federation*, **48**, 1894–1912.

Stirn, J., A. I. Avcin, I. Kerzan, B. M. Marcotte, N. Meith-Avcin, B. Vriser, and S. Vukovic. 1975. Selected methods for assessment of marine pollution. *In*: Marine Pollution and Marine Waste Disposal. Proceedings of the Second International Congress, San Remo, 1973, E. A. Pearson and E. deF. Frangipane (Eds.). Pergammon Press, New York, pp. 307–327.

Swift, D. J. P., J. W. Kofoed, F. P. Saulsbury, and P. Sears. 1972. Holocene evolution of the shelf surface, central and southern Atlantic shelf of North

America. *In*: Shelf Sediment Transport: Process and Pattern, D. J. P. Swift, D. B. Duane, and O. H. Pilkey (Eds.). Dowden, Hutchinson and Ross, Stroudsburg, Pennsylvania, pp. 499–574.

Taylor, D. 1974. Natural distribution of trace metals in sediments from a coastal environment, Tor Bay, England. *Estuarine and Coastal Marine Science*, 2, 417–424.

Welch, C. S., and E. P. Ruzecki. 1979. Physical oceanography and climatology. *In*: Middle Atlantic Outer Continental Shelf Environmental Studies, Vol. IIA, Chemical and Biological Benchmark Studies. Virginia Institute of Marine Science, Gloucester Point, Virginia, pp. 1–36.

Word, J. Q. 1979. The infaunal trophic index. *In*: Annual Report 1978, Southern California Coastal Water Research Project, El Segundo, California, pp. 19–39.

Word, J. Q., B. L. Meyers, and A. J. Mearns. 1977. Animals that are indicators of marine pollution. *In*: Annual Report 1977, Southern California Coastal Water Research Project, El Segundo, California, pp. 199–206.

16

SEWAGE SLUDGE DUMPING IN THE MID-ATLANTIC BIGHT IN THE 1970s: SHORT-, INTERMEDIATE-, AND LONG-TERM EFFECTS

Bernard L. Oostdam

Department of Earth Sciences
Millersville State College
Millersville, Pennsylvania

ABSTRACT

The fate of treated sewage sludge dumped by the City of Philadelphia off the coast of Delaware and Maryland was considered on short-, intermediate-, and long-term time scales. Short-term (minutes to hours) studies of the water column before, during, and after dumping operations show the importance of the thermocline both as a barrier to settling and as a surface enhancing dispersal. Intermediate-term (days to months) events indicate the importance of the net southward bottom drift to the dispersal of sewage sludge. A fine flocculent material consisting of natural aggregates, possibly mixed with settled sewage sludge, disperses widely. Long-term effects (>1 year) deal with the environmental changes at a newly established interim dumpsite and the recovery of an abandoned old dumpsite; attempts to evaluate these effects from a study of changes in trace element concentrations in sediments and in the surf clams, *Spisula solidissima*, were inconclusive.

16.1. INTRODUCTION

The investigations reported here concentrated on the two sewage sludge dumpsites, and surrounding areas (Fig. 16.1), in the Mid-Atlantic Bight off the coasts of Delaware and Maryland, United States (see Muir, Chapter 14), and an extended study area comprising that part of the inner continental shelf bounded between 38° and 39°N.

The so-called "old" sewage sludge dumpsite (Chapter 14) had an area of about 8 km^2 and was about 24 km from the mouth of Delaware Bay (Fig. 16.1). The

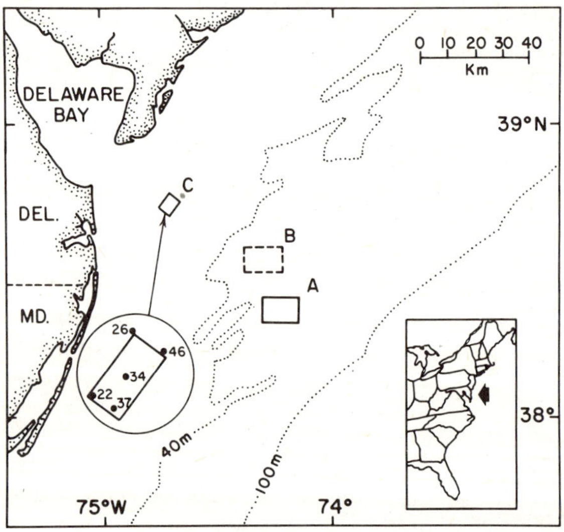

Figure 16.1. Location of dumpsites off the coast of Delaware and Maryland: (A) interim sewage sludge dumpsite; (B) acir-iron waste dumpsite; (C) old sewage sludge dumpsite.

dumpsite was used from 1961 to 1973 when a total of about 557,000 metric tons of sewage sludge solids, on a dry weight basis, were dumped mainly by the City of Philadelphia Water Department (Chapter 14, Fig. 14.2). In June 1973, dumping at this old dumpsite was prohibited, and the U.S. Environmental Protection Agency (EPA) issued its first interim license to dump sludge at a new site (Fig. 16.1). This new or interim sewage sludge dumpsite was 172 km^2 in area and centered about 87 km from the mouth of Delaware Bay. From 1973 until the end of 1980, when sludge dumping was phased out at the interim dumpsite, about 570,000 tons of treated sewage sludge solids were dumped in gradually diminishing annual quantities (Chapter 14).

In addition to the sewage sludge dumpsites, the study area included the acid-iron waste dumpsite (Fig. 16.1) used by the E. I. du Pont de Nemours & Co. (hereafter referred to as DuPont) from 1968 to 1977 (Chapter 14). Lear and Pesch (1975) calculated that in 1974 the City of Philadelphia dumped about 10% more material (including over 90% of the total copper, lead, zinc, and cobalt) than DuPont, whose wastes contained large amounts of iron, titanium, and vanadium.

16.2. METHODS

16.2.1. Field

Experiments to assess short-term effects (minutes to hours) of sewage dumping usually included background measurements and evaluation of natural variations at one or more control stations before a dump, followed by intensive sampling during the dumping event and more widespread sampling for hours after dumping. Short-term effects include (1) changes in turbulent dispersion triggered by the rapid discharge of large quantities of sludge ($\simeq$1000 liters s^{-1}) and enhanced or hampered by ambient current and density distributions (for example, see Chapter 4); (2) rapid chemical transformations forced by the discharge of a barge load of sewage sludge, up to 90–95% freshwater, into seawater with its contrasting high ionic strength; and (3) disturbance of plankton and nekton in the water column by impeding light transmission due to the dumped sludge.

In the present work percent light transmission was determined by a 30-cm path length transmissometer; water samples for the determination of dissolved metals and suspended solids were collected in Teflon-lined Niskin (General Oceanics, Miami, Florida) water bottles.

Current measurements to assess intermediate-term effects were made using drift cards, Woodhead seabed drifters, current crosses suspended from marker floats fitted with radar reflectors, and current meters. Navigation and buoy tracking was done by LORAN C and radar.

Sediments and clams were sampled to assess long-term effects. A Shipek grab and undisturbed fractions were subsampled using sterile gloves. Cores were taken by divers or by piston corer and immediately frozen until analysis. Flocculent material, hereafter referred to as "floc," near or at the bottom was collected by a diver.

Clams for trace element analysis were collected in 10 min hauls by hydraulic dredge and were individually wrapped and kept frozen until analysis.

16.2.2. Laboratory

Concentrations of suspended solids in water samples were determined by vacuum filtration of 1-liter samples through preweighed 0.45-μm Millipore® 47-mm filters (Millipore Corporation, Bedford, Massachusetts) followed by rinsing with distilled water, 48-h drying, and then reweighing. Concentrations of trace elements in sediments, in whole clams or in seawater, were determined (in triplicate) using a Varian 70 (Varian Corporation, Palo Alto, California) atomic absorption spectrophotometer (AAS) equipped with graphite furnace and background corrector. Interlaboratory AAS comparisons with the Annapolis Field Station, EPA, gave satisfactory accord (Oostdam et al., 1975). Total organic carbon and total carbohydrate were determined by a commercial laboratory using a carbon analyzer.

16.3. RESULTS

16.3.1. Short-Term Effects

On 13 November 1971, just before the discharge of sludge in the old sewage sludge dumpsite, two buoys were anchored at about 16 m depth, upwind and downwind of the planned track, and background concentrations of suspended solids were determined (Fig. 16.2). After disposal of the sludge the center of the wake in the barge track was immediately marked by fixed buoy C (Fig. 16.2b); buoy 2 (Fig. 16.2a) was 100 m upwind, and buoy 4 (Fig. 16.2c) was 900 m downwind from buoy C. Wind speeds averaged 20 km h^{-1}. Within 10 min from the passage of the barge, concentrations in the wake had decreased from about 90,000 mg kg^{-1} in the barge itself to 156 mg kg^{-1} at the sea surface near buoy C (Fig. 16.2b). The sludge settled by downward entrainment to mid-depth (8 m) and locally decreased to $\simeq$ 30 mg kg^{-1} in 1 h and to $\simeq$10 mg kg^{-1} in 2 h. Simultaneously, the surface slick moved downwind, reaching and passing buoy 4 about 100 min after discharge from the barge. At buoy 4 (Fig. 16.2c) some fallout of sludge to the bottom was noticeable at $\simeq$100 min after discharge. Currents were in a direction almost opposite to that of the wind and carried the settling sludge past buoy 2 (Fig. 16.2a) where concentrations of suspended solids of at least 25 mg kg^{-1} were observed at mid-depth.

Downward entrainment of sewage sludge was also observed in the interim dumpsite on 11 and 12 June 1974, when suspended solid concentrations were measured in water depths up to 60 m over a period of somewhat more than a day after sludge discharge from a barge (Table 16.1). Background suspended solid concentrations before the dumping event showed a maximum at the surface and a secondary maximum near the thermocline (Table 16.1, column A), a common feature in summer. Within the first 8 h after dumping (Table 16.1, column B), concentrations in the

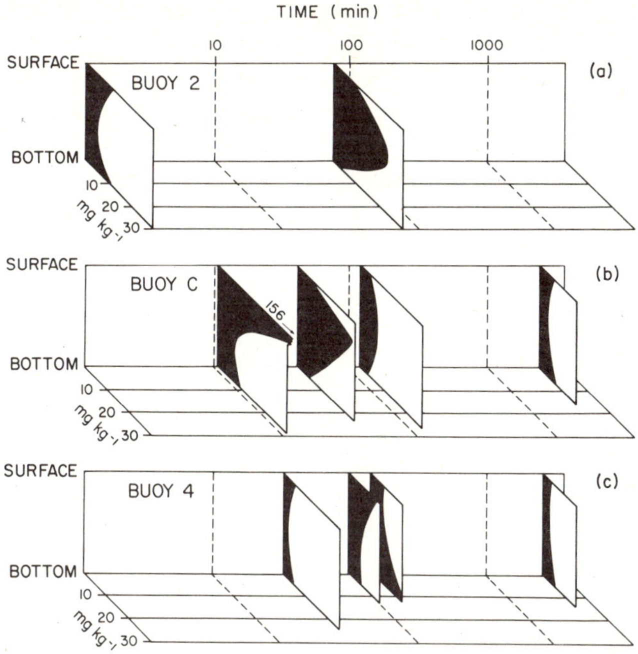

Figure 16.2. Concentrations (mg kg^{-1}) of suspended solids with depth and time after dumping, old sewage sludge dumpsite, 13 November 1971: (a) buoy 2 located 100 m upwind of scheduled dump; (b) buoy C marked time and place of dump; and (c) buoy 4 located 900 m downwind of buoy C.

surface and mixed layers decreased rather dramatically, whereas values for the thermocline and bottom layer increased, demonstrating the considerable downward momentum of the sludge being discharged. This affects not only the particulates in the sludge itself, but also causes a downward drag by entrainment of ambient suspended solids and plankton present in the water column. The results of subsequent downward motion and of the recovery of surface water to predump conditions can be seen in Table 16.1 (see column C).

To evaluate the local and advective changes in the water column following the dump during the June 1974 experiment, four lightweight radar reflecting buoys were deployed: one was anchored, the other three carried current crosses at 3-, 15- (top of thermocline), and 30-m depths. The buoys were tracked by radar, and water samples were collected over 24 h. Although their trajectories (Fig. 16.3) reflect the effect of tides, the buoys did show residual nontidal velocities of 11 and 8 cm s^{-1} southerly at depths of 3 and 15 m, respectively, and of 5.5 cm s^{-1} southeasterly at 30 m. The 3-m buoy followed the surface slick reasonably well during the first 8 h, after which the slick was no longer sighted. Slick displacement, caused by surface currents and by 8–9 m s^{-1} northwesterly winds, was about 25° to the right of the wind and the slick moved at a steady rate of $\simeq$20 cm s^{-1}, or slightly over 2% of the wind speed for both time intervals, 0945 to 1200 and 1200 to 1715 U.S. Eastern Daylight Savings Time (EDT).

Table 16.1. Mean Concentration (mg kg^{-1}) of Suspended Solids Collected before and after Discharge of Sewage Sludge in the Interim Dumpsite

Depth Interval	(A) Before dumping 1700 EDT[a] 10 June to 0300 EDT 11 June 1974	(B) Within 8 h after dumping 0920 EDT to 1720 EDT 11 June 1974	(C) 1720 EDT 11 June to 1300 EDT 12 June 1974
Surface layer	1.87 (7)[b]	0.96 (8)	1.98 (7)
Mixed layer (1–20 m)	1.21 (14)	1.03 (23)	0.95 (17)
Lower region of the thermocline (21-30 m)	1.36 (6)	1.78 (8)	1.79 (6)
Bottom layer (31-60 m)	1.20 (7)	1.44 (9)	1.51 (6)

[a]U.S. Eastern Daylight Time.

[b]Numbers in parentheses represent number of samples collected and analyzed.

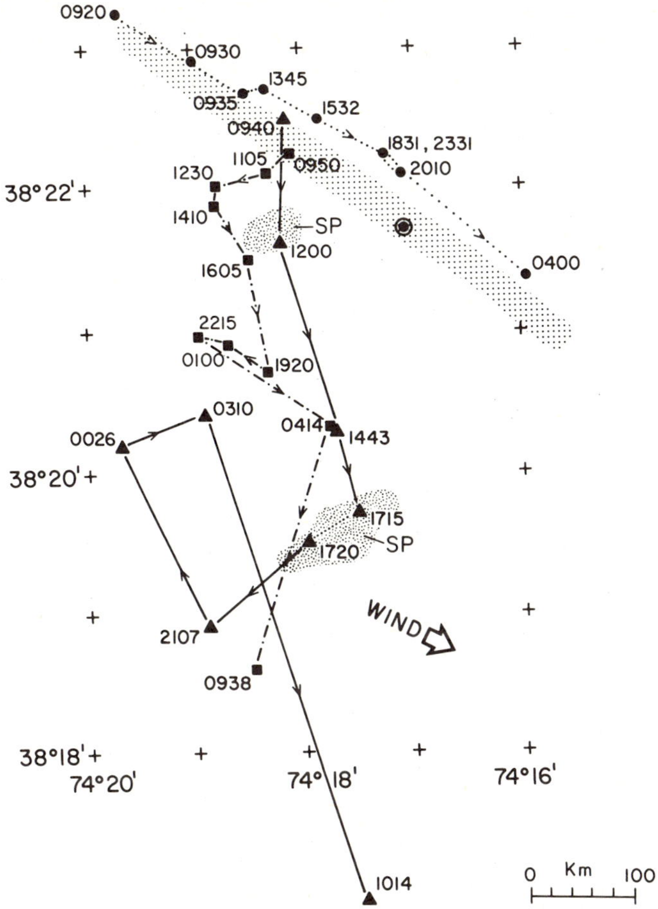

Figure 16.3. Trajectories of current crosses at three depths during sewage sludge disposal experiment, 11–12 June 1974: ⊙, anchored buoy; ▲, current cross at 3 m; ■, current cross at 15 m (thermocline); and ●, current cross at 30 m. Barge track and surface slicks dotted and marked SP. Dotted line at 1715 EDT represents repositioning of current cross after repairs to radar reflector.

In addition to the above physical processes chemical fractionation may have occurred, based on concentrations of dissolved copper, zinc, cadmium, and chromium in seawater before and after discharge of sewage sludge (Table 16.2). After the dumping event some copper may have transformed from dissolved to the particulate phase. The results for cadmium and chromium are less conclusive. Zinc, however, may have been either transformed from the particulate to dissolved phase, possibly responding to the higher chloride concentration of seawater by forming metal-chloride ion complexes, or added to seawater from the dissolved fraction in sewage sludge. Sewage sludge is especially enriched in copper, zinc, cadmium, and chromium (Chapter 1). It is unknown, however, what fraction of these metals in sewage sludge is present in the dissolved or solid phases.

Table 16.2. Trace Elements in Seawater[a] Collected before and after Discharge of Sewage Sludge

	Copper	Zinc	Cadmium	Chromium
Before dumping				
Background median concentration[b] (μg kg^{-1})	1.5	3.0	0.03	0.045
After dumping				
Percentage of samples[c] collected within 24 hours showing filtrate concentrations that are:				
Higher	26	50	38	59
Same	18	25	43	
Lower than the background median concentration	56	25	19	41

[a]Filtered through 0.45 μm-filter.
[b]Based on analysis of 42 samples.
[c]Based on analysis of 128 samples.

In another experiment light transmission was measured on 14 and 15 July 1974 after 7000 m^3 of sludge had been discharged between 2310 and 0050 EDT at the interim sludge dumpsite (Fig. 16.1). Temperatures decreased from 22.8°C at 5 m depth to 8.3°C at 24 m, and an anchored current recorder showed mean currents of about 30–40 cm s^{-1} in direction 033° at 18 m depth. Before the dumping event, practically uniform light transmission of 70–80% per 30 cm occurred throughout the water column (Fig. 16.4, 14 July, 2005 EDT). After the sludge barge passed, the surface slick moved due east at a rate of $\simeq$25 cm s^{-1} until about 0500 EDT. Transmissometer readings taken during the 5 h after dumping showed the top 10 m to alternate between low and high transmission values, with a minimum of less than 10% per 30 cm observed at about 6 m at 0440 EDT. Turbid clouds interspersed with clear seawater are common in the wake of discharging barges. Practically no variation from background values was present between depths of 10–20 m. By contrast, low transmissometer values were noted near the bottom of the thermocline at 24 m. Although these anomalies may have been caused by fallout from the slick immediately above, they may represent fractions derived from waste dumped further along the track of the barge which subsequently were advected by north–east currents occurring at mid-depth.

Detailed transmissometer readings were made on 7 August 1974 in the wake of a discharging barge at the interim sewage sludge dumpsite (Fig. 16.5). It appears that the upper part of the thermocline (at depths from 10 to 12 m) behaved as a barrier to the settling sludge mass. Pronounced variations in the readings were noticeable up to one hour after passage of the barge. A similar phenomenon of internal oscillations in the thermocline has been reported by Proni and Hansen (1981) and by Duedall et al. (1980).

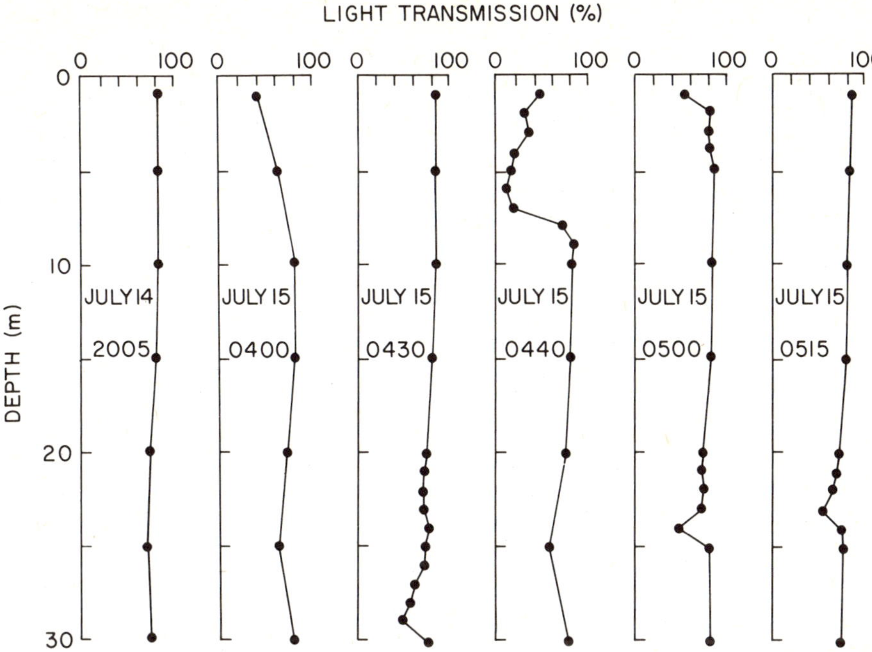

Figure 16.4. Light transmission (% per 30 cm) before and after dumping at the interim sewage sludge dumpsite, 14–15 July 1974. The dumping event occurred 14 July 2310 EDT to 15 July 0050 EDT. Water depth was 54 m. At the top of thermocline at 10 m, temperature was 22.8°C; at the bottom of the thermocline at ≃24 m, the temperature was 8.3°C. Current speed at ≃18 m was ≃40 cm s^{-1} and the direction was toward 033°.

16.3.2. Intermediate-Term Effects

Intermediate-term effects range in time from several hours to a year, and thus include tidal, diurnal, and seasonal cycles. In the summer density stratification may impede for several hours the settling of sludge to the seafloor (Duedall et al., 1977). In the winter the water column approaches uniform density; thus, a greater part of the sludge probably settles directly to the bottom (Callaway et al., 1976). The ultimate spatial distribution of the sludge will, however, depend mainly on prevailing horizontal advective processes, especially those occurring as a result of storm events. Previously published work on surface and bottom currents on the Mid-Atlantic inner shelf indicates the existence of residual currents toward bay mouths and to the south (Bumpus, 1965; Bumpus and Lauzier, 1965; Davey, 1972; Palmer and Lear, 1974). A 60-h current meter record by the Marine Science Consortium in the center of the interim dumpsite during March 1975 (Fig. 16.6) indicates a net southwesterly bottom current of 12 cm s^{-1}.

To determine the degree of dispersal of sludge beyond the boundaries of the dumpsites, observations of the seafloor were made by underwater television, scuba divers, and submersibles (Folger et al., 1979), and bottom samples were collected in

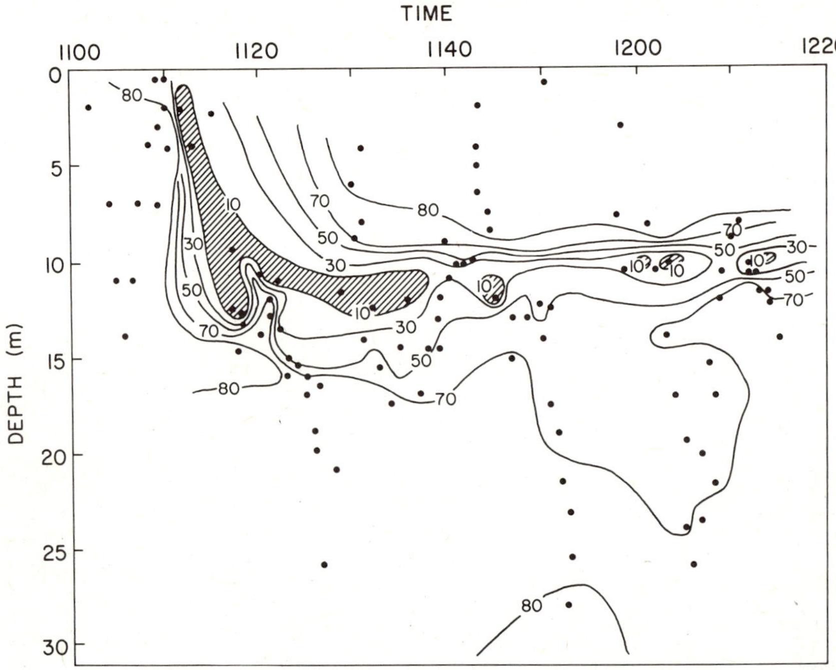

Figure 16.5. Light transmission (% per 30 cm) before, during, and after dumping, at interim sewage sludge dumpsite, 7 August 1974; dumping event occurred at 1110 EDT.

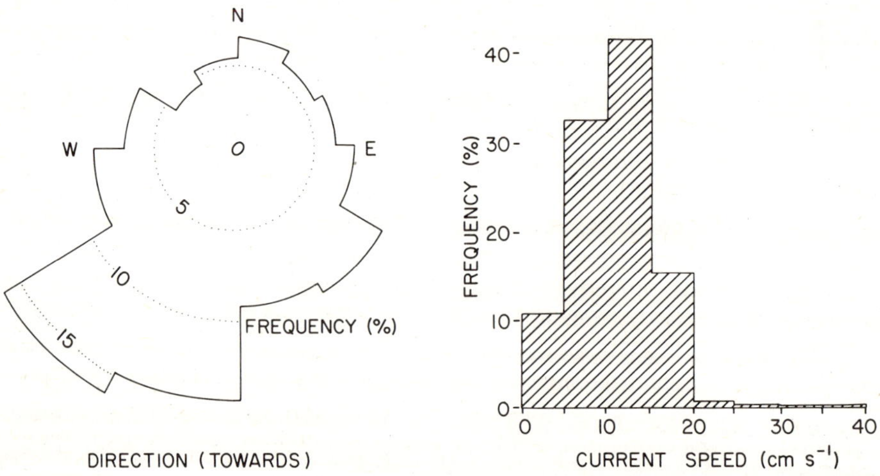

Figure 16.6. Current direction and speeds observed during a 60 h bottom current record at the interim sewage sludge dumpsite, 8 March [0354 Eastern Standard Time (EST)] to 10 March (1551 EST) 1975. The current meter, a General Oceanics (Miami, Florida) Model 2010 film recording type, was anchored 2 m above seafloor; water depth was 56 m. The frequency of recording was once per min.

and near the dumpsites. During dives made in August 1974 the seafloor below the track of a sewage sludge transport barge, which had discharged the day before, was partially covered with patches of brownish floc up to a few millimeters in thickness. The appearance of this floc was similar to that found elsewhere on the continental shelf after a period of quiet weather. On a later dive in October 1974, samples of floc for analysis were collected at a depth of 43 m in the northwest corner of the interim sewage dumpsite. None of the samples contained any fibers typical of sludge (Figs. 16.7 and 16.8), and the floc in the samples was more translucent than sewage sludge. Analysis for selected trace elements in 1-cm sections of floc trapped in a piston core taken 1 km south of the interim dumpsite in December 1975 is shown in Table 16.3.

A floc sample taken in April 1976 at 5 km from the interim dumpsite consisted of (based on 500 particle count): 49% mineral, 33% biogenic, 14% aggregate, 2% fiber, and 2% unknown. Concentrations of total organic carbon (TOC) and total carbohydrate (TCH) were $\simeq$250 and $\simeq$190 g kg^{-1} (dry wt), respectively. For comparison, the TOC and TCH in sewage sludge are in the range of 250–500 g kg^{-1} and 100–350 g kg^{-1}, respectively (Charnell, 1975; Duedall et al., 1975). In sediment TOC and TCH concentrations measured in 1-cm intervals of cores collected from troughs and crests (Table 16.4) shows considerable spatial variations over very short horizontal distances, but no significant changes with depth.

Although other investigators of floc in the study area report that floc was widespread (Lear and Pesch, 1975; Lear et al., 1981; Raytheon Corporation, 1975), no material on the seafloor has conclusively been identified as sewage sludge based on direct measurements. Indirect evidence of possible sludge on the seafloor are elevated trace element concentrations in surficial layers (Table 16.3), high TOC in floc (Table 16.4), and the presence of some fibrous material.

16.3.3. Long-Term Effects

Long-term effects of sewage sludge dumping over periods of years exert their main impact on the benthic environment. They include changes in bathymetry, in sediment grain size, and in trace element concentrations in bottom sediments. These effects can cause chronic disturbances to benthic biota and produce changes in population structure. Sediments and filter-feeding organisms are indicators of changes in the chemistry of bottom waters because of their high adsorption and bioconcentration capacities, respectively.

Of particular interest in assessing long-term effects are the rates of environmental changes at new sewage sludge dumpsites and recovery rates of abandoned dumpsites. EPA's closure of the old sewage sludge dumpsite (Fig. 16.6, see also Chapter 14) and the subsequent initiation of dumping at the interim dumpsite in June 1973 afforded the opportunity to assess both recovery and deterioration rates. Sampling of sediments and surf clams was conducted in and around both the interim and abandoned old sewage sludge dumpsites in December 1974, using bottom grabs, sediment corers, and commercial clam dredge (Oostdam et al., 1975). The bottom sediments, cores, and surf clams were analyzed for concentrations of aluminum,

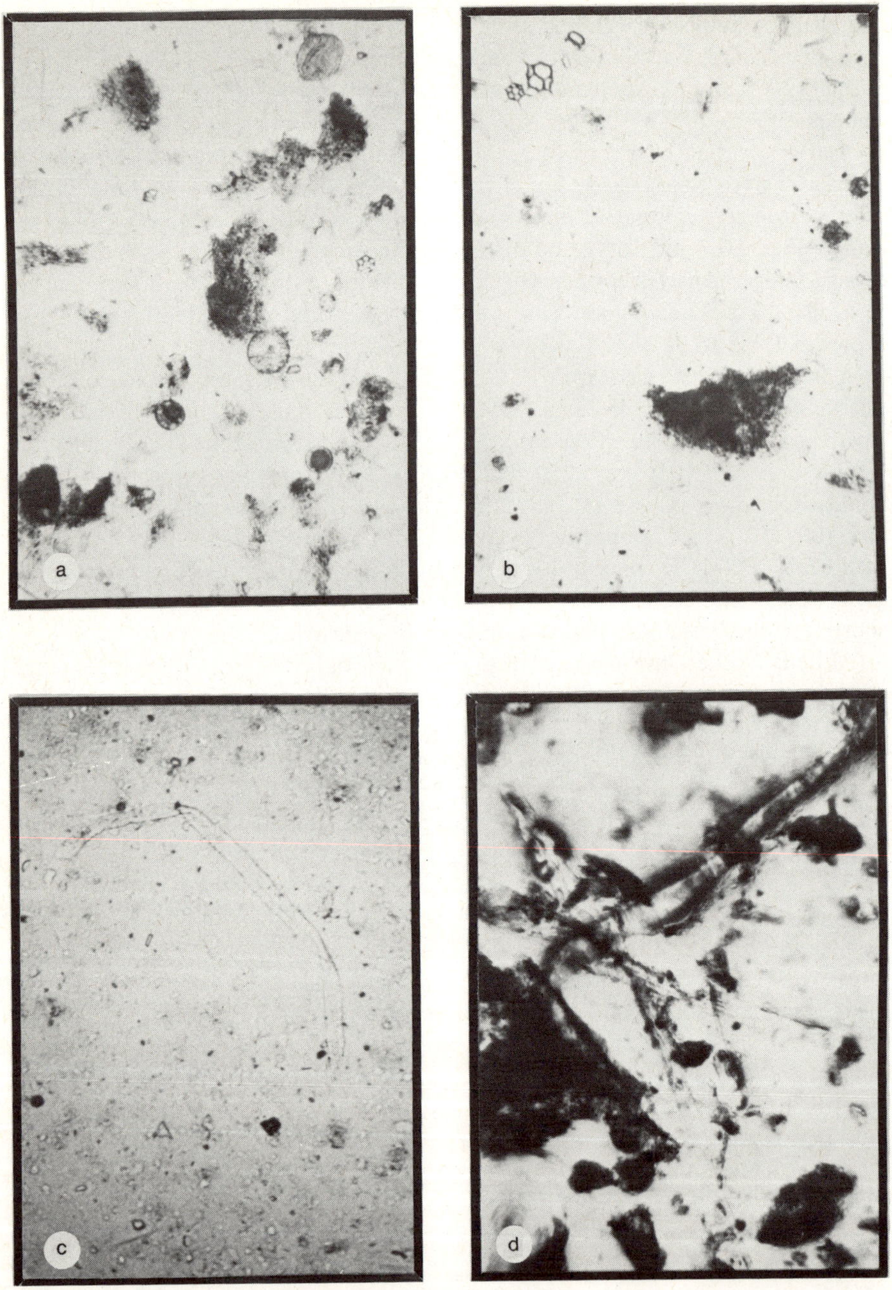

Figure 16.7. Photomicrographs (X 100) of Philadelphia sewage sludge and of diver collected "floc" overlying seafloor sewage sludge dumpsite, 27 October 1974: (a) digested sewage sludge from City of Philadelphia; (b) sample collected 1 km north of the old sewage sludge dumpsite; (c) and (d) samples of floc collected by diver in the NW corner of the interim sewage sludge dumpsite.

Figure 16.8. Floc, shells, and sand dollar on seafloor below barge tract at the interim sewage sludge dumpsite, 27 October 1974. (Photo credit: Lee Robinson)

Table 16.3. Concentrations (mg kg^{-1}, dry wt) of Trace Elements in Floc and in Underlying Sediment

	Cadmium	Copper	Iron
5-cm layer of floc	0.37–0.17	100–15	11,000–9,000
5 to 24 cm of sediment beneath floc layer	<0.01	7.4 to 2.0	3,700–1.600

cadmium, chromium, iron, mercury, nickel, lead, titanium, vanadium, and zinc, which were compared with other available and similar data for the study area and for an extended area (for clams only) on the inner continental shelf (Tables 16.5 and 16.6).

16.3.3a. Elemental Concentrations in Sediment

Inspection of Table 16.5 shows apparent increases by one order of magnitude or more in the time series from 1970 to 1974 for mean concentrations of chromium, iron, nickel, lead, and zinc in surface sediments in the old sewage sludge dumpsite [Table 16.5, see old sewage sludge dumpsite results: Oostdam et al. (1975) versus Davey (1972) and Clem (1975)]. The least apparent change occurred for cadmium; the greatest major change was for iron. However, replicates of Davey's 1971 sediment samples (Davey, 1972) analyzed later by the Marine Science Consortium in 1974 showed concentrations of one or even two orders of magnitude higher, suggesting variations in methodology. T. D. Meyers (personal communication)

Table 16.4. Concentrations (mg kg^{-1}, dry wt) of Total Organic Carbon (TOC) and Total Carbohydrate (TCH) in Sediments[a] in and near Interim Sewage Sludge Dumpsite

Depth in Core (cm)	Site 1[b] Core 01[c] TCH	Site 1[b] Core 01[c] TOC	Site 1[b] Core 03[d] TCH	Site 1[b] Core 03[d] TOC	Site 2[b] Core 01 TCH	Site 2[b] Core 01 TOC	Site 2[b] Core 02 TCH	Site 2[b] Core 02 TOC
0–1	78	300	200	1300	130	800	150	650
1–2	67	560	120	860	140	700	60	530
2–3	91	640	110	1000	72	340	40	540
3–4	89	550	100	1500	90	460	160	370
4–5	—	660	—	1100	90	470	70	510
5–6	—	—	—	—	110	740	—	350
6–7	—	—	—	—	140	500	—	—

[a]In cores collected by scuba divers on 19 April 1976; the dashes indicate no information available.

[b]Site 1, 5 cm south of the interim sewage sludge dumpsite, depth 52 m. Site 2, 2 km north of the center of the dumpsite, depth 46 m.

[c]Core taken in ripple trough.

[d]Core taken on ripple crest.

Table 16.5. Concentration (mg kg^{-1}, dry wt) of Elements in Surface Sediments in Vicinity of the Mid-Atlantic Dumpsites[a]

Location, Year Sampled Reference	Number of Samples	Range and Mean	Element									
			Aluminum	Cadmium	Chromium	Iron	Mercury	Nickel	Lead	Titanium	Vanadium	Zinc
Center of old sewage sludge sludge dumpsite (within 2 km radius)												
November, 1970, Clem (1975)	138	Min.	–	0.00	0.00	0.09	–	0.01	0.02	–	–	0.00
		Max.	–	0.30	0.04	2.0	–	0.14	0.14	–	–	0.06
		Mean	–	0.01	0.02	0.75	–	0.03	0.06	–	–	0.02
1971, Davey (1972)	9	Min.	–	0.00	0.00	–	–	0.0	0.02	–	–	0.02
		Max.	–	0.01	0.80	–	–	1.2	0.40	–	–	2.5
		Mean	–	0.005	0.14	–	–	0.2	0.08	–	–	0.50
1974, Oostdam et al. (1975)	72	Min.	24	<0.01	0.01	400	<0.01	0.2	0.3	1.1	0.1	1.0
		Max.	370	0.26	11	3,000	0.11	3.1	7.2	4.3	12	8.9
		Mean	200	0.09	1.1	1,500	0.04	1.6	2.9	2.5	0.5	4.1
1974, T. D. Meyers (personal communication)	3	Min.	–	–	–	520	–	–	–	5	1.3	–
		Max.	–	–	–	1,200	–	–	–	10	3.3	–
		Mean	–	–	–	810	–	–	–	8	2.2	–
Acid-iron waste dumpsite												
1974, T. D. Meyers (personal communication)	2	Min.	–	–	–	880	–	–	–	3.0	–	–
		Max.	–	–	–	2,300	–	–	–	8.0	–	–
		Mean	–	–	–	1,600	–	–	–	5.5	–	–
Interim sewage sludge dumpsite												
1974, Oostdam et al. (1975)	6	Min.	56	0.02	0.59	880	0.01	1.0	0.2	1.9	0.3	3.0
		Max.	440	0.10	1.9	2,400	0.26	3.4	5.4	5.7	1.3	9.4
		Mean	250	0.05	1.2	1,500	0.05	2.2	2.8	3.7	0.8	6.0
1974, U.S. Environmental Protection Agency (in Oostdam et al., 1975)	11	Min.	–	<1	1.0	1,300	0.01	<1	3	3	–	3
		Max.	–	<1	3.0	2,600	0.04	2.0	10	8	–	9
		Mean	–	<1	2.4	2,200	0.02	1.4	7	5.5	–	7.4

[a]The dashes indicate no information available.

Table 16.6. **Concentrations (mg kg^{-1}, dry wt) of Elements in the Surf Clams (*Spisula solidissima*) in the Vicinity of the Mid-Atlantic Dumpsites**[a]

Location, Month/Year Sampled, Reference	Number of Samples	Range and Mean	Element: Aluminum	Cadmium	Chromium	Iron	Mercury	Nickel	Lead	Titanium	Vanadium	Zinc
Off Delaware Bay												
May–June 1967, Buelow (1968)	9	Min.	–	<0.15	0.62	43	–	<0.10	<0.15	–	–	9.9
		Max.	–	<0.15	7.5	140	–	3.7	<0.15	–	–	13
		Mean	–	<0.15	2.8	94	–	1.3	<0.15	–	–	12
Old sewage sludge dumpsite												
February 1970, Clem (1975)	6	Min.	–	0	0.65	24	–	2.2	0	–	–	12
		Max.	–	0	2.3	76	–	4.3	0	–	–	14
		Mean	–	0	1.3	49	–	3.3	0	–	–	13
February 1970, Davey (1972)	6	Min.	–	0	0.01	–	–	0.03	0.02	–	0.04	–
		Max.	–	0.09	0.19	–	–	0.09	0.11	–	0.65	–
		Mean	–	0.04	0.07	–	–	0.06	0.05	–	0.29	–
December 1974, Oostdam et al. (1975)	28	Min.	200	0.24	0.2	350	0.12	3.6	5.6	7.2	1.7	34
		Max.	440	1.7	73	1,100	0.99	160	38	36	9.1	110
		Mean	260	0.76	32	670	0.36	24	12	14	3.3	68
Interim sewage sludge dumpsite												
November, 1973, Lear (1974)	7	Min.	68	0.22	1.1	93	–	1.2	1.6	–	1.4	46
		Max.	170	0.58	29	580	–	8.1	6.4	–	4.4	430
		Mean	130	0.45	5.4	330	–	5.2	3.0	–	2.3	110
Landward of interim sewage sludge dumpsite												
July, 1975, Marine Science Consortium (unpublished results)	40	Min.	33	0.03	9.5	19	0.04	4.6	0.82	10	2.2	36
		Max.	500	3.6	110	1,700	1.0	65	12	31	9.2	150
		Mean	110	0.56	46	60	0.31	26	3.0	22	5.0	100
Extended study area												
June 1975, Reynolds (1979)	16	Min.	–	0.25	0.6	–	–	2.1	–	–	0.8	37
		Max.	–	0.48	9.3	–	–	17	–	–	7.5	71
		Mean	–	–	–	–	–	–	–	–	–	–

[a]The dashes in table denote no information available.

found elevated concentrations of iron, titanium, and vanadium near the center of the old sewage sludge dumpsite (Table 16.5). These three elements are particularly abundant in acid-iron wastes (Chapter 14, Table 14.6) dumped at the acid-iron disposal site (Fig. 16.1).

At the acid-iron dumpsite, T. D. Meyers (personal communication) reported concentrations of titanium (Table 16.5) in the sediment which were similar to those he found at the old sewage sludge dumpsite. Iron in the sediment at the acid-iron site, however, was found similar to that observed by Oostdam et al. (1975) at the old sewage sludge dumpsite. Because of the few data reported by Meyers, too much significance should not be attached to the differences between his data and the other results given in Table 16.5.

At the interim sewage sludge dumpsite (Fig. 16.1), concentrations of the trace elements (excluding cadmium) in sediments collected in 1974 (Table 16.5) and analyzed by Oostdam et al. (1975) and by EPA (in Oostdam et al., 1975) are comparable within a factor or two. Lear and O'Malley (Chapter 15, Fig. 15.2) report concentrations of lead at stations in and south of the interim sewage sludge dumpsite grid that are within the range given by Oostdam et al. (1975) in Table 16.5.

In an attempt to study the long-term temporal variations in elemental concentrations, ten sediment cores taken in and around the old sewage sludge dumpsite in December 1974 were analyzed at 25-cm intervals. For most elements only a minor decrease in trace element concentrations was noted from the sediment surface to a depth of 50 cm; both cadmium and mercury displayed nonsystematic variations with depth in cores in the dumpsite (Fig. 16.9) which may or may not

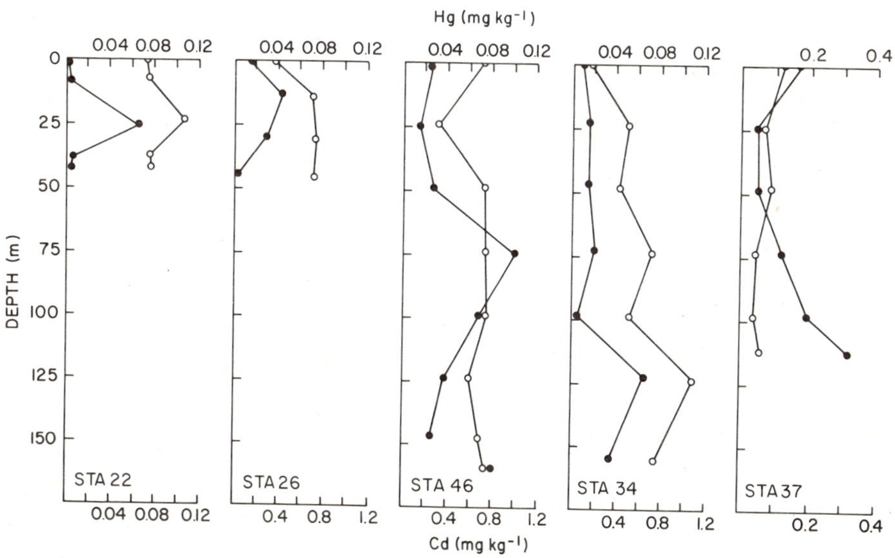

Figure 16.9. Concentrations of cadmium (•) and mercury (○) in sediment cores taken from the old sewage sludge dumpsite, December 1974. (Note different scales for mercury and cadmium.) The location of the stations is shown in Fig. 16.1.

be due to dumping activity. Inside the dumpsite all ten elements analyzed, except nickel, showed lower concentrations than those found in the surroundings although the difference was only statistically significant for mercury.

16.3.3b. Elemental Concentrations in the Surf Clam, Spisula solidissima

Trace element concentrations in surf clams (Table 16.6) taken in the period 1967 through 1970 off Delaware Bay away from the old sewage sludge dumpsite showed order of magnitude lower values by previous investigators (Buelow, 1968; Davey, 1972; Clem, 1975) compared to the values found by Oostdam et al. (1975) for surf clams collected in 1974. Because of the increasing sophistication in AAS during the 1967–1974 period, the apparent increase in trace element concentrations observed at the dumpsite cannot be unequivocally attributed to sludge dumping; at best the 1974 data (Oostdam et al., 1975) may represent a valuable base line for future comparisons.

Mean concentrations for some trace elements such as chromium and nickel in clams taken in and near the interim sewage sludge dumpsite (Lear, 1974) were appreciably less than those found in clams taken landward of the dumpsite or those from an extended study area, north and south of the interim dumpsite investigated by Reynolds (1979) (Table 16.6). Thus, it is difficult to establish clearly any cause and effect of sludge disposal on the concentrations of trace elements in the surf clams at this site.

16.4. DISCUSSION OF ENVIRONMENTAL CHANGES AT THE SEWAGE SLUDGE DUMPSITE

To evaluate the rate of environmental change of the interim sewage sludge dumpsite (Fig. 16.1), which was opened in 1973, spatial and temporal trends in trace element concentrations in sediment were compared (see Table 16.5; interim sewage sludge dumpsite; Lear et al., 1981). Results by EPA and the Marine Science Consortium (Table 16.5) for surface sediments collected during 1974 proved most useful in the evaluation of natural and anthropogenic effects because the analytical methods were uniform and intercalibrated with EPA's Annapolis Field Station (Oostdam et al., 1975). The year 1974 was about one year after cessation of dumping in the old sewage sludge dumpsite and the subsequent initiation of dumping in the interim dumpsite. Therefore, if sewage sludge dumping operations were the sole cause of variation in the temporal distribution of trace element, the concentrations in the old dumpsite may still have been decreasing toward background values, while concentrations of trace elements in the interim dumpsite would be on the increase. Consequently, all trace element concentrations for sediment (and biota) are likely to show values above background levels; unfortunately, there are no background data for these two sites, and few results for the interim dumpsite (Table 16.5). To provide a basis for comparison with another highly industrialized country, summary data on trace element concentrations in British coastal areas are compared with results reported here.

The mean cadmium concentration in 1974 for surface sediments sampled within the interim dumpsite (Table 16.5; Oostdam et al., 1975) was well below that for the old dumpsite. Furthermore, the 1974 maximum cadmium value at the interim dumpsite was less than the maximum for the old dumpsite for 1970 and 1974. The highest cadmium concentrations in over 200 sediment samples analyzed by various investigators, including Oostdam et al. (1975), was 0.3 mg kg^{-1}, compared to a range from 0.1 to 9.3 mg kg^{-1} reported for sediments in 11 British coastal areas (Vivian, in Collins et al., 1981). By contrast, mercury showed a considerable increase in maximum concentration for the six samples collected by the Marine Science Consortium within the interim dumpsite compared to the maximum for the old dumpsite (Table 16.5). Because of the relatively low concentrations for both cadmium and mercury, which EPA found in their samples (EPA, Table 16.5), EPA unfortunately discontinued monitoring these two toxic elements in sediments samples after 1974.

Comparison of means and maxima for iron and titanium in the interim sewage sludge dumpsite indicates a possible increase in concentrations of these elements, which are diagnostic of acid-iron waste dumping operations (Chapter 14). Unfortunately, only results for two samples are available for the acid-iron dumpsite (Table 16.5). The fact that zinc is diagnostic of City of Philadelphia sewage sludge (Chapter 14) may account for its increased concentration at the interim and old sewage sludge dumpsite. Nevertheless, highest concentration of zinc [9.4 mg kg^{-1} in Oostdam et al. (1975) or 15.2 mg kg^{-1} reported by Lear et al. (1981)] in over 100 samples for the study area was virtually less than the minimum in the 15-830 mg kg^{-1} range reported for zinc in sediments in British coastal areas. It is possible that nickel in the interim dumpsite may have increased during 1974; the highest nickel concentration in the study area was 3.4 mg kg^{-1} [Lear et al. (1981) report a value of 3.5 mg kg^{-1}]. Again, this value is still below the minimum in the 4-93 mg kg^{-1} range for the British area impacted by the input of sewage sludge. Finally, lead, also diagnostic of sewage sludge (Chapters 1 and 14), may have also increased within the interim sewage sludge dumpsite in 1974; however, the highest lead value in the sediment samples, 10 mg kg^{-1}, lies in the lowest part of the 5-320 mg kg^{-1} range for the British coastal areas.

In an attempt to evaluate effects of wastes on benthic organisms, mean cadmium concentrations were determined in the viscera and muscles of sea scallops collected within the dumpsite and north or south of the dumpsite (Reynolds, 1979) (Fig. 16.10). The northern samples, unlikely to be affected by waste disposal were considered as controls. Seven of eight concentration values inside the dumpsite were lower than the control; concentrations of cadmium in the tissues of organisms collected south of the dumpsite fluctuated around the control levels. The only consistent upward trends were those for the muscle of scallops samples inside the dumpsite and for the viscera in scallops from the control. There was no consistent relation between cadmium in viscera and muscles. Consequently one cannot unequivocally attribute the increase in cadmium in the muscle to sludge dumping operations, unless the control were also affected by dumping. Furthermore, inspection of Reynold's (1979) data shows no significant increase in the concentrations of trace elements in the muscle and viscera of the sea scallop or in whole (less shell)

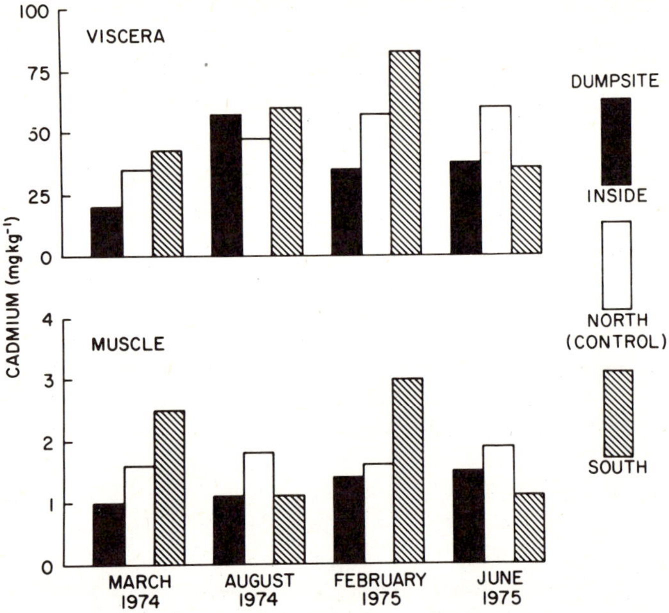

Figure 16.10. Mean concentrations of cadmium in the sea scallop, *Placopecten magellanicus* in and near the interim sewage sludge dumpsite off the coast of Delaware, Maryland (data from Reynolds, 1979). Within the dumpsite, 2 and 5 stations were sampled and 6 to 23 scallops analyzed for each station; south of the dumpsite 4 to 5 stations were sampled and 8 to 27 scallops analyzed for each station; and north of the dumpsite 3 to 5 stations were sampled and 8 to 15 scallops for each station.

quahogs. On the contrary, in many samples analyzed by EPA for trace elements there were decreasing concentrations with time, suggesting bioaccumulation was not occurring in these organisms.

16.5. CLOSING REMARKS

One of the major problems in the assessment of the impact of sewage sludge dumping is the proper evaluation of all possible natural and anthropogenic sources of variation in each parameter measured. The lack of consistent increase with time of trace element levels in sediments and biota sampled in the interim sewage sludge dumpsite, and the lack of correlation between changes in trace element levels and changes in inputs from dumping, cast considerable doubt on statements about progressive deterioration of the dumpsite.

It is my belief that EPA's closure of the old sewage sludge dumpsite and the opening of the interim dumpsite entailed substantial extra costs and was only partially warranted. Unfortunately, at least from EPA's viewpoint, the interim sewage sludge dumpsite proved to be a dispersal site rather than a containment site. Expensive changes imposed in wastewater treatment procedures resulted in such doubtful practices as reducing the cadmium content of sludge and concurrently increasing the cadmium level of effluent to the Delaware River.

The U.S. Congress decision to make ocean dumping of sewage sludge illegal in 1981 appeared unnecessary and premature. The option of ocean disposal of sewage sludge should have been kept available as a valuable alternative until it has been conclusively proved that land-based alternatives are both economically and environmentally acceptable (Oostdam, 1975). Recently, the National Advisory Committee on the Oceans and Atmosphere (1981) expressed a similar view.

ACKNOWLEDGMENTS

The cooperation of many faculty members, staff, and students of the Marine Science Consortium, and the support of the U.S. National Aeronautics and Space Administration, the Pennsylvania Science and Engineering Foundation, the City of Philadelphia, Millersville State College, and the U.S. Environmental Protection Agency are gratefully acknowledged. Dr. F. K. Szucs performed all trace element analyses, and Robert Swift supervised most of the research cruises. William "Smokey" Roberts and Lee Robinson provided capable diving support. Special recognition is given to Dr. Michael A. Champ who, by his example and enthusiasm as chief scientist of several major marine pollution research cruises, guided and inspired dozens of students.

REFERENCES

Buelow, R. W. 1968. Ocean disposal of waste material. *In*: Ocean Sciences and Engineering of the Atlantic Shelf, Transactions, National Symposium Marine Technology Society, Philadelphia, Pennsylvania, pp. 311–337.

Bumpus, D. F. 1965. Residual drift along the bottom on the continental shelf in the Middle Atlantic Bight area. *Limnology and Oceanography*, **10** (Suppl.), R50-53.

Bumpus, D. F., and L. M. Lauzier. 1976. Surface circulation on the continental shelf off eastern North America between Newfoundland and Florida. Folio 7, Serial Atlas of the Marine Environment. American Geographical Society, New York, 4 pp.

Callaway, R. J., A. M. Teeter, D. W. Browne, and G. R. Ditsworth. 1976. Preliminary analysis of the dispersion of sewage sludge discharged from vessels in New York Bight waters. *In*: Middle Atlantic Continental Shelf and the New York Bight, Special Symposia, Vol. 2, M. G. Gross (Ed.). American Society of Limnology and Oceanography, Allen Press, Lawrence, Kansas, pp. 199–211.

Charnell, R. L. 1975. Assessment of Offshore Dumping, Technical Background: Physical Oceanography, Geological Oceanography, Chemical Oceanography. NOAA Technical Memorandum ERL MESA-1, National Oceanic and Atmospheric Administration, Boulder, Colorado, 83 pp.

Clem, J. D. 1975. Letter dated 14 October 1975 from Chief, Shellfish Sanitation Branch, Food and Drug Administration, Department of Health, Education and Welfare, Washington, D.C.

Collins, M. B., F. T. Banner, P. A. Tyler, S. J. Wakefield, and A. E. James (Eds.). 1981. Industrialized Embayments and their Environmental Problems. Pergamon Press, London, 616 pp.

Davey, C. T. 1972. Assessment of the Effects of Digested Sewage Sludge Disposal off the Mouth of Delaware Bay. Technical Report F-C2970, Franklin Institute Research Laboratories, Philadelphia, Pennsylvania, 129 pp.

Duedall, I. W., H. B. O'Connors, and B. Irwin. 1975. Fate of wastewater sludge in the New York Bight apex. *Journal Water Pollution Control Federation*, **47**, 2702–2706.

Duedall, I. W., H. B. O'Connors, S. A. Oakley, and H. Stanford. 1977. Short-term water column perturbations caused by wastewater sludge dumping in the New York Bight apex. *Journal Water Pollution Control Federation*, **49**, 2074–2080.

Duedall, I. W., S. A. Oakley, J. H. Parker, R. Dayal, H. B. O'Connors, K. W. Jones, H. W. Kraner, and J. Foehrenbach. 1980. Second Sewage Sludge Tracking Acoustic Experiment (STAX II): SUNY Results. Unpublished Report, Marine Sciences Research Center, State University of New York, Stony Brook. Submitted to Marine EcoSystems Analysis (MESA), National Oceanic and Atmospheric Administration, Stony Brook, New York, 59 pp.

Folger, D. W., H. D. Palmer, and R. A. Slater. 1979. Two waste disposal sites on the continental shelf off the Middle Atlantic states: Observations made from submersibles. *In*: Ocean Dumping and Marine Pollution, H. D. Palmer, M. G. Gross (Eds.). Dowden, Hutchinson and Ross, Inc., Stroudsburg, Pennsylvania, pp. 163–184.

Lear, D. W. 1974. Environmental Survey of Two Interim Dumpsites. Middle Atlantic Bight–Supplemental Report. EPA-903/9-74-010a, U.S. Environmental Protection Agency, Region III, Philadelphia, Pennsylvania, 53 pp. + appendices.

Lear, D. W., and G. B. Pesch. 1975. Effects of Ocean Disposal Activities on Mid-Continental Shelf Environment off Delaware and Maryland. EPA-903/9-75-015, U.S. Environmental Protection Agency, Region III, Philadelphia, Pennsylvania, 175 pp.

Lear, D. W., M. L. O'Malley, and S. K. Smith. 1981. Effects of Ocean Dumping on a Temperate Midshelf Environment. *In*: Ocean Dumping of Industrial Wastes, B. H. Ketchum, D. R. Kester, and P. K. Park (Eds.). Plenum Press, New York, pp. 485–502.

National Advisory Committee on Oceans and Atmosphere (NACOA). 1981. The Role of the Ocean in a Waste Management Strategy. National Advisory Committee on Oceans and Atmosphere, U.S. Governmental Printing Office, Washington, D.C., 103 pp. + 6 appendices.

Oostdam, B. L. 1975. Statement to the Committee on Science and Technology of the U.S. House of Representatives concerning environmental effects of the dumping of wastes into the ocean waters and the Great Lakes. Washington, D.C., 12 October 1975.

Oostdam, B. L., F. K. Szucs, and R. N. Swift. 1975. Heavy Metals in Sediments and Surf Clams (*Spisula solidissima*) in the Former Philadelphia Treated Sludge

Dump Area Off Delaware Bay. Unpublished Report (21 March 1975), The Marine Science Consortium, Lewes, Delaware, Erie and Millersville, Pennsylvania, and Wallops Island, Virginia. Submitted to the City of Philadelphia Water Department, Pennsylvania, 181 pp.

Palmer, H. D., and D. W. Lear. 1974. Low Recovery Rates from Seabed Drifter Study, Chesapeake Bight, Offshore Maryland, Unpublished Report, Submitted to Westinghouse Ocean Research Laboratory, Annapolis, Maryland, 10 pp.

Proni, J. R., and D. V. Hansen. 1981. Dispersion of particulates in the ocean studied acoustically: The importance of gradient surfaces in the ocean. *In*: Ocean Dumping of Industrial Wastes, B. H. Ketchum, D. R. Kester, and P. K. Park (Eds.). Plenum Press, New York, pp. 161–173.

Raytheon Corporation. 1975. Preliminary Oceanographic Survey, Cruise 1, City of Philadelphia Water Department Ocean Dispersal Site. Unpublished Report, Raytheon Corporation. Submitted to Water Department, City of Philadelphia, Pennsylvania. 45 pp.

Reynolds, B. H. 1979. Trace Metals Monitoring at Two Ocean Disposal Sites. EPA-600/3-79-037, U.S. Environmental Protection Agency, Region III, Philadelphia, Pennsylvania, 64 pp.

17

MARINE AMOEBAE (PROTOZOA: SARCODINA) AS INDICATORS OF HEALTHY OR IMPACTED SEDIMENTS IN THE NEW YORK BIGHT APEX

Thomas K. Sawyer and Shawn M. Bodammer

U.S. Department of Commerce
National Marine Fisheries Service
Northeast Fisheries Center
Oxford, Maryland

ABSTRACT

Thirty-two species of marine amoebae were identified from the northwest Atlantic Ocean in the New York Bight apex near active sewage, dredge, and acid-waste dumpsites. Twenty-seven of the 32 species were present in surface waters, 25 in bottom waters, and 20 in sediment core samples. Bottom cores from the sewage site yielded 11 of the 32 species, acid-waste cores 11, dredge spoil cores 9, and control station cores 13. Only 3 of the 32 species were present in cores from all of the collection sites: *Paramoeba pemaquidensis*, *Clydonella vivax*, and *Platyamoeba langae*. The recovery of the three species from all collection sites indicated that they might be useful indicators for monitoring ocean sediments for effects more serious than those brought about by existing disposal practices. The presence of 11 amoeba species from the New York sewage dumpsite, already depleted of most macrofaunal species, indicated that the polluted sediments still support microfaunal species which contribute to organic decay and nutrient regeneration. The 20 species present in sediments are bactivorous and thrive at the lowest level of the food web. Environmental modifications, which might disrupt protozoan/bacterial interactions in surficial sediments, can be measured qualitatively by monitoring the seabottom for changes in protozoan species diversity. Preliminary studies on species diversity showed that, when seabottom sediments were cultured on distilled-water agar to test for physiologically adaptable species, well-known freshwater or soil species of *Acanthamoeba* were common at stations harboring sewage-associated bacteria. Sewage sites yielded species of *Acanthamoeba* from 20 to 80% of the sediment samples, and sites that were not affected by sewage sludge yielded the amoebae from 0 to 5% of the samples.

17.1. INTRODUCTION

One of the most important roles played by protozoans in fresh and seawater environments is to consume bacteria at a rate that sustains their respective population densities. Environmental changes that depress protozoan growth are known to support unchecked bacterial growth and may sometimes cause aerobic environments to become anaerobic with a negative Eh. In contrast, changes which depress bacterial growth may deprive protozoans of an adequate food supply and effectively eliminate them from food webs in which they are important decomposers. Laboratory studies (Mitchell and Yankofsky, 1969) have shown that at least one marine amoeba, *Vexillifera telmathalassa* Bovee, consumes fecal bacteria, *Escherichia coli*, and increases in numbers with a corresponding decrease in bacterial numbers. Protozoan populations thus may be manipulated experimentally by varying the added food supply, and naturally by differences in the availability of food organisms in the sea. The small size of marine protozoa and the bactivorous feeding habits of many of them provides a means by which their abundance and species diversity can be used to monitor changes in the health of ocean sediments.

Efforts to select marine plants or animals as biological monitors of environmental change are not always successful for a variety of reasons, including (1) a lack of historical data on species abundance and diversity, (2) very little knowledge of the factors responsible for natural fluctuations in abundance and diversity, and (3) insufficient data on the rate at which different species respond to environmental changes. Seasonal changes alone are known to influence the numbers of animals available for study at different times of the year, the types of available food organisms, and the life cycle stages that are present. Water temperature, dissolved oxygen, and phytoplankton and zooplankton blooms are also responsible for temporal environmental modifications. When the influences of naturally occurring events are known on a spatial and temporal basis, they may be taken into account when efforts are made to interpret the effects of unnatural events such as oil spills, sewage disposal, and dredge disposal. Preliminary studies (Davis et al., 1978; Sawyer, 1980) have shown that many amoeba species are distributed worldwide and may be collected at any level in the water column from the sea surface microlayer to deep-sea sediments. Since the amoebae do not have complex life cycles and many of them eat only one type of food (bacteria, algae, or other protozoans), there are fewer variables to be taken into account when attempting to measure their response to environmental change than with other marine organisms.

During the past decade, there has been a renewed interest in the distribution of marine protozoans in the sea and in the seabottom. Davis et al. (1978) surveyed oceanic amoebae in surface waters of the Northwestern Atlantic Ocean and the Florida Straits. Sawyer (1980) published an extensive account of the distribution of marine amoebae in clean and stressed bottom sediments of the Northwest Atlantic Ocean and the Gulf of Mexico. Fenchel (1969) published extensively on changes in ciliate species composition and diversity in marine sediments with different redox potentials. In the New York Bight apex Soukup (1976) noted profound influences by bacteria associated with sewage sludge particles, and seasonal changes in the thermocline on ciliate abundance and species composition.

Approximately one-third of the known species of marine amoebae have been described during the last 10 years, and a key to the identity of most of them has been published by Bovee and Sawyer (1979).

The design and development of deep-sea submersibles and photographic equipment have led to the descriptions of many species of protozoans that inhabit the ocean floor. New protozoans from the ocean depths probably will be discovered with each advance in the design of collecting devices. At present improved techniques are needed for making quantitative measurements on protozoan abundance regardless of their source in the sea.

We have chosen marine amoebae as valuable tools for the development of a marine monitoring program in waters of the Northwest Atlantic Occan. Published accounts of species diversity, identifying characteristics, and research methodology that have appeared during the past 10 years provide a foundation from which specific answers to important questions may now be explored. Potentially pathogenic species of *Acanthamoeba* Volkonsky, 1931, have been discovered in ocean sediments which harbor sewage-associated bacteria (Sawyer et al., 1977; Sawyer, 1979, 1980) and already have been used in carefully designed monitoring studies in clean and sewage-impacted environments. Studies presented here provide a brief overview of the current status of our efforts to use several important groups of marine amoebae as environmental monitors.

17.2. METHODS

Methods used for the isolation, cultivation, and identification of marine amoebae were developed according to the type of study in progress. During 1973 and 1975, collections were made in the New York Bight apex (Fig. 17.1) as part of the Marine EcoSystems Analysis (MESA) program conducted by the National Oceanic and Atmospheric Administration (NOAA). The methods of isolation, cultivation, and identification were those described by Bovee and Sawyer (1979) and Sawyer (1980). All of the principal genera of amoebae that appeared in culture were identified using the descriptions and illustrations provided in a recent key to the marine amoebae (Bovee and Sawyer, 1979). Two collections of sediments made in the Gulf of Mexico (November 1977 and March 1978) were studied by procedures cited above (Sawyer, 1980). The first extensive surveys for pathogenic and non-pathogenic species of *Acanthamoeba* were conducted in the New York Bight and Philadelphia-Camden sewage sludge dumpsites during 1978–1979 in cooperation with scientists from the Food and Drug Administration, Davisville, Rhode Island, who provided data on most probable numbers (MPNs) for sewage-associated bacteria present in the same sediments that were used for amoeba isolations. Procedures for the selective cultivation of euryhaline species of *Acanthamoeba* employed distilled-water or brackish-water agar media (Sawyer et al., 1977; Sawyer, 1980). Observations from the sewage sites were then tested for reproducibility in 1980 by sampling stations in Narragansett Bay, Rhode Island, located from near the city of Providence to 43 km (25 nautical miles) to sea. The final results from ocean surveys are in preparation for publication, but brief summaries are presented in the Results section here. Parts of the data from work in Baltimore Harbor, Maryland, and from a previously published report also are included (Sawyer et al., 1977).

References to the methods employed in this study appear in a number of different publications; therefore, a brief review is presented below.

17.2.1. Sample Collection

Sediment and water samples were collected by standard methods using Smith-McIntyre or Shipek bottom grabs, and Niskin (General Oceanics, Inc., Miami, Florida) or van Dorn water samplers. (Reference to trade names does not imply endorsement by the National Marine Fisheries Service.) Sediments were subsampled with plastic cores, capped, and refrigerated until returned to the laboratory, or subsampled with sterile wooden applicators, transferred to sterile plastic dishes, and refrigerated. Each sediment was subsampled once during 1973–1977. Culture results indicated, however, that replicate subsamples yielded more frequent positive results and the number of replicates was increased to three beginning with

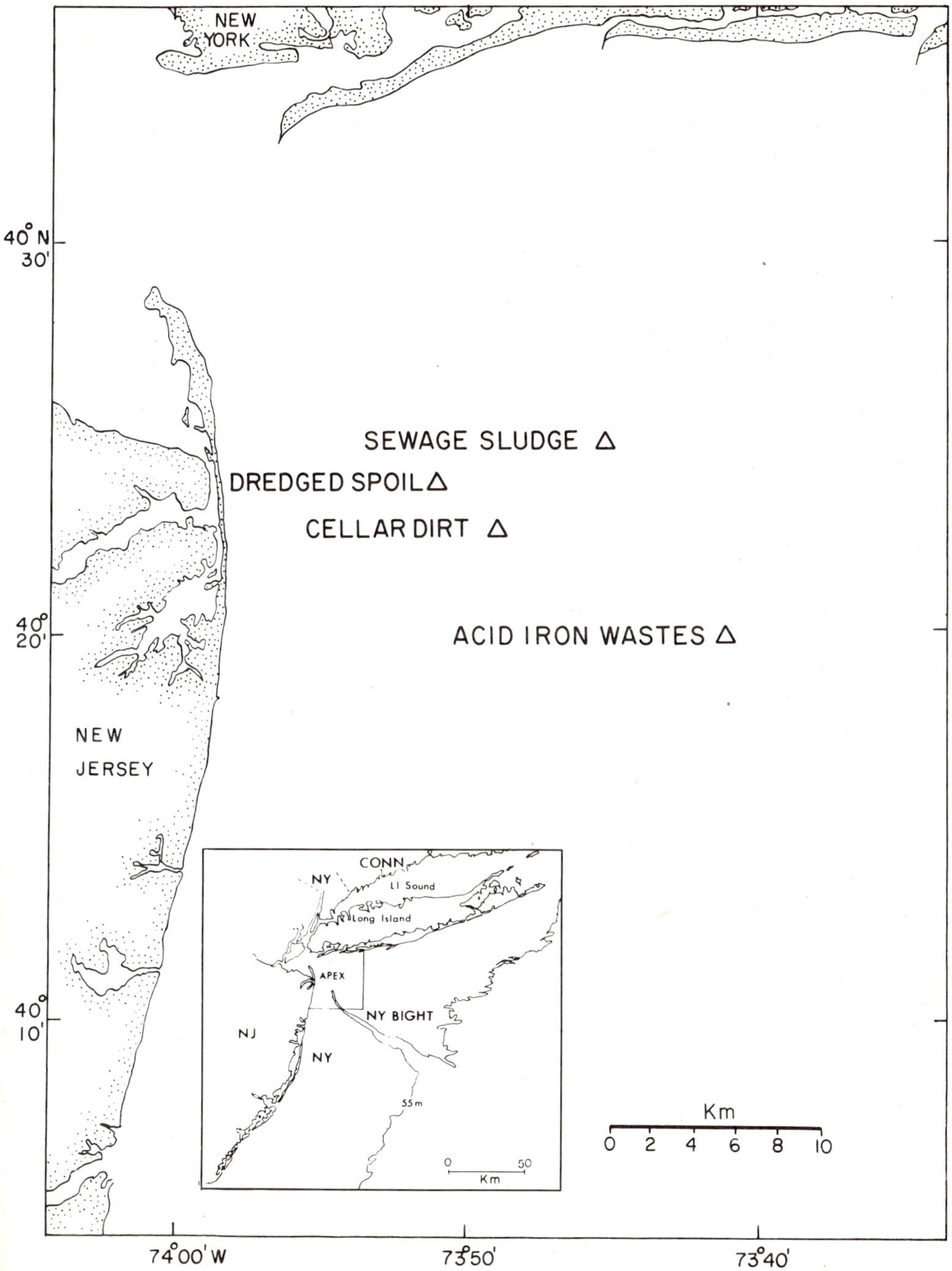

Figure 17.1. Location of ocean dumpsites sampled for marine amoebae. Sewage disposal, 25–30 m deep; dredged spoil disposal, 22–30 m deep; cellar dirt, ~25 m deep; acid-iron waste disposal, 26–28 m deep.

collections made after 1977. Surface waters were sampled with water sampling devices but bottom water was collected as it drained from bottom samplers immediately after retrieval (grab water). Control sites were considered to be those which were not in proximity to the dumpsites and which were consistently negative for sewage-associated bacteria. Amoebae were cultured from water and sediment by placing samples on agar media or in plastic dishes containing seawater.

17.2.2. Agar Media

Low nutrient agar was prepared by adding 10 g Difco agar, 0.1 g malt extract, and 0.1 g yeast extract to 1 liter of filtered seawater. The mixture was then heated in hot water until the agar was completely dissolved, poured in 100-ml amounts into glass bottles, and sterilized by autoclaving. Bottles were stored in a refrigerator for use as needed. Plastic dishes, 60 mm in diameter, were inoculated with 6 ml of agar, which was allowed to harden, and were placed in a refrigerator for no more than 2 days prior to use. Each plate was streaked with bacteria, *Klebsiella aerogenes*, immediately before use.

17.2.3. Seawater Media

Several hundred milliliters of seawater from each sample (salinity 25-32‰) were filtered through filters having a pore size of 1.2 μm (Millipore® RA Type, Millipore Corp., Bedford, Massachusetts) and stored in a refrigerator until needed. When cultures were prepared immediately upon returning to the laboratory, the seawater was not filtered. Liquid cultures were prepared by placing 6 ml of filtered or unfiltered seawater in sterile plastic dishes (60-mm diameter) and adding concentrated samples (see below), and were examined with a Leitz inverted microscope.

17.2.4. Culture Preparation and Examination

Sediments were cultured by collecting small amounts (0.1-0.8 g) of the sample on the end of a microbiological loop and streaking it through the center of bacterized agar plates. Cultures were observed frequently and amoebae removed and placed on new plates as they appeared. Amoebae that migrated from the sediment streak and appeared as small colonies often could be removed with a small piece of agar and established as pure cultures after only one transfer. Agar cultures were inverted and stored in plastic boxes containing moist paper towels to provide a humid environment. Seawater samples were cultured by filtering 400-500 ml water samples through 1.2-μm filters (Millipore® RA), leaving a thin layer of water on the filter to minimize pressure damage to concentrated organisms. Water samples were processed by vacuum filtration at no more than 110-200 mm Hg (2-4 lb in.$^{-2}$). Filters were removed with fine-tipped forceps, inverted, and placed on the surface of water from the sample contained in a plastic dish. Filters were removed after overnight storage in the dish. Each culture was examined with a phase-contrast water immersion objective and various organisms photographed to provide a record of delicate forms which would not grow on semimoist agar. Several days after preparation, seawater was removed from the dish with a sterile pipette, leaving about 1 ml behind, and was placed in a sterile plastic tube for subsequent return to the culture. The bottom of the dish was flushed to suspend the organisms and one-drop amounts were transferred to bacterized plates. The aspirated water was then returned to the original dish to maintain the culture for further study. Culture dishes were tilted gently to allow the drop to form a streak through the center of the plate. Culture dishes were then inverted and stored in humid containers as described above. Frequent observations were made and amoebae subcultured as described for sediment cultures.

Living amoebae were observed, measured, and photographed in hanging drops, depression slides, and in dishes of seawater where they were not subjected to coverslip pressure. Species that required further study were stained by the methods described by Page (1967), or examined for their type of rayed or floating stage (Sawyer, 1975c; Bovee and Sawyer, 1979). Lists of the identified amoebae were prepared to compare the distribution of species in the water column and in the sediments of each collection site. Seawater cultures often yielded species that were nearly identical in appearance and difficult to distinguish from one another. One important test for differentiating such species was to place them in media prepared with distilled water, brackish water, half-strength seawater, and full-strength seawater to determine their tolerance to diluted seawater. Distilled-water agar media often yielded cyst-forming amoebae belonging to the genus *Acanthamoeba* almost to the exclusion of other species. Beginning in 1977, all water and sediment samples were cultured on both distilled-water and seawater media, and sediments from sewage-impacted sites often yielded acanthamoebas, amoebas that rarely appeared on seawater cultures. Cooperative studies were initiated with the U.S. Environmental Protection Agency and the U.S. Food and Drug Administration to sample each sediment for sewage bacteria and the amoebae. Surveys were conducted in the New York Bight apex and Philadelphia–Camden sewage dumpsites, in Narragansett Bay, Rhode Island, and in their adjacent waters, which consistently were negative for coliform bacteria. Qualitative tests were initiated in 1979 to provide new information on the frequency with which acanthamoebas could be isolated from bottom sediments. The tests were conducted by preparing six replicate cultures from each sediment sample and comparing the number of positive dishes with bacterial MPNs per 100 g sediment. Cultures of *Acanthamoeba* species were tested further by incubating them at 40°C to test for heat tolerance, and several strains which grew satisfactorily were then inoculated into mice to test for pathogenicity (Sawyer et al., 1977).

17.3. OBSERVATIONS AND RESULTS

17.3.1. Non-Cyst-Forming Marine Species

Cultures prepared from surface water, bottom water, and sediment from the New York sites (Fig. 17.1) yielded 32 species of amoebae (Table 17.1): 27 from the

Table 17.1. Distribution of Marine Amoebae in Water Column and Sediment in Waters of the New York Bight Apex[a]

Species	Surface Water	Bottom Water	Sediment	Frequency
Paramoeba pemaquidensis	+	+	+	3/3
Paramoeba aesturina	+	+	+	3/3
Mayorella corlissi	+			1/3
Mayorella conipes	+	+		2/3
Mayorella smalli	+			1/3
Dactylosphaerium sp.	+			1/3
Vexillifera telmathalassa	+			1/3
Vexillifera ottoi	+	+	+	3/3
Vexillifera browni	+	+	+	3/3
Vexillifera minutissima	+	+	+	3/3
Triaenamoeba jachowskii		+	+	2/3
Stygamoeba polymorpha		+	+	2/3
Rhizamoeba polyura		+	+	2/3
Vannella sensilis	+	+	+	3/3
Platyamoeba mainensis	+	+		2/3
Platyamoeba murchelanoi	+	+	+	3/3
Platyamoeba langae	+	+	+	3/3
Platyamoeba plurinucleolus		+		1/3
Platyamoeba douvresi	+	+	+	3/3
Platyamoeba weinsteini		+		1/3
Clydonella vivax	+	+	+	3/3
Clydonella sindermanni	+	+	+	3/3
Clydonella rosenfieldi	+	+	+	3/3
Clydonella wardi	+	+	+	3/3
Lingulamoeba leei	+	+	+	3/3
Hyalodiscus angelovici	+	+	+	3/3
Gibbodiscus newmani	+			1/3
Unda schaefferi	+	+		2/3
Flabellula sp.[b]	+	+	+	3/3
Cochliopodium sp.[b]	+	+	+	3/3
Pelomyxa sp.[b]	+			1/3
Saccamoeba sp.[b]	+			1/3
Totals	27/32	25/32	20/32	

[a]See Fig. 17.1 for station location and depth.
[b]Amoebae not identified to species; it is probably a new species.

surface, 25 from the bottom, and 20 from the sediments. Nine to 13 species were cultured from sediment cores of the different sites (Table 17.2). Data were analyzed to determine which species were distributed throughout the water column and which were common inhabitants of sewage, dredge, acid, and control stations. Three species were present in bottom sediments from all collection sites: *Paramoeba pemaquidensis* Page, 1970, *Clydonella vivax* (Schaeffer, 1926) Sawyer, 1975, and *Platyamoeba langae* Sawyer, 1975 (Table 17.2). Seven species, *Vexillifera tel-*

Table 17.2. Distribution of Marine Amoebae in Sediment Cores from the New York Bight Apex[a]

Species	Type of station: Control	Sewage	Dredge	Acid	Frequency
Paramoeba pemaquidensis	+	+	+	+	4/4
Paramoeba aesturina	+		+		2/4
Vexillifera ottoi	+	+	+		3/4
Vexillifera browni	+				1/4
Vexillifera minutissima		+			1/4
Triaenamoeba jachowskii	+	+		+	3/4
Stygamoeba polymorpha			+	+	2/4
Rhizamoeba polyura				+	1/4
Vannella sensilis		+	+		2/4
Platyamoeba murchelanoi	+		+	+	3/4
Platyamoeba langae	+	+	+	+	4/4
Platyamoeba douvresi	+	+			2/4
Clydonella vivax	+	+	+	+	4/4
Clydonella sindermanni				+	1/4
Clydonella rosenfieldi	+				1/4
Clydonella wardi				+	1/4
Lingulamoeba leei	+	+			2/4
Hyalodiscus angelovici	+		+		2/4
Flabellula sp.		+		+	2/4
Cochliopodium sp.	+	+		+	3/4
Totals	13/20	11/20	9/20	11/20	

[a]See Fig. 17.1 for station location.

mathalassa Bovee, 1956, *Mayorella corlissi* Sawyer, 1975, *M. smalli* Sawyer, 1975, *Pelomyxa* sp., *Saccamoeba* sp., *Gibbodiscus newmani* Sawyer, 1975, and *Dactylosphaerium* sp., were present in dishes of seawater but not on agar plates. Two species of *Mayorella* found only in the water column survived on agar only when subcultured at frequent intervals. They always had a very vacuolated appearance and failed to reproduce in large numbers. Species that were the first to appear on agar media often consumed bacterial food and produced large populations that may have inhibited or depressed the growth of other species. Thus, amoebae that were not isolated from all stations, or from throughout the water column, may have been present but not recovered with the methods employed. Among the 32 different species of the amoebae that were identified to genus or species, 24 (75%) were discovered during the past 10 years. Thirteen of the species recovered from the nearshore waters of the New York Bight apex were isolated from the surface waters of the open ocean by Davis et al. (1978). The characteristic shapes of six different genera (*Paramoeba, Hyalodiscus, Gibbodiscus, Flabellula, Vexillifera, Mayorella*) are shown in Fig. 17.2. All the genera that were identified from cultures have been described and illustrated in a key to the marine amoebae (Bovee and Sawyer, 1979).

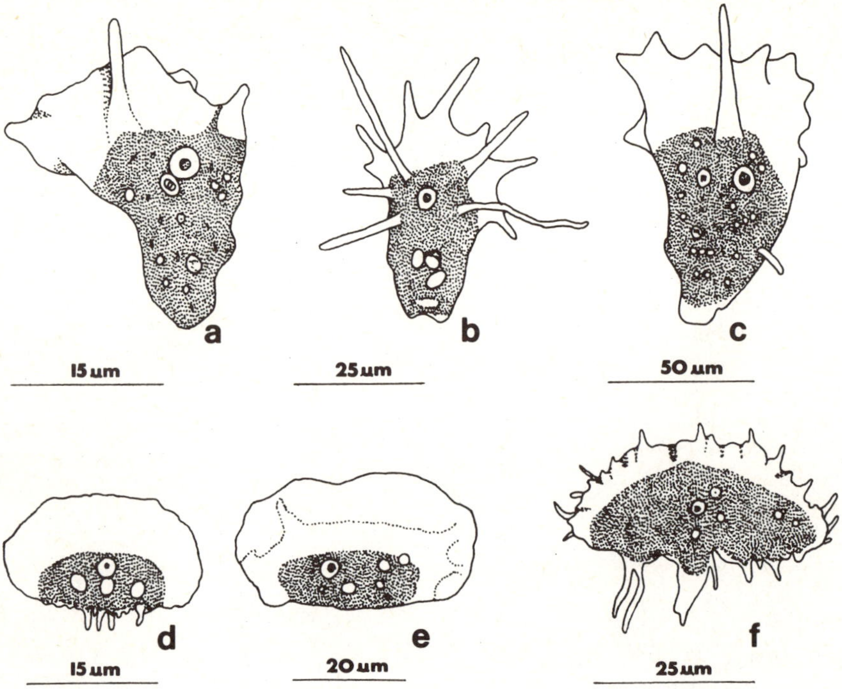

Figure 17.2. Species of marine amoebae that represent six of the principal genera commonly found in the coastal waters of the northeast United States (see Bovee and Sawyer, 1979). (a) *Paramoeba pemaquidensis* Page, 1970 (note oval secondary body adjacent to nucleus). (b) *Vexillifera telmathalassa* Bovee, 1956 (note wavy pseudopod to the right which bends at the base and is resorbed). (c) *Mayorella corlissi* Sawyer, 1975 (note broad base of conical pseudopod and uniformly scalloped anterior margin). (d) *Gibbodiscus newmani* Sawyer, 1975 (note broad oval shape without pointed pseudopods, and short filaments along center of posterior margin). (e) *Hyalodiscus angelovici* Sawyer, 1975 (note oval shape with rippled surface protoplasm shown by stippling). (f) *Flabellula hoguae* Sawyer, 1975 (note short spike-like pseudopods along anterior margin and trailing uroidal filaments).

17.3.2. Cyst-Forming Euryhaline Species

The only amoebae that formed distinctive cysts and grew equally well on distilled-water, brackish-water, and seawater agar were species within the genus *Acanthamoeba*. Four or more of the species were found at all collection sites where sewage-associated bacteria were present, and two of them were found at control sites (Table 17.3). Experiments that employed six replicate cultures per sediment sample showed a direct relationship between the number of positive cultures and the number of fecal bacteria. When fecal coliform MPNs were less than 50 bacteria per 100 g of sediment, zero to two cultures yielded the amoebae, and when MPNs were over 50 bacteria per 100 g of sediment, three to six cultures usually were positive for amoebae. The amoebae were recovered from 20% of the Philadelphia-Camden stations, 60–80% of the New York Bight stations, and at all the Narragansett Bay stations. A report on detailed accounts of studies at each

Table 17.3. Species of *Acanthamoeba* Isolated from Sediments at Sewage Sludge and Control Sites

Species	Locations with Sewage Bacteria				Locations without Sewage Bacteria			
	New York Sewage Site[a]	Philadelphia Sewage Site[b]	Narragansett Bay[c]	Baltimore Harbor[d]	Southeast of New York Sewage Site	East/Northeast of Philadelphia Site	Mouth of Narragansett Bay	Gulf of Mexico
A. castellanii	+	+	–	+	–	–	–	–
A. polyphaga	+	+	+	+	–	–	–	+
A. rhysodes	+	+	+	+	–	–	–	–
A. triangularis	+	+	+	–	–	–	–	–
A. hatchetti	+	+	+	+	–	–	+	–
A. culbertsoni	+	+	–	–	–	–	–	–

[a]60–80% of stations positive for *Acanthamoeba*, high bacterial counts.
[b]20–25% of stations positive for *Acanthamoeba*, low bacterial counts.
[c]100% of bay stations positive for *Acanthamoeba*, low and high bacterial counts.
[d]50% of stations positive for *Acanthamoeba*, counts not done; stations known to have sewage-associated bacteria.

location is in preparation. Sediments collected from control stations ranging from Rhode Island to the Gulf of Mexico rarely yielded cultures of *Acanthamoeba* (Table 17.3). All species of *Acanthamoeba* grew at room temperature on agar media and two of them grew rapidly at 40°C. The two temperature-tolerant species, *A. culbertsoni* and *A. hatchetti*, killed laboratory mice after internasal inoculation (Sawyer et al., 1977). Both strains have been deposited with the American Type Culture Collection, Rockville, Maryland, for use by interested investigators (ATCC nos. 30732 and 30730). Other strains of the two pathogens were identified on the basis of cyst morphology but were not tested in mice because of time and cost limitations.

17.4. DISCUSSION

Paramoeba Schaudinn, 1896, was the first genus to be proposed for which a marine species (*P. eilhardi*) was designated as the type species. Prior to 1896, there were only ten known species of marine amoebae, and all of them were assigned to genera that had been established previously for freshwater or soil species. Today, there are about 23 genera of marine amoebae, and they accommodate about 100 carefully described species. The relative ease with which most of them may be identified is advantageous to studies on species diversity and distribution. Among 20 species cultured from shallow water in Chincoteague Bay, Virginia (Sawyer, 1975a,b,c), 14 were readily identified in coastal waters or ponds from Virginia to Massachusetts (Bovee, 1979), and 7 were found in surface waters of the open Atlantic Ocean (Davis et al., 1978).

Lighthart (1969), Sieburth (1976), and Davis et al. (1978) have published accounts of the numbers of bacteria present in association with marine protozoa. Their observations suggest that the presence and abundance of protozoa in seawater or sediment samples may depend more on the numbers of available microbial food organisms than on the size of the sample or subsample that is examined. In this respect, there is convincing evidence that planktonic amoebae may serve as useful indicators of sporadic bacterial blooms. Soukup (1976) found that the number of bactivorous ciliates in the water at a given time increased significantly during periods when barge-delivered sewage sludge was being discharged in the New York Bight. Berk et al. (1976) showed that the growth of two ciliates, *Uronema* and *Potomacus*, was inhibited when they were fed less than 10^6 to 10^7 bacterial cells ml^{-1}. Sawyer (1980) found that in the Atlantic Ocean and in the Gulf of Mexico seasonal blooms of minute heliozoans which occurred during sampling periods yielded agar cultures in which the organisms covered the surface and exhausted food sources before other protozoans could become established. Soukup (1976) similarly observed that sporadic blooms of a dinoflagellate, *Gymnodinium*, retarded the growth of marine bactivorous ciliates. The importance of the "bacterial-protozoan partnership" proposed by Sieburth (1976) is emphasized by the fact that it is essential to know when bacterial or protozoan "blooms" occur naturally and when they occur as a result of ocean dumping or other pollutant additions. The value of marine amoebae as biological monitors is greatly enhanced when samples collected for study are used in companion microbiological studies, especially when sewage-associated bacteria are present.

The recovery of six species of *Acanthamoeba* from ocean sewage sludge dumpsites and inland waters impacted by sewage outfalls was of particular interest since all of them originally were reported from nonmarine sources. Only one species in the family Acanthamoebidae Sawyer and Griffin, 1975, was isolated originally from seawater. The amoeba *A. griffini* Sawyer, 1971, was isolated from a shallow-water bottom sampler near New London, Connecticut, and has not been reisolated since its discovery in 1962. New studies to determine if certain amoeba species have preferential food requirements and others do not might provide useful information on which of them are useful for ocean monitoring. Two species that are heat tolerant, *A. culbertsoni* and *A. hatchetti*, are known to kill laboratory mice, and are now undergoing biochemical (Daggett et al., 1980) and immunological (Visvesvara and Sawyer, 1980) tests to determine if they are identical at the molecular level to the nonmarine-type species. Studies in progress in our laboratory clearly demonstrate that the frequency with which *Acanthamoeba* may be recovered from sewage-impacted environments is related statistically to the density of bacterial populations.

Deep-sea protozoans that might be useful as indicators of ocean health are yet to be discovered. Several major groups of protozoans have received attention in recent years largely because of the development of improved collecting devices and submersibles. Problems that remain, however, include breakage due to the fragile nature of certain species, and excessive shrinkage of preserved specimens (Burnett, 1973). Burnett (1973) illustrated several proteomyxan-like organisms, flagellates, ciliates, and testaceans on stained slides collected from the San Diego Trench; amoebae and flagellates were among the most commonly occurring protozoans. Tendal and Lewis (1978) found delicate, large foraminiferans (xenophyophores), some up to 25 cm diameter, in the subbathyal and abyssal depths around New Zealand. Sawyer (1980) reviewed some of the literature on the occurrence of slime mold-like organisms reported from the seabottom in the mid-1800s. Thus, many free-living protozoans, probably primitive, have been catalogued from considerable depths in the sea, and they can be used as part of the historical record from which temporal changes can be measured. New efforts to conduct surveys of species diversity and density in selected sites now are practical. Such efforts would provide an important data base from which changes, or the lack of changes, could be measured as the open ocean becomes utilized for disposal practices.

Ruddell (1979) compiled a very useful list of benthic invertebrates and the symbionts and epibionts that infested their body surfaces and cavities. Further studies of these often specialized, dual associations might serve as highly quantitative and qualitative indications of changes in water quality or suspended particulates. Wolff (1979) reviewed the literature on the deposition of wood, leaves, fruits, and seagrass in many areas of the deep sea and listed a rich fauna that was associated with plant remains. He suggested that in the deep sea, as in shallow waters, bacteria and fungi associated with organic decomposition are available as food for predators and as dead remains for scavengers. Fecal pellets produced by scavengers may also be utilized by deposit feeders and bactivorous protozoans. Deep-sea counterparts to shallow and surface-water food webs are known to exist and should be worthy of investigation. Scavenger amoebae have been isolated and identified from the surface film of seabottom sediments, from fouled gills of fish and shellfish, and from the body cavities of marine invertebrates–all sources of abundant bacterial food. The "bacterial-protozoan" relationship appears to be worthy of consideration in future efforts to select tools for ocean monitoring.

It is suggested that the potential impacts of ocean mining, explorations for sources of oil, and ocean-dumping activities probably will be difficult to monitor in the deep ocean for an unknown number of years. During the interim, however, a unique opportunity is presented to obtain a historical data base. Without such a base, it is not likely that long-term effects will be recognized until modifications of the seabottom are severe. Recent ocean research indicates that single-celled animals, especially amoebae and flagellates at or near the bottom of the food web, are ideally suited to be ocean monitors.

17.5. CONCLUSIONS

Approximately one-third of the well-described marine amoebae known from the east coast of the United States have been found in the New York Bight apex; 27 from surface water, 25 from bottom water, and 20 from sediment cores. Sediment from the New York sewage sludge dumpsite yielded a minimum of 11 amoeba species and they may be considered tolerant of the chemical and biological components of sewage. About one-half of the species occur throughout the water column and in the sediment, and all of them utilize bacteria as a food source. Potentially pathogenic species of *Acanthamoeba* grow on agar media prepared with distilled water and may be recovered in up to 80–100% of the sediment samples collected at the dumpsite. The frequency with which *Acanthamoeba* species are isolated correlates statistically with the abundance of sewage-associated bacteria.

ACKNOWLEDGMENTS

The authors gratefully acknowledge the MESA New York Bight Project, Stony Brook, New York, and the NOAA Office of Marine Pollution Assessment, Rockville, Maryland, for the interest and support that made our studies possible. The officers and crews of NOAA's vessels *Ferrel, George B. Kelez, Oregon II*, and *Albatross IV* provided the high degree of seamanship and cooperation that made each cruise a pleasant and successful experience. We especially thank Dr. Suzanne Suchard and Miss Cady Soukup (former students, University of Maryland, College Park) who made collections for us during extended cruises. Mr. Mark Galasso kindly prepared the line drawings of amoeba shown in the text figure.

REFERENCES

Berk, S. G., R. R. Colwell, and E. B. Small. 1976. A study of feeding responses to bacterial prey by estuarine ciliates. *Transactions of the American Microscopical Society*, **95**, 514–520.

Bovee, E. C. 1979. A preliminary report on amebas found in marine coastal waters of Virginia, New Jersey and Massachusetts. (Abstract). *Journal of Protozoology*, **26** (3), Pt. 1, 26A.

Bovee, E. C., and T. K. Sawyer. 1979. Marine Flora and Fauna of the Northeastern United States. Protozoa: Sarcodina: Amoebae. Circular 419, U.S. Dept. of Commerce, National Marine Fisheries Service, 56 pp.

Burnett, B. R. 1973. Observation of the microfauna of the deep-sea benthos using light and scanning electron microscopy. *Deep-Sea Research*, **20**, 413–417.

Daggett, P.-M., T. A. Nerad, T. K. Sawyer, and E. J. Lewis. 1980. Preliminary observations on the possible use of starch-gel electrophoresis as an effective method for separation of *Acanthamoeba* species. (Abstract). *Transactions of the American Microscopical Society*, **99**, 230.

Davis, P. G., D. A. Caron, and J. M. Sieburth. 1978. Oceanic amoebae from the North Atlantic: Culture, distribution, and taxonomy. *Transactions of the American Microscopical Society*, **97**, 73–88.

Fenchel, T. 1969. The ecology of marine microbenthos. IV. Structure and function of the benthic ecosystem, its chemical and physical factors and the microfauna communities with special reference to the ciliated protozoa. *Ophelia*, **6**, 1–182.

Lighthart, B. 1969. Planktonic and benthic bacteriovorous protozoa at eleven stations in Puget Sount and adjacent Pacific Ocean. *Journal of the Fisheries Research Board of Canada*, **26**, 299–304.

Mitchell, R., and S. Yankofsky. 1969. Implication of a marine ameba in the decline of *Escherichia coli* in seawater. *Environmental Science and Technology*, **3**, 574–576.

Page, F. C. 1967. Taxonomic criteria for limax amoebae, with descriptions of 3 new species of *Hartmannella* and 3 of *Vahlkampfia*. *Journal of Protozoology*, **14**, 499–521.

Ruddell, C. 1979. Histopathological studies. *In*: Special Report in Applied Marine Science and Ocean Engineering No. 200. (Contract no. AA550-CT6-62 with the U.S. Bureau of Land Management.) Virginia Institute of Marine Science, Gloucester Point, Virginia, Chapter 12, pp. 125 (+ 140 figures).

Sawyer, T. K. 1975a. Marine amoebae from surface waters of Chincoteague Bay, Virginia: Two new genera and nine new species within the families Mayorellidae, Flabellulidae, and Stereomyxidae. *Transactions of the American Microscopical Society*, **94**, 71–92.

Sawyer, T. K. 1975b. Marine amoebae from surface waters of Chincoteague Bay, Virginia: One new genus and eleven new species within the families Thecamoebidae and Hyalodiscidae. *Transactions of the American Microscopical Society*, **94**, 305–323.

Sawyer, T. K. 1975c. *Clydonella* n.g. (Amoebida: Thecamoebidae), proposed to provide an appropriate generic home for Schaeffer's marine species of *Rugipes*,

C. vivax (Schaeffer, 1926) n. comb. *Transactions of the American Microscopical Society*, **94**, 395–400.

Sawyer, T. K. 1979. Species diversity among marine protozoa in sediment from a sewage disposal site. (Abstract). *Journal of Protozoology*, **26** (3), Pt. 1, 22A.

Sawyer, T. K. 1980. Marine amebae from clean and stressed bottom sediments of the Atlantic Ocean and Gulf of Mexico. *Journal of Protozoology*, **27**, 13–32.

Sawyer, T. K., G. S. Visvesvara, and B. A. Harke. 1977. Pathogenic amoebas from brackish and ocean sediments, with a description of *Acanthamoeba hatchetti*, n. sp. *Science*, **196**, 1324–1325.

Sieburth, J. M. 1976. Bacterial substrates and productivity in marine ecosystems. *Annual Review of Ecology and Systematics*, **7**, 259–285.

Soukup, C. A. 1976. Seasonal population fluctuations of planktonic ciliates of the New York Bight. M.Sc. Thesis, University of Maryland, College Park, Maryland, 65 pp.

Tendal, O. S., and K. B. Lewis. 1978. New Zealand xenophyophores: Upper bathyal distribution, photographs of growth position, and a new species. *New Zealand Journal of Marine and Freshwater Research*, **12**, 197–203.

Visvesvara, G. S., and T. K. Sawyer. 1980. Antigenic analysis of pathogenic, euryhaline, and eurythermal *Acanthamoeba hatchetti* (Amoebida: Acanthamoebidae). (Abstract). *Transactions of the American Microscopical Society*, **99**, 238–239.

Wolff, T. 1979. Macrofaunal utilization of plant remains in the deep sea. *Sarsia*, **64**, 117–136.

PART VI: STABILIZED COAL WASTE

Physical and Chemical Properties in Seawater

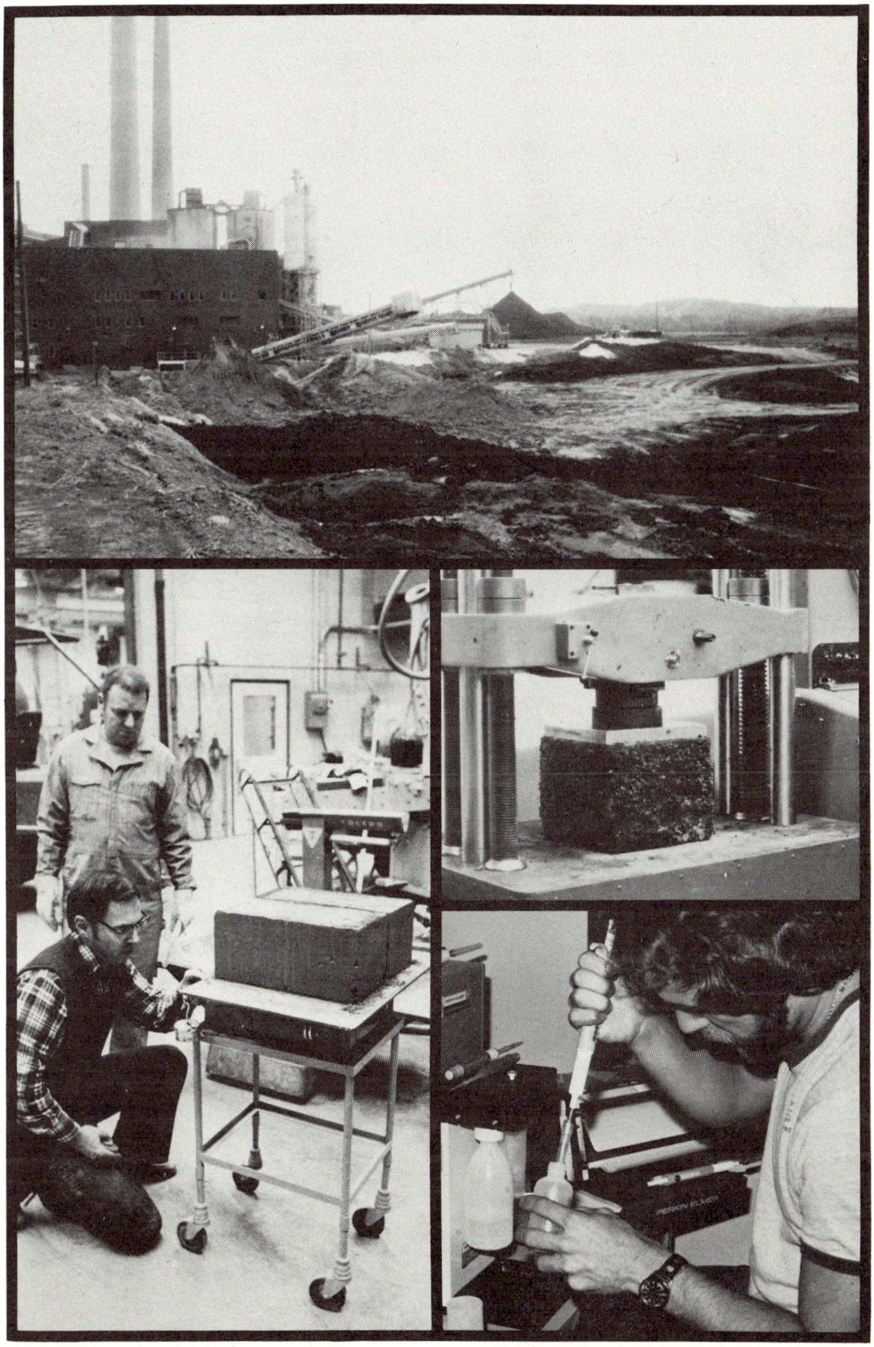

18

TESTING AND EVALUATION OF STABILIZED COAL WASTES FOR OCEAN DISPOSAL

Herbert R. Carleton and Franklin F. Y. Wang

College of Engineering and Applied Sciences
State University of New York
Stony Brook, New York

ABSTRACT

A variety of compositions of fly ash and filtercake scrubber sludge stabilized with lime were evaluated in the laboratory and at an offshore site over a period of two years. The primary purpose of this work was to establish the suitability of stabilized coal wastes for ocean disposal through examination of the physical properties of this material over an extended period, and to determine those properties that adequately reflect physical integrity.

Groups of test blocks were prepared from fly ash and scrubber sludge obtained from two eastern coal-burning power plants. Blocks were divided into groups for placement at an Atlantic Ocean test site (at 25 m depth) and for laboratory tests in seawater tanks. The apparent elastic modulus of these materials was evaluated using a time-of-flight ultrasonic intervalometer operating at 162 kHz developed for this application. All blocks were ultimately tested for compressive strength in the laboratory to obtain correlation between compression strength and elastic modulus. Results showed that the ultrasonic method was a sensitive indicator of internal physical changes. The accuracy of the method was sufficient to differentiate block-to-block variations in density and to reflect changes in water absorption and cementation.

18.1. INTRODUCTION

18.1.1. Background

The projected conversion of oil-fired powerplants to coal and the construction of coal burning units involves many new problems, particularly those introduced in making coal burning environmentally acceptable. In addition to recovering fly ash from flues, it is necessary to carry out flue gas desulfurization (FGD) to minimize air pollution by sulfur dioxide. One method uses a scrubber process in which lime or limestone reacts with the flue gases to produce calcium sulfite hemihydrate and calcium sulfate (gypsum) (see Duedall et al., Chapter 19). These by-products appear as a sludge filtercake, and when combined with fly ash produce a thioxotropic suspension.

Because the combined fly ash and sludge filtercake constitute 5–15% of the mass of coal consumed by the power plant, an appropriate method of disposal of this residue is of great importance. One process realizes the consolidation of the fly ash and dewatered sludge filtercake residues with the addition of lime to obtain a stabilized composite of high mechanical strength (Duvel et al., 1978). Although consolidation of coal-waste residues may facilitate the transportation of the large quantity of material, characteristics of the composite limit its use to landfill and similar applications. Lunt et al. (1977) have pointed out that ocean dumping may be an attractive option in coastal areas. This alternative is further enhanced by the opportunity to construct fishing reefs if the waste material is ecologically compatible with the marine environment (Duedall et al., 1981; Woodhead et al., 1981; see also Chapter 19).

A primary requirement for any material to be eligible for ocean disposal in large quantities is that it possesses adequate physical rigidity not to "smother" the bottom (for example, as a dumped slurry). A measure of physical rigidity is the compressive strength of the solid residue. Compressive strength is widely used as a measure of quality in load-bearing materials such as concrete (Orchard, 1973), and extensive testing methods have been developed (Whitehurst and Malhotra, 1978).

We evaluated the compatibility of stabilized coal-waste composites for ocean disposal and reef construction and monitored the physical properties of these materials for two years. In addition to periodic sampling and measurement of compressive strengths, we have developed an ultrasonic testing procedure to monitor the ultrasonic velocity of block compositions in the laboratory and at sea. The ultrasonic properties of velocity and modulus were found to be a sensitive indication of the evolution of the physical properties of cementation and water absorption within these materials. Correlation is obtained between the compressive strength and modulus of these compositions, although the precision of this comparison is limited by the large statistical variation in compressive strength data. Results indicate that ultrasonic evaluation can be applied to the continuous monitoring of block integrity in the marine environment. The sensitivity of this technique warrants our further understanding of the physical properties of cemented composites as they relate to ultrasonic measurement.

18.1.2. Compressive Strength and Elastic Modulus

A solid is stressed to its compressive limit when the applied load causes an irreversible rearrangement of the elements of its structure. This leads to crack formation. The structure loses its ability to withstand the applied stress because of a radical decrease in the effective elastic modulus which, until the point of failure, limits the relative displacement of substructural elements. The role of displacement in the failure mechanism establishes the qualitative relation between the compressive strength and the elastic modulus of the material. Because the local displacement is proportional to the strain induced in a solid under stress, the compressive strength of a solid can be expected to be proportional to the elastic modulus (Barrett et al., 1973).

The fracture mechanism in a given material depends on the presence of any flaws, voids, or structural aberrations that may be present in the material. A stress multiplication process creates high stress values at the flaws, and causes fracture at average values well below the elastic modulus. This process is very complicated in blocks of stabilized fly ash and sludge filtercake, which are comparatively inhomogeneous. If incomplete cementation occurs, the granular properties of the mix play a dominant role in determining its compressive strength (Sadasivan and Raju, 1977).

Whitehurst (1966) reported on the sonic velocity and flexual strength of 180 prisms of concrete of various compositions and found correlation between these parameters. The correlation for samples of a given composition and curing treatment was substantially better than for those of a different formulation. This indicates that the dependence of compressive strength on modulus is not a single-valued function. Tomsett (1980) reported on extensive studies of structural concrete and developed a series of functional relations between the compressive strength and pulse velocity based on experimental data. His studies showed that ultrasonic measurements may be used as an indication of material integrity, particularly for measurement of differences between samples of material of a fixed composition.

18.1.3. Ultrasonic Velocity and Elastic Modulus

The theory of sound propagation in homogeneous materials is well developed (Auld, 1973). Three fundamental velocities of propagation exist for any direction of propagation in isotropic solids. These modes are given by

$$V_l^2 = \rho M_l \qquad 1$$

and

$$V_{\sigma_1}^2 = V_{\sigma_2}^2 = \rho M_\sigma, \qquad 2$$

where V_l and V_{σ_1}, V_{σ_2} are the longitudinal and (degenerate) shear mode velocities, respectively; ρ is the density; and M_l and M_σ are the appropriate elastic moduli for these modes of propagation, respectively. Modes σ_1 and σ_2 represent shear mode propagation with particle motion perpendicular to the direction of propagation.

Equations 1 and 2 are idealizations of actual material behavior and do no include the effect of defects and irregularities that can be expected to play a significant role at short wavelengths of ultrasound. These concepts and their relation to coal-waste composites will be discussed in Section 18.3.

18.2. EXPERIMENTAL

18.2.1. Experimental Design

The fly ash and sludge filtercake used in the block compositions evaluated in this study were obtained from electric power plants in Elrama, Pennsylvania (Duquesne Light), and Coshocton, Ohio (Columbus and Southern Ohio Electric). The composition of a sludge filtercake varies from day to day, particularly in water content. Also, the Elrama filtercake contains large and variable quantities of fly ash whereas the Conesville filtercake is essentially calcium sulfite hemihydrate and gypsum (see Chapter 19).

The test blocks were made by IU Conversion Systems, Inc. (IUCS, Horsham, Pennsylvania), by adding lime to a scrubber sludge and fly ash mixture using the Poz-o-Tec® process (IUCS). This material was then hand-compacted into 0.027-m^3 plywood forms. After the blocks set, the forms were removed and the blocks and allowed to cure at outdoor temperatures (summer) while covered with plastic. The water content and curing time will strongly affect the physical characteristics of the material.

The composition of blocks varied in their fly ash to sludge ratio and the lime used. Blocks were identified by a classification system consisting of four symbols: the first symbol indicates the power plant of origin (C = Conesville, E = Elrama); the second is an ordinal number denoting the production cycle; the third symbol is the type of lime used in the mix (C = coarse lime, F = fine lime); and the fourth symbol is the nominal ratio of fly ash to scrubber sludge in the mix by weight. For example, the designation C2F3 represents second Conesville production of blocks; fine lime; and 3:1 ratio of fly ash to sludge (by dry weight). In all cases the lime content was approximately 5% by weight. Eight compositions were evaluated.

After curing for 30 days, the blocks were transported to Stony Brook, New York, where a group were selected for testing. The 0.027-m^3 size blocks were too large for practical testing. Also, it was desirable to evaluate the possible effects of seawater intrusion after the blocks were subsequently retrieved from the sea. Thus, a number of larger blocks were cut on a band saw into 24 subblocks of about 6.5 X 9.5 X 14 cm. These blocks were then divided into groups and subgroups to be placed under various experimental conditions in the laboratory and at a test site in the ocean. For convenience of discussion, the three major experimental categories reported here are designated A, B, and C as discussed below. A summary of the test sites, compositions, and test results reported here is indicated in Table 18.1.

18.2.1a. Group A

A group of six E3F1 and four C3F3 blocks was kept immersed in covered seawater tanks in the laboratory for 575 days (Table 18.1). Tank waters were replenished with distilled water when required. At appropriate intervals, more frequent in the beginning, each block was removed from the tank and its ultrasonic velocity determined by the method described in Section 18.2.2. Each block was also weighed and its density determined by the immersion method using an electronic balance with a resolution of 10 mg. At the end of the test period, each block was tested for compressive strength while wet as described in Section 18.2.3.

18.2.1b. Group B

A complementary test of 64 blocks of type E2C1 was undertaken under the same laboratory conditions as Group A in order to obtain replication of compressive strength data. Each block was carefully cut on a band saw to a 7-cm cube. In this

Table 18.1. Summary of Three Test Groups of Coal-Waste Blocks Tested

Group	Number Tested	Block Identification	Seawater Immersion (days)	Environment	Tests
A	4	C3F1	575	Laboratory seawater tanks at ambient temperature (~19°–22°C)	Ultrasonic velocity, density, compressive strength
	6	E3F3	575		
B	64	E2C1	Various to 199	Same as Group A	Final ultrasonic velocity, density, compressive strength
C	57	C1F1	Various to 114	Atlantic Ocean, 3.2 km offshore at 25 m depth	Final ultrasonic velocity, density, compressive strength
		C1C1			
		C3F3			
		C3C3			
		C3C1.6			
		E2C1			
		E2C3			
		E3F1			

case ultrasonic and density data were taken on subgroups of four blocks at a time at selected time intervals; these blocks were then tested for compressive strength.

18.2.1c. Group C

A series of blocks of productions 1 to 3 were prepared in the same manner as Group A and placed in plastic racks at an ocean test site located about 3.2 km off Long Island's south shore at 25 m depth during June and August of 1979. Each set of block compositions was placed in a separate plastic rack that was open to the sea. Blocks were taken by divers and returned to the laboratory for testing. Every effort was made to prevent loss of water from the blocks until testing. Where possible, these blocks were immersed in seawater in the laboratory and pretested before placement in the ocean; these tests are reported as one-day tests in data of Group C.

18.2.2. Ultrasonic Method

To obtain accurate and replicable measurements of the ultrasonic velocity in the blocks under test, a system was designed to measure the time-of-flight of an acoustic pulse over a path between two ultrasonic transducers immersed in a seawater test tank (Fig. 18.1). This system consisted of a gated modulator that selected a preset number of sinusoidal cycles from an independently adjustable continuous wave oscillator (which established the center frequency of transmission). The output of the pulse generator was amplified by a linear amplifier that can produce up to 250 V peak-to-peak output to the source transducer. The received signal was amplified by a preamplifier and displayed on an oscilloscope that was triggered

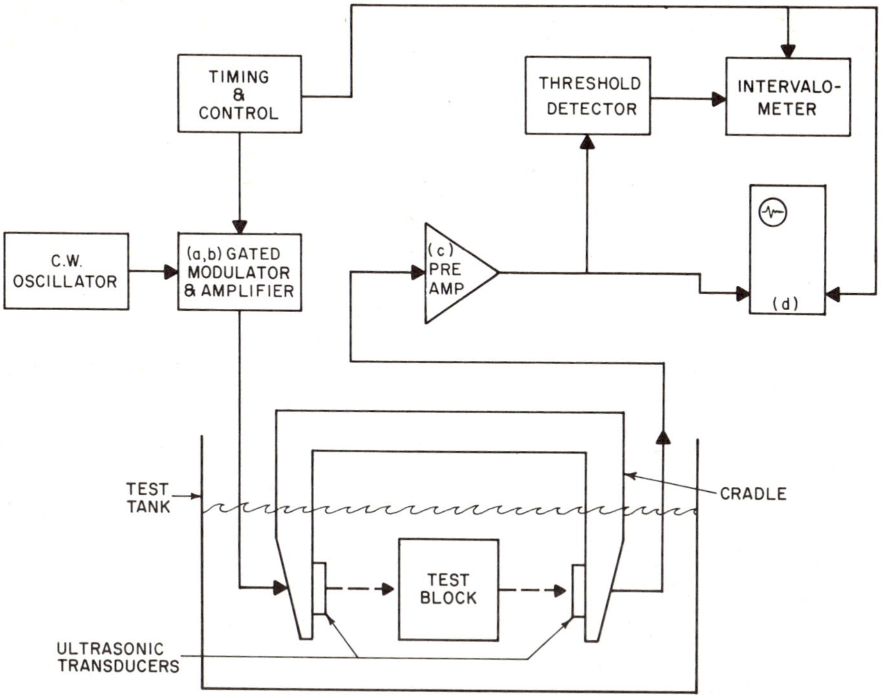

Figure 18.1. System for ultrasonic velocity measurement: (a) General Radio (Concord, Massachusetts) gated modulator No. 1396-B; (b) Electro Navigation Industries (Rochester, New York) linear amplifier, No. 240-L; (c) Tektronix (Beaverton, Oregon) amplifier, No. 1101; and (d) Tektronix dual beam oscilloscope, No. 585A. The differential time delay of a gated single pulse of 163 kHz is measured with and without the test block present. Measurements are taken directly from the oscilloscope or can be electriconically determined with an intervalometer/counter.

at transmit time. The signal was not filtered and the pulse delay was measured directly by using the expanded sweep feature of the oscilloscope. Two 6.25-cm diameter PZT-5 transducers, resonanting at 162 kHz, were used as transmitter and receiver.

The change in pulse arrival time, when a test block was in the acoustic path, was used to calculate the ultrasonic velocity (V) from the equation:

$$V = \left(\frac{1}{V_t} - \frac{\Delta T}{L} \right)^{-1} \qquad 3$$

where ΔT is the observed time differential; L is the block thickness in the acoustic path; and V_t is the velocity of sound in the immersion liquid. The ultrasonic velocity of the liquid was sensitive to specific gravity and temperature variations and was checked frequently by measuring the time delay of a plastic reference block. Accuracies of better than 0.5% in the sound velocities were readily obtained by this method.

The use of a seawater tank eliminated the necessity of bonding transducers to each block to be tested. We found that good bonds were difficult to obtain, and failures in bonding introduced unacceptable uncertainties in the velocity measurements. This method was also more practical when large number of blocks had to be tested.

18.2.3. Compressive Strength

Measurements of the compressive strengths of test blocks were taken on selected samples of each group. All blocks were tested on a Reihle (East Moline, Illinois) test system, No. FS-160; the loading rate was 6 mm min^{-1}. As far as possible, the testing was consistent with ASTM standard C-39 (American Society for Testing and Materials, 1974) with exceptions as discussed below.

Generally, the blocks did not have perpendicular faces and some degree of erosion in the blocks taken from the ocean was visible along edges. The compressive strength was obtained by dividing the load by the measured cross-sectional area of the block. A large variation in compressive strengths within groups containing three or four blocks was found. Although the fracture mechanism itself is statistical in nature, irregularities in the blocks made it difficult to achieve uniform loading and thus contributed to the variability of compressive strengths.

All blocks were ultrasonically tested prior to compressive tests. Hence, to maintain consistency, all compressive tests for "initial strength" were carried out after an immersion time of one day.

18.3. RESULTS AND DISCUSSION

18.3.1. General Considerations

Composite materials are characterized by the existence of two or more different materials or phases with significant grain boundaries; in the case of a sintered or cemented material, the second phase consists of voids embedded in a matrix of the primary material. Although a statistical distribution is required for an exact description, such materials can be characterized by a volume-fraction porosity, The scattering from grain boundaries or voids must be included in any consideration of ultrasonic propagation, because they produce significant attenuation of the wave, particularly in the region where the ultrasonic wavelength approaches the grain size (Christensen, 1979).

The material in coal-waste block compositions is coarse and nonhomogenous. The fly ash and sludge components generally form a quasi-uniform base material with granules not exceeding a few millimeters in diameter. When the block is formed, complete mixing is rarely achieved, and point-to-point variations in fly ash to sludge ratio, porosity, density, and water content can be expected. The microstructure of this base material is complex (Fig. 18.2).

Statistical variations in composition due to poor mixing can be expected to cause variations in sound velocity. In addition, the rapid increase in weight, which occurs during the first day of submersion, is a direct indication that these materials

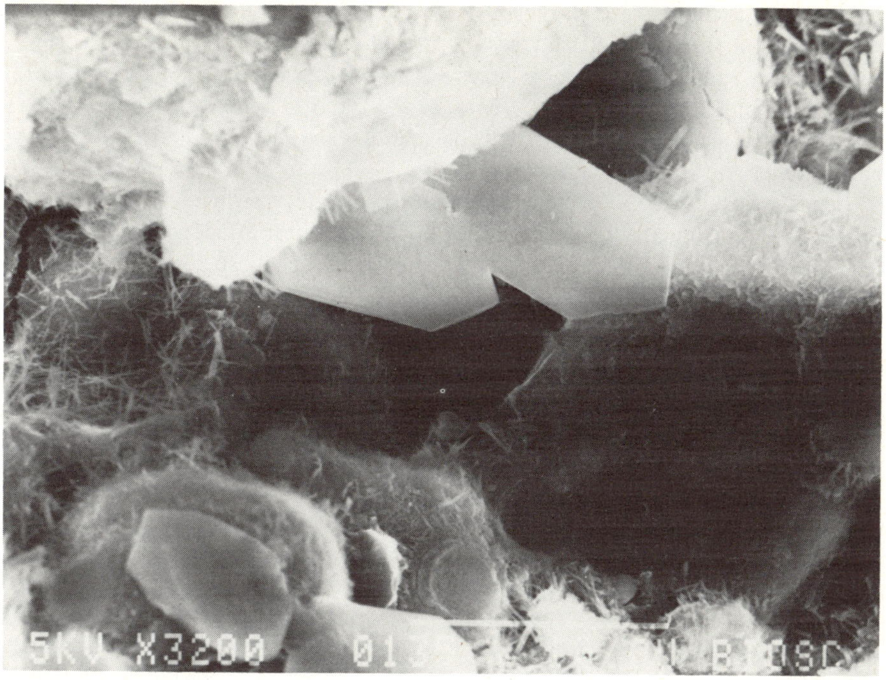

Figure 18.2. Scanning electron micrograph of a sample taken from interior of an E2C1 block after crushing. The open structure in this composition is partially filled with prisms of gypsum ($CaSO_4 \cdot 2H_2O$) and needles of ettringite ($3CaO \cdot Al_2O_3 \cdot 3CaSO_4 \cdot 31H_2O$) which is a product of hydration (see Chapter 19); the spherules are the fly ash component. (Figure used with permission of Frank Roethel.)

are porous after the initial curing. The hydration process, which occurs during cementation, incorporates available water in the formation of new compounds, leaving microvoids in the surrounding area.

Composition and porosity variations have a direct effect on the density of the material. In addition, the hydration process produces chemical consitituents of higher molecular weight thus resulting in increased block density. Any mechanism affecting the density of the block can be expected to produce a change in ultrasonic velocity (equation 1). If the elastic molulus does not change, then water absorption and hydration would be expected to produce a continuous reduction in sound velocity. In fact, the effect of these processes is greater on the modulus, and a net increase in velocity was observed, as discussed below.

The absorption of water by each block produces an accompanying increase in ultrasonic velocity through modification of the scattering mechanism. This is primarily due to the reduction in the number of air-filled pores. These pores act as scattering centers, which result in a reduction of the ultrasonic velocity compared to that of a nonporous matrix. The effect of the pores is complex, but can be understood qualitatively on the basis of scattering theory (Carleton, 1961) or mixture theory (Biot, 1962; Stoll, 1974; Christensen, 1979). If the micromorphology of the block material remains fixed, any physical process that reduces the magnitude of scattering from voids or irregularities will change the ultrasonic velocity. In stabilized coal-waste materials, the voids are of lower density when dry than when filled with water. Because the acoustic match is better to water, the effect of scattering by voids will be reduced and the ultrasonic velocity increased when the blocks are wet.

Cementation can be expected to have a direct effect on the elastic modulus, providing greater rigidity between particles. The greatest effect will be in increasing the shear modulus of the solid. If the material is locally isotropic, the modulus of the intrinsic matrix can be described by equations 1 and 2.

Any of the above mechanisms can be expected to produce variations in the observed ultrasonic velocity within a coal waste block and between blocks of varying composition. Because the diameter of the acoustic beam used in our measurements was 6 cm, the measured velocity was an average for a volume of about 80 cm^3. Variations of velocity were observed when the position of the beam was changed. However, when care was taken to locate the beam accurately, measured velocities were repeatable to better than 2 m s^{-1}.

Three each of E2Cl and of E2C3 blocks (Fig. 18.3) that were immersed offshore for 30 days were selected from Group C and photographed after compression test measurements were made. These blocks typify the morphology and size of all of the test blocks reported in this series. E2C3 blocks appear to be more eroded, as evidenced by more rounded edges. This appearance is due partly to the falling away of softer material during compression and is an additional reflection of the lower integrity of blocks of high fly ash content.

A large sludge inclusion is evident in block number 7, and there is significant fracture in this area of the block (Fig. 18.3). The lighter areas of block number 4 are marine growth, which was found on blocks that have been taken from the ocean. Block 2 is a control block of type E2C1 that was immersed in seawater in the laboratory for four days; it exhibited sharper edges than those blocks immersed for 30 days in seawater.

Initial measurements of the ultrasonic velocity of test blocks provided direct qualitative correlation between higher velocities and block integrity. Blocks that were firm and retained distinct edges also possessed high sonic velocities (2000–2400 m s^{-1}) while those that appeared soft displayed velocities of 1600–2000 m s^{-1} and low compressive strengths. In cases of extreme softness, velocities were difficult to measure because of the high acoustic attentuation. None of the blocks of this series exhibited any statistically significant degradation in integrity over the test periods.

18.3.2. Laboratory Control Blocks

18.3.2a. Group A: Qualitative behavior, Error analysis, Compressive Strength, and Modulus

It was possible to evaluate precisely the velocity evolution of those blocks in Group A over 575 days. Because each block could be measured at intervals that were

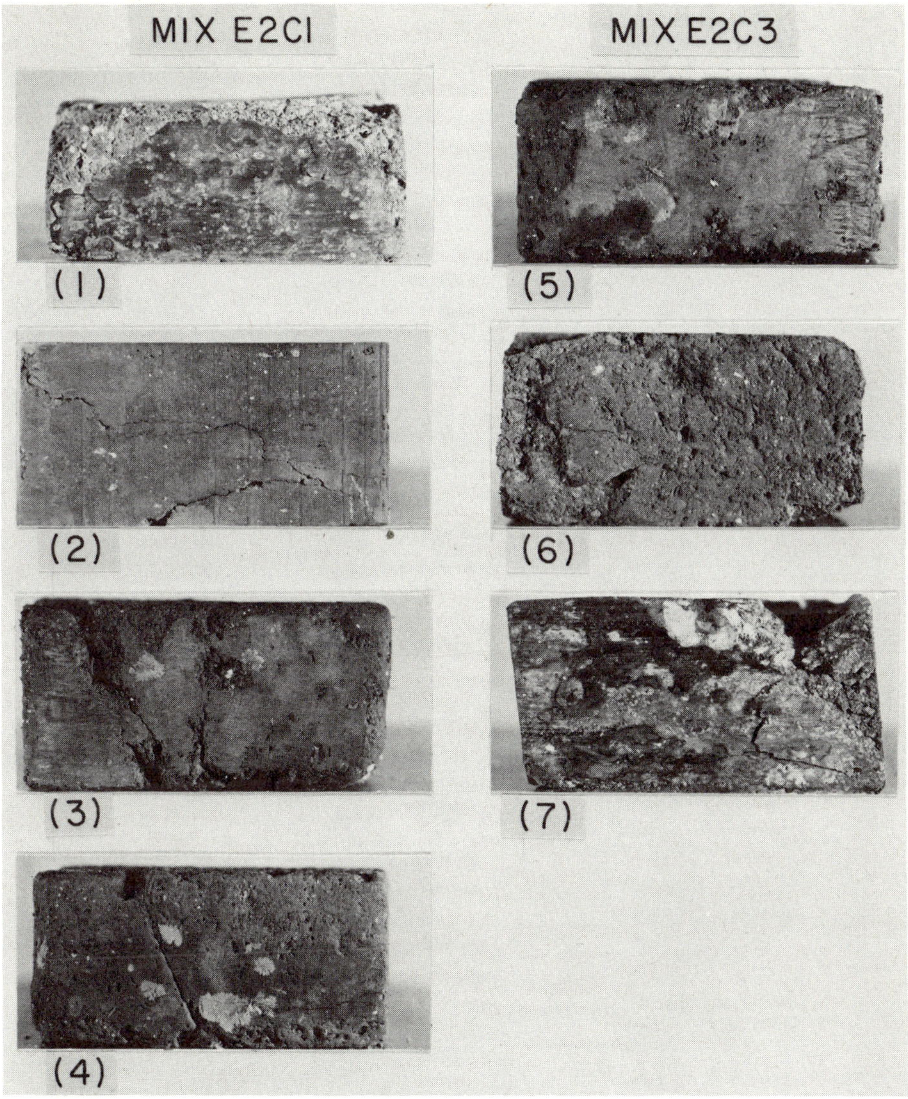

Figure 18.3. Photographs of block erosion after 30 days immersion in the ocean. These blocks were tested for compressive strength prior to photographing; fractures are evident from this test.

appropriate to the rate of change of the observed velocity, this laboratory work provided an important base for the validation of the ultrasonic method as well as for the analysis of control samples.

The velocity evolution of the two groups of blocks from Group A showed an increase in the velocities of all blocks in the test sample over the test period (see Fig. 18.4). With few exceptions, the accuracy and precision of the velocity determination made it possible to identify each block in the set from its velocity profile.

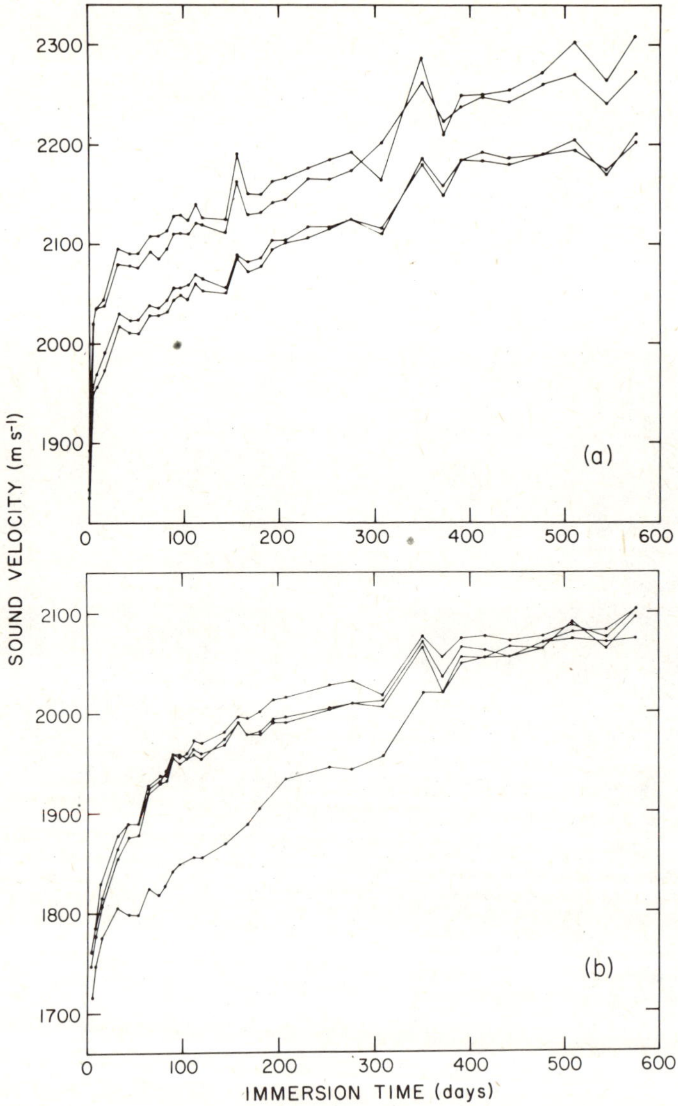

Figure 18.4. The velocity evolution of four replicate blocks of test group A over the period of immersion in seawater in the laboratory: (a) E3F1 composition and (b) C3F3 composition.

The growth in velocity of these blocks was characterized by an initial period of rapid growth for a period of five to ten days, and then continued growth at a rate about tenfold lower for the remainder of the observation period. There was no significant evidence for any decrease in the rate of growth in velocity with time, even after 575 days.

Consistency and replicability of data for each block are strong evidence that differences in block composition were responsible for the block-to-block variations in sound velocity profile. In this context, it is significant that the block exhibiting a lower profile in Fig. 18.4a assumed the same profile as others in the group after 350 days. The implication here is that chemical and physical processes within the block are evolving favorably through the period of the test.

Because the blocks tested contained different amounts of fly ash, the velocity range of each set was different, as reflected in the vertical axes of Figs. 18.4a and 18.4b. The velocities of the E3F1 blocks were higher over the test cycle than those of the C3F3, even though the density of the 1:1 mix design was higher.

The data given in Fig. 18.4 were corrected for all known sources of error, which are primarily those associated with the time-base calibration and corrections in velocity for temperature changes of the seawater tank. Examination of the velocity profiles of the blocks of Group A indicate that the random error in reading the differential time delay is equivalent to 4 m s^{-1}. Except in one instance, all the significant deviations of velocity from the general growth curve showed movement systematically in the same direction. Although the deviations might be a reflection of physical processes occurring in the blocks at that time, we are unable to postulate a mechanism. It is probable that these effects were not physical, but were bias errors from the ultrasonic measurement technique. Such an error is likely attributable to variations in the time-base calibration reference or to observation error.

At 575 days the compressive strength of each block in this group was measured and recorded. The average compressive strength for six E3F1 blocks was 1.9 MPa (279 lb in.$^{-2}$) [1 megapascal (MPa) = 10^6 newton m^{-2}], while that for four C3F3 blocks was 1.29 MPa (188 lb in^{-2}). The greater strength of 1:1 block mix design is evident here and was observed in all comparisons of blocks from the same production.

Figure 18.5 compares the compressive strengths of this series with the elastic

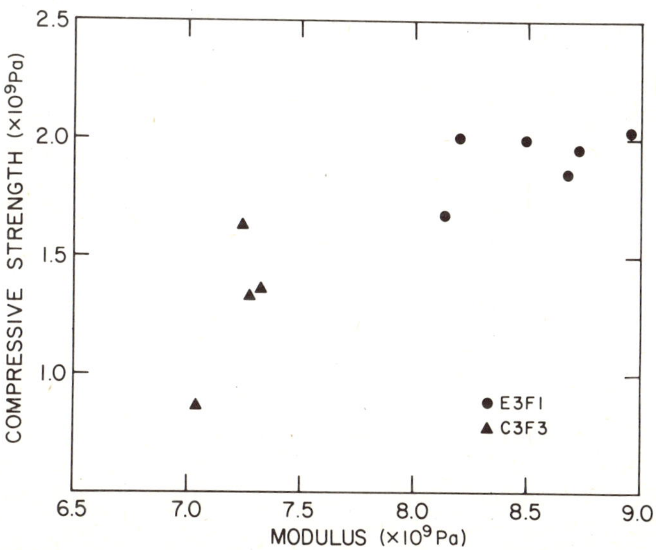

Figure 18.5. The correlation of compressive strength of blocks from test group A to the ultrasonic modulus.

moduli derived from the ultrasonic data. A general increase in compressive strength (S) with elastic modulus (M) is evident. A linear regression analysis of these data leads to the expression

$$S = K (M - M_0) \quad 4$$

with $K = 3.61 \times 10^{-4}$ and $M_0 = 3.2 \times 10^3$ MPa. This result has a correlation coefficient of 0.87, which implies a confidence level of greater than 99% that this data set was not picked from an uncorrelated sample (Bevington, 1969).

A similar analysis of these data for each block type was less definitive. The correlation coefficient of the C3F3 group was 0.79 (80% confidence) and that of the E3F1 group was 0.48 (85% confidence). These results suggest that a much larger data set will be required to achieve a reliable analytical curve for the dependence of compressive strength on apparent elastic modulus, unless the error in compressive strength measurements can be reduced in future work.

18.3.2b. Group B: Time Variation, Correlation of Moduli to Compressive Strength

A summary of data obtained on the 64 blocks of Group B is shown in Table 18.2. The purpose of this experiment was to determine any evidence of systematic changes in compressive strength with time of submersion in laboratory tanks containing seawater. The sound velocities and densities reported were obtained on each group of four blocks immediately prior to strength testing. The standard deviations in velocity were greater than the experimental error and reflect the variation in composition from block to block as discussed in Section 18.3.2a. For the same reason, the standard deviations in density were also larger than the estimated density error of 1 kg m^{-3}.

A linear regression analysis of the variation of compressive strength S with time of immersion yields the relation S (MPa) = 5.95 - 0.00148T, where T is in days. Although this relation indicates a decrease in block strength with time, this behavior was not observed in blocks of Group A for the same mix. The correlation coefficient was -0.24, that is, about 75% of the variation in compressive strength is unexplained. This is undoubtedly due to the large variation in compressive strength for the four-block sets (typically about 1 MPa), which is a factor of 10 greater than the 100-day variation predicted from linear regression.

More meaningful results were obtained from a linear regression analysis of the sound velocity results. The relation V_l (m s^{-1}) = 2367 + 0.83T, with a correlation coefficient of 0.85, was obtained. This result is consistent with those blocks of Group A in which a nonlinear growth in velocity with time was observed, especially in the initial months. Because velocity data of Table 18.2 were obtained on different blocks at the end of the immersion period, individual velocity profiles are not available and further regression analysis is of limited value.

The ratio of compressive strength to acoustic modulus is listed for each group of blocks in Table 18.2. If there is a proportional dependence of the compressive strength on the modulus, this ratio should remain constant; values vary from

Table 18.2. The Physical Properties of 64 Laboratory Control Blocks of Group B[a]

Duration in Seawater (days)	Sound Velocity V_l ($m\ s^{-1}$)	Compressive Strength S (10^6 Pa)	Wet Density ρ ($kg\ m^{-3}$)	Modulus M^b (10^9 Pa)	Modulus Ratio r^c (10^{-3})
1	2300 (36)	6.579 (1.1)	1605 (5)	8.490 (0.17)	0.777 (0.16)
5	2361 (36)	5.516 (0.8)	1608 (9)	8.966 (0.31)	0.615 (0.09)
14	2370 (10)	6.015 (1.4)	1603 (4)	9.014 (0.09)	0.668 (0.15)
26	2437 (56)	5.790 (1.0)	1648 (6)	9.791 (0.46)	0.588 (0.09)
37	2466 (15)	4.968 (1.0)	1628 (4)	9.900 (0.12)	0.503 (0.11)
52	2396 (43)	6.218 (1.0)	1616 (8)	9.280 (0.46)	0.670 (0.01)
64	2425 (10)	6.234 (0.7)	1656 (2)	9.736 (0.08)	0.640 (0.07)
84	2418 (42)	5.826 (1.2)	1658 (8)	9.697 (0.37)	0.598 (0.10)
101	2444 (15)	5.777 (1.2)	1655 (5)	9.885 (0.11)	0.585 (0.12)
122	2472 (18)	5.817 (1.2)	1624 (4)	9.920 (0.14)	0.586 (0.12)
177	2529 (41)	5.971 (0.4)	1620 (5)	10.370 (0.34)	0.578 (0.06)
177	2535 (34)	5.909 (0.3)	1617 (8)	10.380 (0.25)	0.570 (0.03)
199	2490 (47)	5.173 (0.08)	1630 (5)	10.100 (0.36)	0.512 (0.08)

[a]Numbers in parentheses are standard deviations.
[b]$M = \rho V_l^2$.
[c]$r = M/S$.

0.503×10^{-3} to 0.777×10^{-3}. It is more reasonable to assume that the compressive strength will depend on the modulus by the relation of equation 4.

A linear regression analysis of compressive strength versus the moduli (Table 18.2) leads to the values $K = -0.43$ and $M_0 = 30.94$ MPa (confidence level of 85%). This result is unreasonable because it implies a large compressive strength at zero modulus; thus, an alternative explanation must be sought. An argument can be made for negative correlation between S and M; a porous substance will exhibit an elastic modulus and velocity that are lower than that of the same material without porosity. If water enters the pores, the compressive strength will remain approximately the same, but the ultrasonic velocity will increase through the mechanism mentioned earlier. Hence, a negative correlation could be obtained in a set of blocks with varying porosity. This is important in interpreting block behavior from ultrasonic measurements and must be developed further.

18.3.3. Offshore Reef Blocks

Field studies of the effect of immersion in the ocean were undertaken for a variety of compositions and time periods. A representative data set for 57 blocks of eight different compositions is summarized in Table 18.3. In each case, the data presented are the mean for three blocks within each subset. All data were obtained after seawater immersion for the period indicated. Blocks in different sets were not all placed at the same time but all were immersed during the summer and fall months when water temperatures were between 10° and 20°C.

18.3.3a. Velocity and Density

The density and velocity data for these block compositions indicate that higher velocities were associated with higher densities. This trend was particularly evident for subgroups within the same composition. The density of 1:1 (fly ash to sludge filtercake) compositions was higher than that of 3:1. The density of materials within a given composition will vary depending on the composition of the sludge filtercake and the degree of dewatering and compaction obtained during block formation. [The density of the Elrama blocks is particularly variable owing to the large and variable amount of carry over fly ash in Elrama sludge filtercake (R. Hilton, personal communication).] Because the density increased rapidly in the first few days because of the uptake of water, densities after one day of immersion were lower than the final densities.

The quality of the C1C1 blocks was significantly better than that of C1F1 blocks. The densities and velocities were higher in the former material, and compressive strengths were an order of magnitude greater. No similar result was found in a comparison of C3C3 and C3F3, where the densities obtained were almost identical and the modulus and compressibility data were within a standard error.

The 1.6:1 composition (C3C1.6) in group C also displayed the higher values of density, modulus, and compressive strength associated with lower fly ash content (Table 18.3). The differences in physical properties between E2C1 and E2C3 blocks are not as great as between other block sets in group C.

Table 18.3. **The Physical Properties of 57 Blocks of Group C Immersed in the Ocean at a Depth of 25 m for the Indicated Time**

Block Type	Duration in Seawater (days)	Sound Velocity V_l ($m\ s^{-1}$)	Compressive Strength S (10^6 Pa)	Wet Density ρ ($kg\ m^{-3}$)	Modulus M (10^6 Pa)
C1F1	71	1825	0.22	1612.0	5.37
C1F1	75	1758	0.12	1612.7	4.98
C1C1	32	2330	2.0	1679.3	9.12
C1C1	71	2350	2.4	1681.0	9.28
C3F3	30	2196	0.92	1647.3	7.94
C3F3	75	1981	0.31	1643.3	6.45
C3F3	114	2028	0.33	1644.0	6.76
C3C3	32	2164	1.49	1648.7	7.72
C3C3	71	2199	0.70	1641.0	7.94
C3C1.6	32	2245	2.22	1678.7	8.46
C3C1.6	71	2319	3.23	1685.0	9.06
E2C1[a]	1	2034	3.40	1544.0	6.39
E2C1[a]	1	2031	1.93	1733.8	6.11
E2C1	30	2349	2.72	1608.8	8.88
E2C3[a]	1	2013	1.94	1554.0	6.30
E2C3[a]	1	2018	1.29	1654.6	6.03
E2C3	30	2242	2.02	1599.8	8.04
E3F1	30	2112	1.24	1611.6	7.19
E3F1	30	1990	1.42	1633.0	6.47

[a]Blocks tested in the laboratory after one day immersion in seawater.

18.3.3b. Compressive Strengths

The compressive strengths of most compositions in Group C were greater than 1 MPa (Table 18.3), a strength that our observations indicate is adequate for extended survival in the marine environment. The mixes C1C1, C3C1.6, and E2C1 exhibited more than twice this compressive strength, which was also reflected in the more distinct edges of these blocks after retrieval. Types C1C1 and C3C1.6 also exhibited an increase in compressive strength over the 39 days of immersion.

Two compositions did not follow this pattern and had very poor compressive strengths when tested. Blocks of mix C1F1 were soft on retrieval and exhibited an average compressive strength of 0.17 MPa. The compressive strength of blocks retrieved after 75 days was almost half that of the group retrieved after 71 days, but this result cannot be considered a significant time sequence since each subgroup was from a different test sequence placed months apart. Mix C3F3 displayed a distinct decrease of compressive strength with time to a value that is probably unacceptable for marine disposal. Because both of these mixes used fine lime, the quality of compositions of this type is questionable, even though the compressive strength of E3F1 was more acceptable.

Materials with low compressive strength again exhibited a pattern of lower density and lower acoustic velocities and moduli than those compositions with compressive strengths greater than 2 MPa. These properties are consistent with a mix that had higher intrinsic porosity and for which the process of cementation did not progress to fill a high fraction of void volume.

18.3.3c. Comparison to Laboratory Results

Direct comparison can be made between the physical properties of blocks placed in the ocean and in seawater in the laboratory for mixes E2C1, E3F1, and C3F3. The compressive strengths of E2C1 blocks immersed for 30 days in the ocean were 50% of those from Group B for the same period (2.72 versus 5.38 MPa). The values of acoustic moduli and densities were also lower for blocks in the ocean (10 and 2%, respectively). There was greater disparity in time between sampling of blocks of the other two types. Compressive strengths of type E3E1 taken at 575 days in the laboratory and 30 days in the ocean were 1.93 and 1.33 MPa, respectively. Again, for these blocks the moduli and densities in the ocean were lower by 20 and 3%, respectively. For type C3F3 the compressive strength of blocks removed from the ocean was reduced by about 75% and the moduli by 7% although the density was the same. These results consistently show less complete consolidation of the blocks in the ocean environment. Possible processes causing this difference are block erosion and diffusion of major components from the block (Duedall et al., Chapter 19).

18.4. CONCLUSIONS

Our results show that blocks of stabilized fly ash and scrubber sludge filtercake are capable of maintaining adequate physical integrity over periods of years in the ocean environment. Adequately formulated compositons can be expected to maintain compressive strengths greater than 3 MPa. The most favorable compositions are those with higher scrubber-sludge content in the range of 1:1 by weight that use coarse lime for stabilization.

The nondestructive evaluation of consolidated coal-waste composites by ultrasonic techniques is feasible and is a sensitive measure of physical process occurring in the material. A direct measurement of the ultrasonic velocities of these materials reflects processes of water absorption and cementation. There is a correlation between the compressive strength and apparent modulus of the block material. The large error in compressive strength measurement makes it difficult to establish any exact analytical relation, even on a sample as large as 64 blocks.

Because the ultrasonic velocity is also related to other factors such as water content and the size and density of inhomogeneities of the material, the correlation of the compressive strength of different compositions to apparent modulus is not definitive. This result emphasizes the necessity of developing a more quantitative understanding of the relationships among porosity, cementation, density, and the elastic and ultrasonic properties of these materials.

ACKNOWLEDGMENTS

The authors wish to acknowledge the contributions of Andrew Hamilton for the construction and assembly of the ultrasonic immersion system, Joseph Muratore for the preparation and ultrasonic measurement of test Groups A and C, and Thomas Baum and Sarit Bhaduri for the measurements of Group B blocks. We appreciate the continuous effort of Jeffrey Parker and the divers and crew of *RV Onrust* in the placement and retrieval of test blocks in the sea. Rose Bouleris and Ingeborg Schulz typed the manuscript and assisted in the preparation of figures.

This work has been funded by New York State Energy Research and Development Authority, the Electric Power Research Institute, the Power Authority of the State of New York, the U.S. Department of Energy, and the U.S. Environmental Protection Agency.

REFERENCES

American Society for Testing and Materials (ASTM). 1974. Compressive strength of cylindrical concrete specimens. *In*: Annual Book of ASTM Standards Part 14, ASTM Philadelphia, Pennsylvania, pp. 26–29.

Auld, B. A. 1973. Acoustic Fields and Waves in Solids, Vol. 1. Wiley, New York, 350 pp.

Barrett, C. R., W. D. Nix, and A. S. Tetelman. 1973. The Principles of Engineering Materials. Prentice-Hall, Englewood Cliffs, New Jersey, pp. 193–218.

Bevington, P. R. 1969. Data Reduction and Error Analysis for the Physical Sciences. McGraw-Hill, New York, pp. 119–127.

Biot, M. A. 1962. Generalized theory of acoustic propagation in porous dissipative media. *Journal of the Acoustic Society of America*, **34**, 1254–1264.

Carleton, H. R. 1961. Theoretical development of volume reverberation as a first-order scattering phenomenon. *Journal of the Acoustic Society of America*, **33**, 317–323.

Christensen, R. M. 1979. Mechanics of Composite Materials. Wiley, New York, pp. 258–265.

Duedall, I. W., F. J. Roethel, J. D. Seligman, H. B. O'Connors, J. H. Parker, P. M. J. Woodhead, R. Dayal, B. Chezar, B. K. Roberts, and H. Mullen. 1981. Stabilized power plant scrubber sludge and fly ash in the marine environment. *In*: Ocean Dumping of Industrial Wastes, B. H. Ketchum, D. R. Kester, and P. K. Park (Eds.). Plenum Press, New York, pp. 315–346.

Duvel, Jr., W. A., W. R. Gallagher, R. G. Knight, C. R. Kolarz, and R. J. McKlaren. 1978. State-of-the-Art of FGD Sludge Fixation. EPRI-FP-671, Electric Power Research Institute, Palo Alto, California, 10 sections (paginated separately) + appendices.

Lunt, R. R., C. B. Cooper, S. L. Johnson, J. E. Oberholtzer, G. R. Schimke, and W. I. Watson. 1977. An Evaluation of the Disposal of Flue Gas Desulfurization Wastes in Mines and the Ocean: Initial Assessment. EPA 600/7-77-051, U.S. Environmental Protection Agency, Research Triangle Park, North Carolina, 308 pp.

Orchard, D. F. 1973. Concrete Technology. Halsted Press, Essex, England, 375 pp.

Sadasivan, S. K., and V. S. Raju. 1977. Theory of shear strength of granular materials. *Journal of the Geotechnical Engineering Division, American Society of Civil Engineers*, **6T8**, 851–861.

Stoll, R. D. 1974. Acoustic waves in saturated sediments. *In*: Physics of Sound in Marine Sediments, L. Hampton (Ed.). Plenum Press, New York, pp. 19–39.

Tomsett, H. N. 1980. The practical use of ultrasonic pulse velocity measurements in the assessment of concrete quality. *Magazine of Concrete Research*, **32**, 7–16.

Whitehurst, E. A. 1966. Evaluation of Concrete Properties from Sonic Tests. American Concrete Institute Monograph No. 2, Iowa State University Press, Ames, Iowa, 94 pp.

Whitehurst, E. A., and V. M. Malhotra. 1978. Significance of Tests and Properties of Concrete and Concrete-Making Materials. STP 169B. American Society for Testing and Materials, Philadelphia, Pennsylvania, pp. 200–225.

Woodhead, P. M. J., I. W. Duedall, J. H. Parker, and H. R. Carleton. 1981. Acceptability of coal-fired power plant wastes for ocean disposal in New York. *In*: Proceedings of the Governor's Conference on Expanding the Use of Coal in New York State: Problems and Issues (May 21–22, 1981, Albany). Research Foundation of the State University of New York, Albany, pp. 31–36.

19

DIFFUSION OF CALCIUM AND SULFATE IONS IN STABILIZED COAL WASTES

Iver W. Duedall, Jeffrey S. Buyer, Monteith G. Heaton, Sue A. Oakley, Akira Okubo, Ramesh Dayal, Mark Tatro, Frank J. Roethel, Richard J. Wilke, J. Peter Hershey

Marine Sciences Research Center
State University of New York
Stony Brook, New York

Dr. Duedall's present address: Department of Oceanography and Ocean Engineering, Florida Institute of Technology, Melbournc, Florida 32901. Mr. Buyer's present address: 7785 Caminito Glorita, San Diego, California 92122. Mr. Heaton's present address: Inerstate Electronics Corp. 707 E. Vermont Avenue, P.O. Box 3117, Anaheim, California 92803. Ms. Oakley's and Mr. Wilke's present address: Oceanographic Sciences Division, Brookhaven National Laboratory, Upton, New York 11973. Dr. Dayal's present address: Department of Nuclear Energy, Brookhaven National Laboratory, Upton, New York 11973. Mr. Tatro's present address: Perkin-Elmer Inc., Elmwood Park, New Jersey 07407. Dr. Hershey's present address: Rosenstiel School of Marine and Atmospheric Science, University of Miami, Miami, Florida 33149.

ABSTRACT

Fly ash and scrubber sludge wastes (mainly calcium sulfite hemihydrate and gypsum) from coal-burning power plants were stabilized with lime to produce solid, brike-like forms. The flux of calcium (Ca^{2+}) and sulfate ($SO_4{}^{2-}$) ions from blocks placed in test tanks containing seawater or estuarine water was measured for 168 days for one waste and 147 for another. Initially, fluxes of Ca^{2+} and $SO_4{}^{2-}$ leaving the blocks were about 10^{-7} mole mm^{-2} d^{-1} for two different wastes with approximately the same fly ash:sludge ratio of 1:1; at the end of the experiment, fluxes had decreased to 2×10^{-8} to 3×10^{-8} mole mm^{-2} d^{-1}. A one-dimensional model based on diffusion was developed to describe the fluxes. The model predicts diffusivities of 1.2×10^{-9} to 3×10^{-9} cm^2 s^{-1} and the depth (x_c) of penetration of the diffusion process: in 10 days, x_c = 0.06–0.10 cm; in 365 days x_c = 0.39–0.61 cm; and in 30 years, x_c = 2.1–3.3 cm. The results of the model suggest that stabilized coal waste blocks will have a long life in seawater if erosion and biological processes do not have a major effect on block properties.

19.1. INTRODUCTION

The present national energy policy encourages a rapid increase in the use of coal. Coal-fired power plants equipped with fly ash precipitators and flue gas desulfurization scrubbers generate large amounts of solid wastes that must be removed from the plant site. Typically, a 1000 megawatt plant will consume 2.4 million tons (metric) of 15% ash–3.5% sulfur coal annually and generate about 644,000 tons of ash and scrubber sludge. The disposal of these wastes must be accomplished in an environmentally acceptable manner.

For four years a multidisciplinary team of scientists and engineers from the Marine Sciences Research Center (Woodhead and Duedall, 1979), the Department of Materials Science at the State University of New York (Carleton and Wang, Chapter 18), IU Conversion Systems, Inc. (IUCS, Horsham, Pennsylvania), and the Horn Point Environmental Laboratory of the University of Maryland have been investigating the composition, physical and chemical behavior, and biological acceptability of blocks of stabilized fly ash and scrubber sludge in seawater. The scrubber sludge is a mixture of calcium sulfite hemihydrate ($CaSO_3 \cdot \frac{1}{2}H_2O$) and gypsum ($CaSO_4 \cdot 2H_2O$), the relative proportions of which depend on the power plant characteristics (that is, the scrubber process and boiler efficiency). Stabilization is accomplished by adding lime to fly ash and scrubber sludge. The lime reacts chemically with the fly ash to form compounds possessing cementitious properties; the resultant mixture forms a hard mass. Preliminary activities of this investigation were reported by Duedall et al. (1981); other reports include Seligman and Duedall (1979), Heaton et al. (1982); Humphries (1982); Roethel et al. (1980), and Parker et al. (1981).

One area of research has been determination of the rate of release in seawater of major block components, principally the ions calcium (Ca^{2+}), sulfite ($SO_3{}^{2-}$), and sulfate ($SO_4{}^{2-}$) from phases including $CaSO_3 \cdot \frac{1}{2}H_2O$, $CaSO_4 \cdot 2H_2O$, calcium carbonate ($CaCO_3$, mainly calcite), lime [$Ca(OH)_2$], and ettringite [$3CaO \cdot Al_2O_3 \cdot 3CaSO_4 \cdot 32H_2O$, a synthetic mineral (Faust, 1977)] present in the blocks. These phases are only sparingly soluble in aqueous systems. However, over long time periods (years), the slow release of Ca^{2+}, $SO_3{}^{2-}$, and $SO_4{}^{2-}$ from the solid waste could lead ultimately to block failure through increased porosity and/or chemical dissolution.

Experiments were conducted using solid blocks of coal waste suspended in tanks of seawater to determine the rate of release of Ca^{2+} and $SO_3{}^{2-}$ + $SO_4{}^{2-}$ and to study the kinetic processes involved. Because $SO_3{}^{2-}$ rapidly oxidizes to $SO_4{}^{2-}$, only the combined release of $SO_3{}^{2-}$ + $SO_4{}^{2-}$ could be determined. We report here results of the experiments and also preliminary results from a mathematical model in which the flux of Ca^{2+} and $SO_3{}^{2-}$ + $SO_4{}^{2-}$ (hereafter referred to as $SO_4{}^{2-}$) from the block is described as a diffusion process. Calculated from the model are the diffusivity (diffusion coefficient) and the depth of penetration, as a function of time, of the diffusion zone in the block. Knowledge of the depth of the diffusion zone permits an estimation of the lifetime of the block.

19.2. METHODS

The work was conducted on blocks from two different coal wastes. The Elrama blocks were made from fly ash and scrubber sludge obtained from the Elrama power plant (Duquesne Light) located near Pittsburgh, Pennsylvania; the blocks had a nominal fly ash:scrubber sludge ratio of 1:1. The Conesville blocks were made of coal waste from the Conesville power plant (Columbus and Southern Ohio Electric) located near Coshocton, Ohio; two fly ash:scrubber sludge ratios (1:1 and 3:1) were tested.

The coal wastes were stabilized using Poz-o-Tec® process (IUCS, Horsham, Pennsylvania). Fine hydrated lime was added to all fly ash–scrubber sludge mixes to stabilize them into block forms. The bulk mineralogy (Figs. 19.1 and 19.2) was determined by x-ray diffraction analysis of unoriented mounts of powdered waste-block materials. Copper Kd Ni-filtered radiation was used and the scan was from 5° to 65° 2Θ, where Θ is the half angle between the incident and diffracted x-rays. Major components of the blocks are given in Table 19.1.

The Elrama test blocks were cut from a larger (0.028 m^3) block using a band saw. The blocks had nominal geometric surface areas of 130, 260, and 520 cm^2. For the Conesville blocks, the samples were cut from 7.6 × 10.2 cm proctor cylinders; the final shape tested was a solid cylinder rather than a cube and only the 260 cm^2 size block was investigated.

The Elrama test blocks were suspended by monofilament line in plastic tanks (Fig. 19.3) that were filled with 1.3, 2, or 4 liters of Chesapeak Bay estuarine water ($S \simeq 7‰$) from the Choptank River. The Conesville blocks were also suspended in plastic tanks containing 2 liters of seawater ($S \simeq 31‰$) from the Atlantic Ocean south of Long Island. Each tank-block system was duplicated. The seawaters

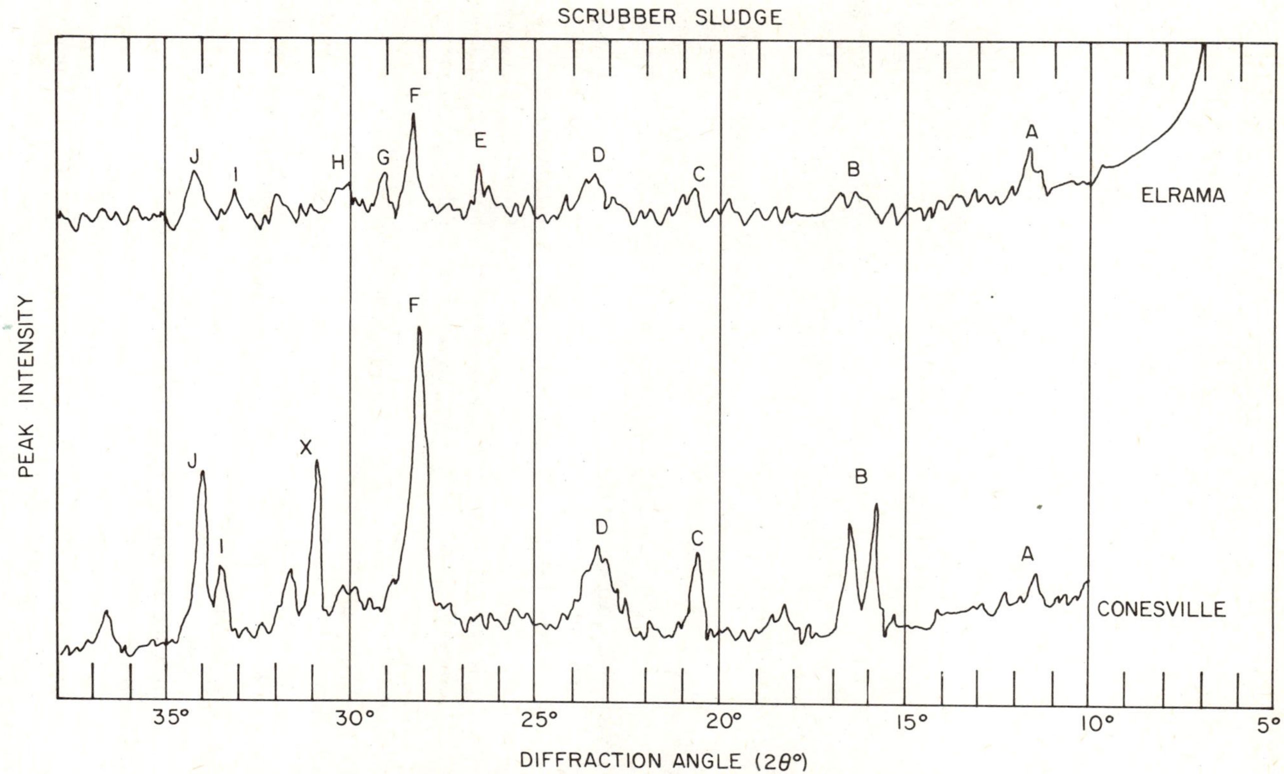

Figure 19.1. X-ray diffractograms for scrubber sludges obtained from Elrama and Conesville power plants. A, C, G, X gypsum; B, D, F, H, I, J calcium sulfite hemihydrate; E quartz.

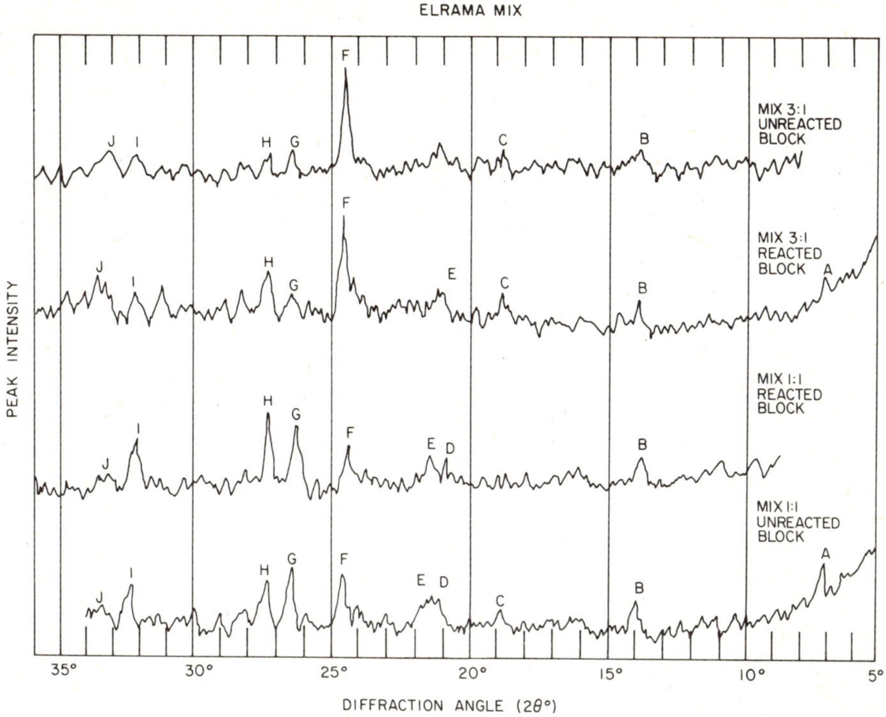

Figure 19.2. X-ray diffractograms of unreacted (with seawater) and reacted samples of blocks of stabilized fly ash and scrubber sludge. A, D ettringite; B, E, G, I calcium sulfite hemihydrate; C, F quartz; H calcite; J magnetite plus hematite.

Table 19.1. Composition Percentage (Dry Weight) of Major Components in Blocks of Stabilized Fly Ash and Scrubber Sludge from the Elrama and Conesville Power Plants

Fly ash:scrubber ratio[a]	Fly Ash	$CaCO_3$	$CaSO_3 \cdot \frac{1}{2}H_2O$	$CaSO_4 \cdot 2H_2O$	$Ca(OH)_2$	Total
Elrama[b]						
1:1	53	17.4	18.4	4.3	–	93
Conesville						
1:1	48	2.0	33.6	–	5.9	90
3:1	71	1.1	16.9	–	5.7	94

[a]Nominal ratio; the actual ratios varied as can be seen from the analysis of fly ash.

[b]These particular analyses were performed on another set of Elrama 1:1 blocks stabilized with coarse lime (CaO) rather than fine hydrated lime.

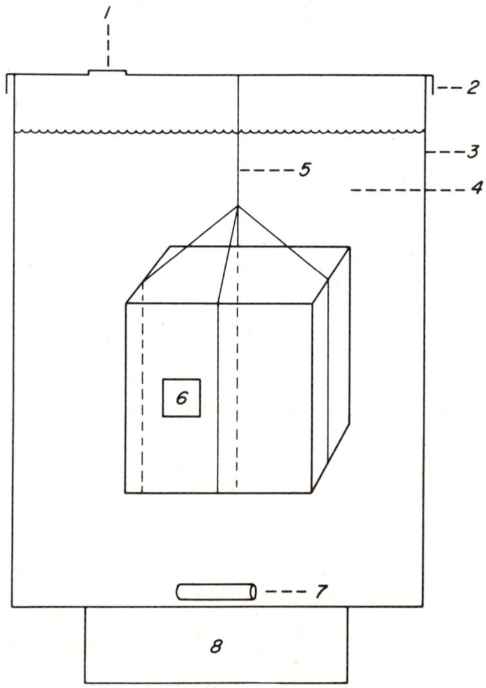

Figure 19.3. Tank configuration: (1) 0.45-μm filter; (2) cover; (3) tank; (4) filtered seawater (from the Atlantic or the Choptank River); (5) monofilament fishing line; (6) coal waste block; (7) magnetic stirring bar; (8) magnetic stirrer.

chosen for these studies were taken from field sites where quantities of the solid coal wastes were placed on the seabed for *in situ* studies of block colonization by marine organisms (Parker et al., 1981; Heaton et al., 1982; Humphries, 1982). The ratio of surface area of the test blocks to volume of seawater in the test tank was held constant at about 130:1 (cm^2 $liter^{-1}$) for most of the systems studied (Table 19.2). Duplicate tanks with 2 liters of seawater, or estuarine water, were set up as controls. Each tank had a 0.45-μm membrane filter placed over an opening in its cover to insure aeration. All tanks were placed on individual magnetic stirrers and an outlet tube was attached to the bottom of each tank for sampling. The experiments were conducted at room temperature (24 ± 2°C) and lasted 147 and 168 days for the Conesville and Elrama blocks, respectively. To avoid saturation with respect to $CaSO_4 \cdot 2H_2O$, with a potential reduction in the rate of release of Ca^{2+} and $SO_4{}^{2-}$ to tank waters, the seawater in the tanks was replaced at 14 day intervals for the Elrama tanks and 14 to 19 day intervals in the first 81 days and every 27 to 39 days in the last 66 days for the Conesville tanks. Results showed that except for the initial leaching period, the solutions were greatly undersaturated with respect to gypsum, except in the one instance where the water was not replaced for 39 days in the Conesville tanks, resulting in 88% of saturation.

At 2 to 3 day intervals during the initial 14 to 19 days between tank water replacements, 5 ml of seawater were sampled from each tank for Ca^{2+} determination using a syringe equipped with a 0.45-μm membrane filter. Just before the tank

Table 19.2. Experimental Design for the Dissolution Study

Sample Designation	Fly Ash: Sludge Filtercake	Surface Area of Blocks (cm^2)	Volume of Seawater[a] (liter)	Surface Area: Volume Seawater (cm^2 $liter^{-1}$)
Convesville Blocks				
C4F1-1	1:1	258	2	129
C4F1-2	1:1	265	2	133
C4F3-1	3:1	268	2	134
C4F3-2	3:1	259	2	130
Elrama Blocks				
E3F1-1	1:1	130	1.3	100
E3F1-2	1:1	125	1.3	100
E3F1-3	1:1	280	2	130
E3F1-4	1:1	265	2	130
E3F1-5	1:1	516	4	129
E3F1-6	1:1	526	4	129

[a]S = 6.6–8.0‰ (Choptank River water) of water in tanks containing the Elrama blocks; S = 30.9‰ of seawater in tanks containing Conesville blocks. Experiments conducted at 25° ± 2°C.

water was replaced, a 100-ml aliquot was taken for SO_4^{2-} analysis and immediately frozen for later analysis in a complete batch. The frequency of Ca^{2+} sampling decreased during the latter period of the experiments because of the decreasing amounts of Ca^{2+} being released to the solution. Only the Ca^{2+} (and SO_4^{2-}) concentrations measured just before water replacement are reported here. The incremental increase in Ca^{2+} measured at 2 to 3 day intervals during the leaching period will be reported elsewhere.

pH, Eh, and the concentration of dissolved oxygen in the control water and in the test tanks were also measured just prior to water replacement. pH and Eh were determined with a glass and platinum electrode, respectively, and oxygen was determined using the micro-Winkler method (Carpenter, 1965).

The concentration of Ca^{2+} was analyzed by flame atomic absorption spectrophotometry after dilution with 1% La^{3+} solution of both samples and standards. The standard deviation (σ) of replicate analyses ($n = 4$) was 0.096 millimolar (mM). The concentration of SO_4^{2-} was determined using the barium sulfate gravimetric method. Picric acid was used to complex the major cations in seawater so that their interference in the precipitation of barium sulfate would be minimized (Bather and Riley, 1954). For 31‰ salinity, we added 50 ml of seawater to 235 ml of distilled water; 10 ml of a saturated solution of picric acid and 50 ml of concentrated reagent grade HCl were then added. For those samples with a salinity of about 7‰ (Choptank River), 5 and 2.5 ml of picric acid and HCl, respectively, were used. The recovery of SO_4^{2-}, based on Copenhagen water as a reference, was 99.43%, $\sigma = 0.55\%$ ($n = 22$), for a 7‰ salinity standard. For salinity 31‰ the recovery was 99.99%, $\sigma = 0.67\%$ ($n = 37$).

The concentration of SO_3^{2-} was measured on fresh tank water samples and found to be negligible (less than 0.025 m*M* demonstrating that the released SO_3^{2-} was oxidized to SO_4^{2-}, or alternatively the SO_3^{2-} in the solid phase $CaSO_3 \cdot \frac{1}{2}H_2O$ in the block was oxidized and released as SO_4^{2-}. The oxidation of SO_3^{2-} was also indicated (see Section 19.3.2) by the depression in dissolved oxygen concentrations observed in the tank waters containing the test blocks.

The chlorinity of the blank tank waters was determined by titration with silver nitrate using the low precision Mohr method (Strickland and Parsons, 1965) and salinity was calculated. The precision, as an average of several standard deviations, was ±0.2‰ in salinity units. The Ca^{2+}/Cl^- ratios by weight for the Elrama and Conesville blank waters were 0.0254 and 0.0212, respectively. The Ca^{2+}/Cl^- for standard seawater is 0.0210 (Kester et al., 1967); thus, the Choptank River water used to leach the Elrama blocks was enriched in Ca^{2+}. This is typically observed when river water is diluted with seawater (Berner, 1971).

19.3. RESULTS

19.3.1. Laboratory Work

The results were tabulated in terms of the blank Ca^{2+} and SO_4^{2-} concentrations (that is, the control tanks), the Ca^{2+} and SO_4^{2-} concentrations in the test tanks, the total millimoles of Ca^{2+} and SO_4^{2-} released to the tank waters, the Ca^{2+}/SO_4^{2-} mole ratio (in the solution and also for the released Ca^{2+} and SO_4^{2-}), and the ion product $(Ca^{2+}) \times (SO_4^{2+})$, expressed as molar^2 ($M^2$).

Table 19.3 is an example of these results for two runs, from the duplicate series, using 260 cm^2 Conesville and Elrama test blocks in 2-liter tanks. Figures 19.4 and 19.5 show the rates of release (normalized flux, J) of Ca^{2+} and SO_4^{2-} to the tank waters for the Elrama and Conesville blocks, respectively. The normalized fluxes [$J(Ca^{2+})$ and $J(SO_4^{2-})$] were calculated by dividing the incremental increase in millimoles of Ca^{2+} and SO_4^{2-} by the dissolution period (14 days for the Elrama blocks and 14 to 39 days for the Conesville Blocks) and by the geometric surface area (mm^2) of the block. The fluxes reported, therefore, are average values over a finite period rather than instantaneous values.

The similarity of $J(Ca^{2+})$ and $J(SO_4^{2-})$ for the different size Elrama blocks (Fig. 19.4) demonstrates the importance of surface area in controlling the release of Ca^{2+} and SO_4^{2-} to the tank waters. The larger values of $J(Ca^{2+})$ relative to $J(SO_4^{2-})$ are probably due to the presence of minor amounts of $CaCO_3$ and unreacted lime (Table 19.1). Ettringite ($3CaO \cdot Al_2O_3 \cdot 3CaSO_4 \cdot 32H_2O$) may also contribute to the $J(Ca^{2+})$ and $J(SO_4^{2-})$; however, nothing is known about its solubility in seawater (Faust, 1977). The larger $J(Ca^{2+})$ and $J(SO_4^{2-})$ (Fig. 19.5) observed for the 1:1 blocks versus those for the 3:1 blocks are probably due to the correspondingly greater concentrations of Ca^{2+}, SO_3^{2-}, and SO_4^{2-} in 1:1 blocks (Table 19.1).

The exponential drop in $J(Ca^{2+})$ and $J(SO_4^{2-})$ with increasing time demonstrates that the surface of the blocks becomes fairly rapidly depleted in Ca^{2+} and

Table 19.3. Sample of Results for Ca^{2+} and SO_4^{2-} Leached during the Tank[a] Dissolution Experiments

Days	Blank (mM)		Solution (mM)		Leached (mM)		Ca^{2+}/SO_4^{2-} (Mole Ratio)		Ion Product (M^2)
Elapsed	$[Ca^{2+}]$	$[SO_4^{2-}]$	$[Ca^{2+}]$	$[SO_4^{2-}]$	Ca^{2+}	SO_4^{2-}	Solution	Leached	$[Ca^{2+}]\times[SO_4^{2-}]$
Elrama 1:1 Block (265 cm^2)									
14	2.30	5.40	20.5	21.4	36.4	32.1	0.96	1.13	4.39×10^{-4}
28	2.30	5.40	16.3	17.6	28.0	23.9	0.93	1.17	2.87×10^{-4}
42	2.30	5.40	12.7	16.8	20.8	22.5	0.76	0.93	2.13×10^{-4}
56	2.74	6.20	12.9	13.9	20.3	15.4	0.93	1.32	1.79×10^{-4}
70	2.74	6.20	12.4	13.1	19.3	13.9	0.95	1.39	1.62×10^{-4}
84	2.74	6.20	10.7	12.0	16.0	11.8	0.89	1.36	1.28×10^{-4}
98	2.74	6.20	10.3	11.6	15.2	10.9	0.89	1.39	1.19×10^{-4}
112	2.87	6.56	9.7	11.4	13.8	9.7	0.85	1.42	1.06×10^{-4}
126	2.87	6.56	9.1	11.1	12.4	9.1	0.82	1.36	1.01×10^{-4}
140	2.87	6.56	8.8	10.8	11.8	8.5	0.81	1.39	9.5×10^{-5}
168	2.87	6.56	7.4	10.0	9.5	6.9	0.74	1.38	7.4×10^{-5}
Conesville 1:1 Block (258 cm^2)									
15	9.26	24.4	31.9	38.3	45.3	27.8	0.83	1.63	1.22×10^{-3}
31	9.47	23.7	31.7	36.5	44.5	25.6	0.87	1.74	1.16×10^{-3}
48	9.36	24.9	26.2	34.9	33.7	20.0	0.75	1.69	9.14×10^{-4}
67	9.40	25.1	24.8	34.6	30.8	19.0	0.72	1.62	8.58×10^{-4}
81	9.36	24.9	22.5	33.2	26.3	16.6	0.68	1.58	7.47×10^{-4}
120	9.68	25.6	30.8	38.7	42.2	26.2	0.80	1.61	1.19×10^{-3}
147	9.73	26.3	24.1	35.3	28.7	18.0	0.68	1.59	8.51×10^{-4}

[a]Volume of tank = 2 liters. These data are from single tanks; typically, duplicate tanks for each block design were studies and mean values were used in the preparation of Figs. 19.4, 19.5, 19.9, and 19.10.

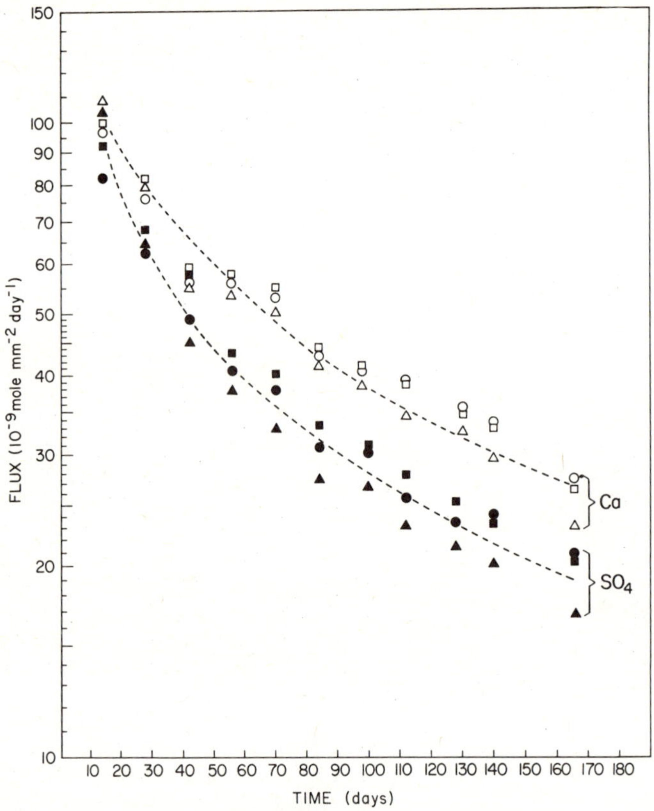

Figure 19.4. Flux of Ca^{2+} (open symbols) and $SO_4{}^{2-}$ (closed symbols), in tank waters ($S \simeq$ 7‰) containing Elrama test blocks. Nominal surface areas of blocks: (△, ▲) 130 cm^2; (□, ■) 260 cm^3; (○, ●) 520 cm^2.

$SO_4{}^{2-}$ because of the dissolution process. Diffusion of Ca^{2+} and $SO_4{}^{2-}$ from the block sublayer to the tank solution is apparently retarded by the increase in path length (tortuosity) of the diffusing ions, with the net result that $J(Ca^{2+})$ and $J(SO_4{}^{2-})$ decrease with increasing time.

Figure 19.6 shows a plot of the percent saturation, defined as $[Ca^{2+}] \times [SO_4{}^{2-}]/K_{sp} \times 100$; K_{sp} is the solubility product of gypsum. Gypsum was chosen for the saturation calculation rather than $CaSO_3 \cdot \frac{1}{2}H_2O$ because $SO_3{}^{2-}$ (from $CaSO_3 \cdot \frac{1}{2}H_2O$) is rapidly oxidized to $SO_4{}^{2-}$ (Templeton et al., 1963) resulting in a solution which is highly undersaturated with respect to $CaSO_3 \cdot \frac{1}{2}H_2O$. As previously mentioned $SO_3{}^{2-}$ was at the detection limit of our analytical method. The K_{sp} for gypsum used to compute the percent saturation was obtained from the measured concentrations of Ca^{2+} and $SO_4{}^{2-}$ in the two test seawaters (31‰ and 7‰) saturated with reagent grade gypsum. For salinity 31‰, K_{sp} was 0.00139 M^2; for the Choptank River water (S = 7‰), K_{sp} was 0.000593 M^2. These values agree within 7-10%, respectively, with the K_{sp} values reported by Marshall and Slusher (1966) and Culberson et al. (1978) for sea salt solutions and NaCl solutions of the same ionic strength as seawater. Before the coal waste blocks were

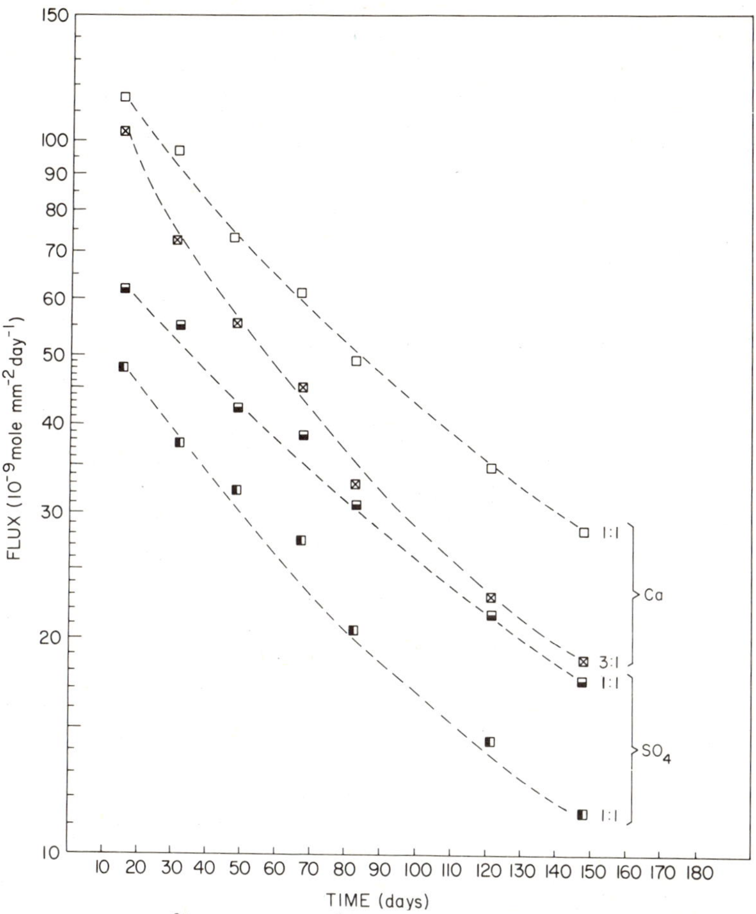

Figure 19.5. Flux of Ca^{2+} (□, ⊠) and $SO_4{}^{2-}$ (⊟, ◧) in $S \simeq 31‰$ tank waters containing Conesville test blocks. 1:1 and 3:1 refer to fly ash: scrubber sludge ratio. Nominal surface area of blocks 260 cm^2.

placed in the tanks the two test waters (Choptank River and Atlantic Ocean) were 2–3% and 18% of saturation, respectively, with respect to gypsum. In the tanks containing the Conesville blocks, where the initial salinity was 31‰, the solutions reached only partial saturation (70–80%) during the initial experimental period when the flux was greatest. In Elrama tanks, which contain 7‰ salinity seawater, the degree of saturation dropped off rapidly in the first 28 days of the experiment even though the blank waters were only 2–3% of saturation.

The Eh of the solutions in Elrama and Conesville tanks were 210 and 430 mV, respectively, with no apparent systematic variation that could be attributed to block size. The pH of the blank tank was close to 7.7, but during initial dissolution the pH dropped to as low as 6 and 7 for the Conesville and Elrama tanks, respectively. With time the pH for both systems graduallly increased; the greatest pH (~8.2) occurred in the tanks containing the 520 cm^2 Elrama blocks.

Figure 19.7 shows a plot of the average depression (test tank minus blank tanks)

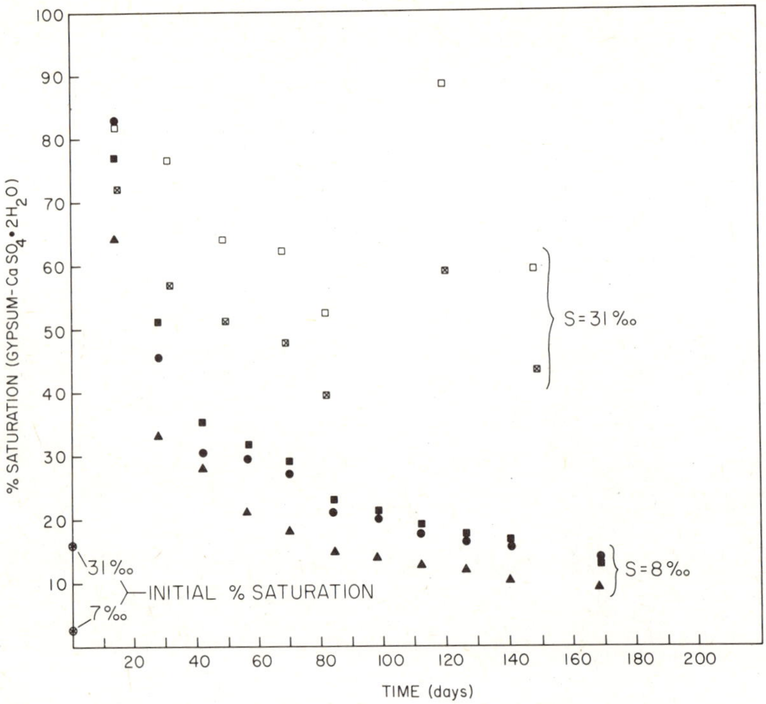

Figure 19.6. Percent saturation in solution with respect to gypsum: closed symbols, Elrama test blocks in tank waters, $S \simeq 7‰$; open symbols, Conesville test blocks in tank waters, $S \simeq 31‰$. Nominal surface area of blocks: (▲) 130 cm^2; (■, ⊠, □) 260 cm^2; (●) 520 cm^2.

of dissolved oxygen concentrations for the Elrama test tanks. Apparently, the SO_3^{2-} leached from the blocks decreased the concentration of dissolved oxygen in the test tanks. The oxygen depletion increased with increasing surface area of the test blocks (Table 19.2, Fig. 19.7). [The anomalous dissolved oxygen depression recorded for day 84 (Fig. 19.7) was caused by the magnetic stirrer accidently being shut off five days prior to sampling.] The rate of oxygen depletion showed a decreasing trend (Fig. 19.7), demonstrating a diminution in the rate of dissolution of sulfite with time. After about 160 days, the rate of oxygen depletion in the tanks was approximately balanced by the rate of oxygen replenishment. The Conesville blocks produced depressions in dissolved oxygen that averaged between -2.31 ml $liter^{-1}$ for the 1:1 blocks to -1.2 ml $liter^{-1}$ for the 3:1 blocks. There was no systematic variation with time in the results obtained for the Conesville blocks.

19.3.2. A Preliminary Model for the Diffusion of Ions in the Test Blocks

The results obtained from these experiments were used to develop a model for the diffusion of major components from the coal waste blocks to the surrounding water. The model is based on $J(Ca^{2+})$ and $J(SO_4{}^{2-})$ although we realize that $J(SO_4{}^{2-})$ is the net result of a combination of processes including dissolution of $CaSO_3 \cdot \frac{1}{2}H_2O$ and $CaSO_4 \cdot 2H_2O$, oxidation of $SO_3{}^{2-}$ to $SO_4{}^{2-}$, and diffusion.

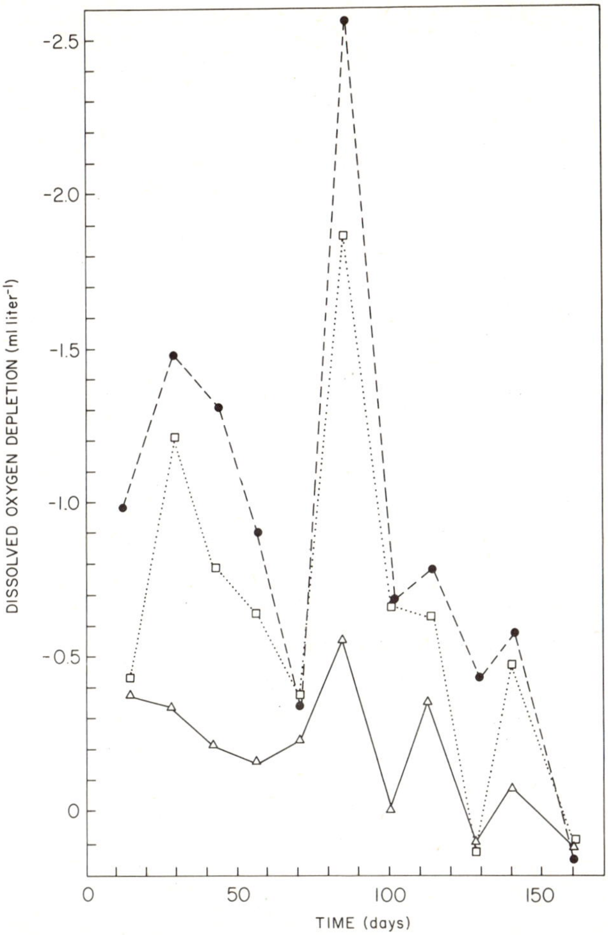

Figure 19.7. The depletion (control tank O_2 minus test tank O_2) of dissolved oxygen (ml liter^{-1}) in tank waters $S \simeq 7‰$. Nominal surface area of blocks: (△) 130 cm^2; (□) 260 cm^2; (•) 520 cm^2.

Ultimately our aim is to develop a model which will include the oxidation rate of $SO_3{}^{2-}$ to $SO_4{}^{2-}$ as well as diffusion processes in order to predict the flux of Ca^{2+}, $SO_3{}^{2-}$, and $SO_4{}^{2-}$ alone from a finite coal waste block comprised of fly ash and calcium sulfite hemihydrate to an infinite ocean (see Roethel, 1981).

For simplicity the model is one-dimensional diffusion for ions in a coal waste block placed in a well-stirred aqueous system (Fig. 19.8). Let x be the one-dimensional coordinate and let the block extended from $x = 0$ to $x = +\infty$. The diffusion of ions in the coal waste block can be studied by the equation

$$\frac{\partial S}{\partial t} = D \frac{\partial^2 S}{\partial x^2}, \qquad 0 \le x \le \infty, \qquad 0 \le t \tag{1}$$

where S is concentration (either Ca^{2+} or $SO_4{}^{2-}$), D is the diffusion coefficient (diffusivity), and t is time. Equation 1 is subjected to the initial condition, that is,

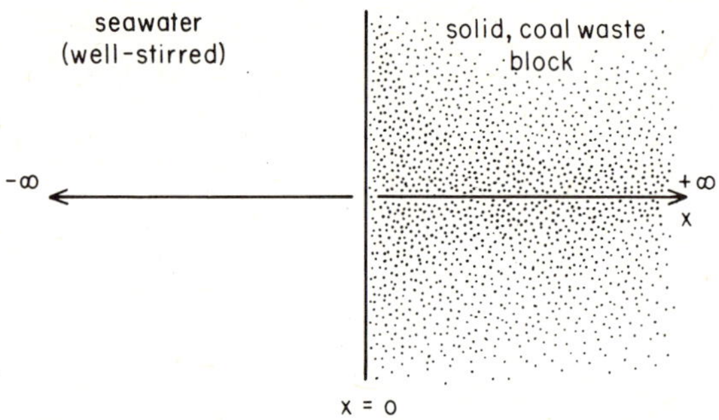

Figure 19.8. Schematic of one-dimensional diffusion.

initially uniform concentration in a block S_0,

$$\text{at } t = 0, \quad S = S_0, \quad 0 \le x \le \infty \tag{2}$$

and to the boundary condition

$$\text{at } x = 0, \quad D\frac{\partial S}{\partial x} = hS \tag{3}$$

where h is a transfer coefficient.

Boundary condition 3 states that flux of ions from solid to the tank water is proportional to the concentration at the interface. Since the water is well stirred and renewed at frequent intervals, we can assume that the excess ion concentration in water surrounding the block is negligibly small; thus

$$\phi \equiv S - \frac{D}{h} \; \frac{\partial S}{\partial S} \tag{4}$$

Equations 1 to 3 can be rewritten as

$$\frac{\partial \phi}{\partial t} = D\frac{\partial^2 \phi}{\partial x^2} \tag{5}$$

$$t = 0, \quad \phi = S_0 \tag{6}$$

$$x = 0, \quad \phi = 0 \tag{7}$$

The solution of equation 5 subject to conditions 6 and 7 is obtained by

$$\phi(x,t) = S_0 \text{ erf } \frac{x}{2\sqrt{Dt}} \tag{8}$$

where

$$\text{erf}(y) \equiv \frac{2}{\sqrt{\pi}}\int_0^y e^{-v^2}\, dv \tag{9}$$

Hence

$$\frac{\partial S}{\partial x} - \frac{h}{D}S = -\frac{h}{D}\phi = -\frac{2h}{\sqrt{\pi D}}S_0\int_0^{x/2\sqrt{Dt}} e^{-v^2}\, dv \tag{10}$$

Solving equation 10 for S and using the boundary condition at $x = \infty$, we obtain

$$S(x,t) = S_0 \operatorname{erf}\left(\frac{x}{2\sqrt{Dt}}\right) + S_0 \; e^{(hx/D + h^2 t/D)} \operatorname{erfc}\left(\frac{x}{2\sqrt{Dt}} + h\sqrt{\frac{t}{D}}\right) \qquad 11$$

where $\operatorname{erfc}(y) = 1 - \operatorname{erf}(y)$. The flux of ion through the interface is then given by

$$J = hS(x = 0) = hS_0 \, e^{(h^2/D)\,t} \operatorname{erfc}\left(h\sqrt{\frac{t}{D}}\right) \qquad 12$$

where J is the flux, that is, the amount of ion transferring through a unit surface area per unit time.

For $t >> D/h^2 \equiv t_c$, the flux is approximated by

$$J = S_0\sqrt{\frac{D}{\pi t}} \qquad 13$$

Thus, J is proportional to $t^{-1/2}$ as time progresses.

If a transfer rate in a thin surface layer is relatively fast (which is expected in this case), then in effect $h \rightarrow \infty$ or $S = 0$ for the boundary condition 3, and $S(x,t)$ is given by

$$S(x,t) = S_0 \operatorname{erf}\left(\frac{x}{2\sqrt{Dt}}\right) \qquad 14$$

and the flux is

$$J = S_0\sqrt{\frac{D}{\pi t}} \qquad 15$$

for all values of t. Thus a plot of $\log J$ against $\log t$ should be a straight line whose slope is $-\frac{1}{2}$.

Figures 19.9 and 19.10 are log-log plots of $J(Ca^{2+})$ and $J(SO_4{}^{2-})$ versus time for the tank results described earlier. Also plotted on the figures is a straight line with a slope $-\frac{1}{2}$. The results show some deviation from the theoretical slope with greater fluxes observed at the beginning of the experiment. These deviations from the $-\frac{1}{2}$ slope line may be due to rapid surface dissolution of Ca^{2+} and dissolution slowed by the tortuous path the ion must travel through the block. Other non-diffusive factors which may have caused deviation from the theoretical diffusion line include the possible rate-limiting role of the oxidation of $SO_3{}^{2-}$ to $SO_4{}^{2-}$ and changes in density of the blocks. Changes in block density have been observed (Carleton and Wang, Chapter 18).

The diffusion coefficient, D, can be estimated using the $J(Ca^{2+})$ values (Figs. 19.9 and 19.10) for test tanks containing coal waste blocks by rearranging equation 15 to give

$$D = \pi t (J/S_0)^2 \qquad 16$$

As an example for Ca^{2+}, from Fig. 19.9 (Elrama 1:1 blocks): t = 100 d; $J = 40 \times 10^{-4}$ millimoles cm^{-2} d^{-1}; S_0 = 2.61 millimoles g^{-1} (Elrama) = 4.44 millimole cm^{-3} [density (wet) = 1.7 g cm^{-3}].

Thus

$$D = 100\pi\,(40 \times 10^{-2}/4.44) = 2.55 \times 10^{-4}\ \text{cm}^2\ \text{day}^{-1} = 2.95 \times 10^{-9}\ \text{cm}^2\ \text{s}^{-1}$$

For the Conesville 1:1 and 3:1 blocks, D equals 2.3×10^{-9} and 1.2×10^{-9} $\text{cm}^2\ \text{s}^{-1}$, respectively.

The effective distance of diffusion x_c, defined as the depth where there is a change in the Ca^{2+} concentration due to diffusion, may be given by

$$x_c = \sqrt{2Dt} \qquad 17$$

Using the above value of D, x_c was calculated for various times (Table 19.4).

The model predicts that the outer 0.5 cm of the block is affected by diffusion in the first year. In 30 years the x_c will vary between 2.1 cm for a 3:1 (fly ash:filtercake ratio) block to about 3 cm for the 1:1 coal waste block.

Roethel (1981) in our laboratory is performing Ca^{2+} analysis of segmented sections of two- to three-year-old coal waste blocks recently retrieved from the Long Island Sound in order to confirm the x_c predicted from the model. It is clear, however, that the calculation of x_c from the model suggests that for a long experiment we must develop a diffusion model in a finite domain, i.e., a three-dimensional problem.

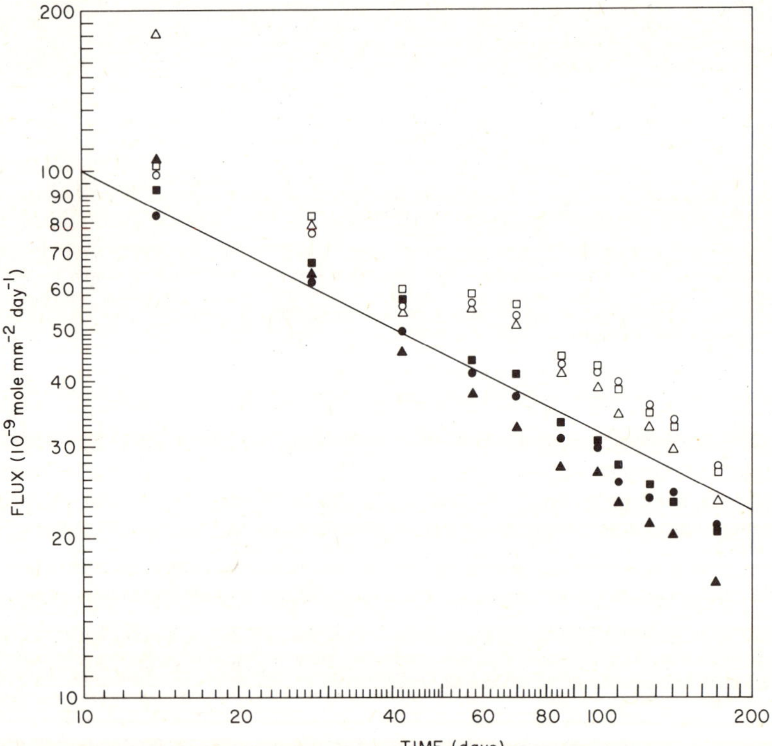

Figure 19.9. The flux, plotted on a log-log scale, of Ca^{2+} (open symbols) and of SO_4^{2-} (closed symbols) from Elrama blocks to tank waters whose $S \simeq 7‰$. Nominal surface area of blocks: (▲, △) 130 cm^2; (■, □) 260 cm^2; (●, ○) 520 cm^2. The straight line is the theoretical slope of $-\frac{1}{2}$ predicted from diffusion model (see text).

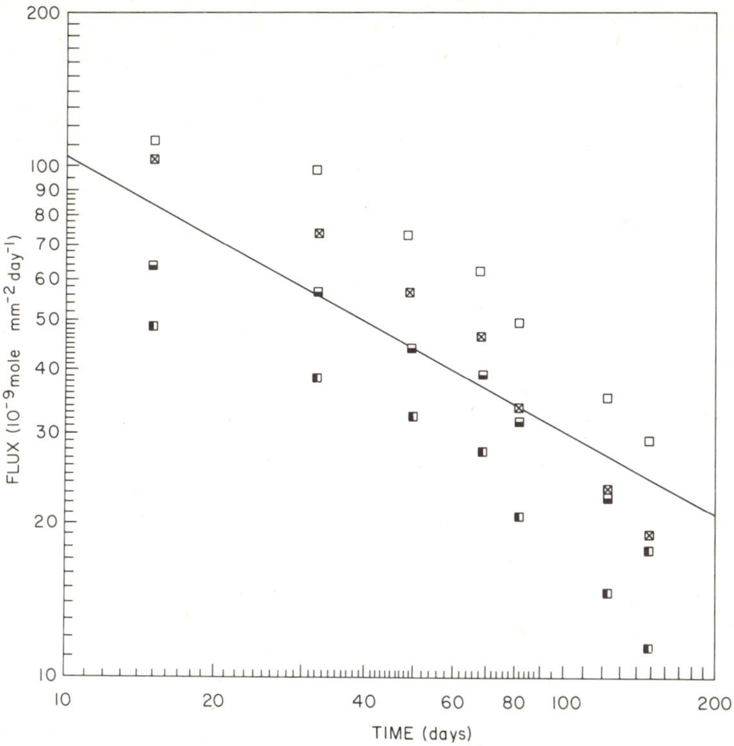

Figure 19.10. The flux, plotted on a log-log scale, of Ca^{2+} (□, 1 : 1; ⊠, 3 : 1 fly ash to scrubber sludge ratio) and of SO_4^{2-} (⊟, 1 : 1, ◧, 3 : 1 ratio) from the Conesville test blocks to tank waters was $S \simeq 31‰$. Nominal surface area of blocks: 260 cm^2. The straight line is the theoretical slope of $-\frac{1}{2}$ predicted from diffusion model (see text).

19.4. DISCUSSION

The results of this work demonstrate that large blocks of stabilized fly ash and scrubber sludge can potentially have a long life in seawater if molecular diffusion is the principal process for the loss of major components. The diffusivity of Ca^{2+} at room temperature (25°C) for a coal waste block with a fly ash to scrubber sludge ratio of 1:1 is 2.3×10^{-9} to 2.9×10^{-9} cm^2 s^{-1} or about four orders of magnitude smaller than the diffusivity of ions in a free medium (that is, seawater). The model predicts that, in 30 years, only the outer 3 cm of 1:1 coal waste blocks at 25°C and at salinity 7-31‰ would be affected by loss of Ca^{2+}. In coastal marine environments, such as the benthic regions south of Long Island, temperatures cycle annually between about 1° and 20°C. Diffusivity decreases with decreasing temperature (Li and Gregory, 1974); thus, the model predictions presented here represent maximum values for the effective depth of the diffusion zone (Table 19.4).

Destructive processes more vigorous than diffusion have been shown, however, to increase significantly the rate of block breakdown. Roethel et al. (1980) found

Table 19.4. Effective Distance (x_c) if the Diffusion[a] Zone in the Coal Waste Blocks

	x_c (cm)		
Time (d)	Elrama 1:1	Conesville 1:1	Conesville 3:1
10	0.10	0.089	0.064
100	0.32	0.28	0.20
200	0.45	0.40	0.29
365 (1 y)	0.61	0.54	0.39
500	0.71	0.63	0.45
1,000	1.01	0.89	0.64
1,095 (3 y)	1.1	0.93	0.67
3,650 (10 y)	1.9	1.7	1.2
5,000	2.3	2.0	1.4
10,000	3.2	2.8	2.0
10,950 (30 y)	3.3	2.9	2.1

[a]Diffusivities calculated from equation 16:

Block	S_0 (mmoles cm^{-3})	D ($cm^2\ s^{-1}$)
Elrama1:1	4.4	2.95×10^{-9}
Conesville 1:1	5.42	2.95×10^{-9}
Conesville 3:1	3.45	1.19×10^{-9}

that some Elrama blocks in Conscience Bay (Long Island Sound) have been bored by the clam, *Zirfaea crispata*. Humphries (1982) reported bioerosion of the Elrama blocks in the Choptank River by the mud crab *Rhithropanopeus harrisii*. Humphries also observed breakdown of the block surfaces by the tube-dwelling amphipod *Corophium lacustre* and the barnacle *Balanus* improvisus. Conversely, Conesville blocks from Conscience Bay examined by Roethel et al. (1980) showed very little evidence of boring. This may be due to increased hardness of the Conesville blocks over those made with Elrama coal waste, the greater concentration of $CaSO_3 \cdot \frac{1}{2}H_2O$ (Table 19.1) leading to locally greater depletions of dissolved oxygen at and very near the surface of a Conesville block, or some combination of these and other unknown factors.

Figure 19.11 compares the condition of coal waste blocks from three different environments: the test tanks (this study), Conscience Bay, Long Island Sound (Roethel et al., 1980), and the Choptank River (Humphries 1982). The tank blocks (after 168 days) used in the present study were only subjected to effects of circulating seawater and the diffusion process, and show virtually no physical degradation. The Elrama blocks retrieved from Conscience Bay adjacent to Long Island Sound show bored holes by *Zirfaea crispata*, as discussed previously. However, Conesville blocks taken from the same area were well preserved after 238 days. Degradation of the Elrama blocks taken from the bottom of the Choptank River in the Chesapeake Bay was caused by various organisms living in and around the site where a number of blocks were placed in *in situ* studies (Humphries, 1982). The relative ease in which Elrama blocks were degraded in Choptank River is probably more a function of the block making procedure. Recent findings show

Figure 19.11. Condition of coal waste test blocks taken from different environments.

that new mixes made with automatic block machines (Parker et al. 1981) and placed near the Choptank River in the Chesapeak Bay are resisting errosion by organisms (E. Humphries, personal communication).

Although some organisms degrade a block, others may stabilize the coal waste material. One of the blocks shown in Fig. 19.11 is coated with a colony of bryozoan (organisms with a calcium carbonate skeleton) that may reduce the $J(Ca^{2+})$ and $J(SO_4{}^{2-})$ and extend the predicted lifetime of the blocks.

19.5. CONCLUSION

The complexity of alteration of the surface of the coal waste block brought about by either bioerosion (Humphries 1982; Roethel et al., 1980) or biostabilization limits further extrapolation of these laboratory results to the marine environment. Recently, Roethel (1981) confirmed the results of the diffusion model by determining the extent Ca^{2+} depletion in coal waste blocks that were in the sea for two to three years. Future work will focus on dissolution studies of coal waste blocks that have been bioeroded or whose surfaces have been otherwise modified or stabilized by marine organisms.

ACKNOWLEDGMENTS

This research was sponsored by the State of Maryland (Power Plant Siting Committee), New York Energy and Research Development Authority, the Department of Energy, Environmental Protection Agency, Electric Power Research Institute, and the Power Authority of the State of New York. We thank IU Conversion Systems, Inc., for their cooperation, in particular Mr. J. Colussi and Mr. Robert G. Hilton for preparation of blocks and helpful discussions. We are especially grateful to Mrs. J. Restivo, Mrs. M. A. Lau, and Mrs. L. Antonacci who assisted in preparation of the typescript.

This work is Contribution 323 of the Marine Sciences Research Center of the State University of New York at Stony Brook.

REFERENCES

Bather, J. M., and J. P. Riley. 1954. The chemistry of the Irish Sea. Part. 1. The sulphate-chlorinity ratio. *Journal du Counseil*, **20**, 145–152.

Berner, R. A. 1971. Principles of Chemical Sedimentology. McGraw-Hill, New York, 240 pp.

Carpenter, J. H. 1965. The Chesapeake Bay Institute technique for the Winkler dissolved oxygen method. *Limnology and Oceanography*, **10**, 141–143.

Culberson, C. H., G. Latham, and R. G. Bates. 1978. Solubilities and activity coefficients of calcium and strontium sulfates in synthetic seawater at 0.5 and 25°C. *Journal of Physical Chemistry*, **82**, 2693–2699.

Duedall, I. W., F. J. Roethel, J. D. Seligman, H. B. O'Connors, J. H. Parker, P. M. J. Woodhead, R. Dayal, B. Chezar, B. K. Roberts, and H. Mullen. 1981. Stabilized power plant scrubber sludge and fly ash in the marine environment. *In*: Ocean Dumping of Industrial Wastes, B. H. Ketchum, D. R. Kester, and P. K. Park (Eds.). Plenum Press, New York, pp. 315–346.

Faust, R. J. 1977. Investigation of the Hydration Reactions of the Calcium-Aluminates with Calcium-Sulfate and with Calcium-Sulfite. Ph.D. Thesis, University of Pittsburgh, Pennsylvania, 201 pp.

Heaton, M. G., I. W. Duedall, S. Oakley, and R. J. Wilke. 1982. Section 1. Chemical and Physical Investigations. *In*: Stabilized Coal Waste in the Chesapeake Bay-Choptank River Estuary: Environmental Investigations, I. W. Duedall and E. M. Humphries (Principal Investigators). Technical Report PPRP-49, Maryland Power Plant Siting Program, Annapolis, Maryland, 2 sections (paginated separately).

Humphries, E. M. 1982. Section 2. Biological Investigations. *In*: Stabilized Coal Waste in the Chesapeake Bay-Choptank River Estuary: Environmental Investigations, I. W. Duedall and E. M. Humphries (Principal Investigators). Technical Report PPRP-49, Maryland Power Plant Siting Program, Annapolis, Maryland, 2 sections (paginated separately).

Kester, D. R., I. W. Duedall, D. N. Connors, and R. M. Pytkowitz. 1967. Preparation of artificial seawater. *Limnology and Oceanography*, **12**, 176–179.

Li, Y-H., and S. Gregory. 1974. Diffusion of ions in seawater and in deep-sea sediment. *Geochimica et Cosmochimica Acta*, **38**, 703–714.

Marshall, W. L., and R. Slusher. 1966. Thermodynamics of calcium sulfate dihydrate in aqueous sodium chloride solutions, 0–110°. *Journal of Physical Chemistry*, **70**, 4015–4027.

Parker, J. H., P. M. J. Woodhead, and I. W. Duedall. 1981. Coal Waste Artificial Reef Program, Phase 3, Volume 2: Comprehensive Report. EPRI CS-2009, Electric Power Research Institute, Palo Alto, California, 4 sections (paginated separately) + 1 appendix.

Roethel, F. J., I. W. Duedall, H. B. O'Connors, J. H. Parker, and P. M. J. Woodhead. 1980. The interactions of stabilized scrubber sludge and fly ash with the marine environment. *Journal of Testing and Evaluation*, **8**, 250–254.

Roethel, F. J. 1981. The Interactions of Stabilized Power Plant Coal Wastes with the Marine Environment. Ph.D. Thesis, State University of New York, Stony Brook, 349 pp.

Seligman, J. D., and I. W. Duedall. 1979. Chemical and physical behavior of stabilized scrubber sludge and fly ash in seawater. *Environmental Science and Technology*, **13**, 1082–1087.

Strickland, J. D. H., and T. R. Parsons. 1965. A Manual of Seawater Analysis. Bulletin No. 125, Fisheries Research Board of Canada, Ottawa, 203 pp.

Templeton, C. C., S. S. Rushing, and J. C. Rodgers. 1963. Solubility factors accompanying oxygen scavenging with sulfite and oil field brines. *Materials Protection and Performance*, **2**, 42–47.

Woodhead, P. M. J., and I. W. Duedall. 1979. Coal Waste Artificial Reef Program, Phase I. EPRI FP-1252, Electric Power Research Institute, Palo Alto, California, 7 sections (paginated separately).

PART VII: CONCLUDING REMARKS

20

SCIENTIFIC STRATEGY FOR INDUSTRIAL AND SEWAGE WASTE DISPOSAL IN THE OCEAN

Iver W. Duedall

Marine Sciences Research Center
State University of New York
Stony Brook, New York

Bostwick H. Ketchum

Senior Scientist Emeritus
Woods Hole Oceanographic Institution
Woods Hole, Massachusetts

P. Kilho Park

Ocean Dumping Program
National Oceanic and Atmospheric Administration
Rockville, Maryland

Dana R. Kester

Graduate School of Oceanography
University of Rhode Island
Kingston, Rhode Island

Dr. Duedall's present address: Department of Oceanography and Ocean Engineering, Florida Institute of Technology Melbourne, Florida 32901

ABSTRACT

Ocean dumping of industrial and sewage wastes will continue into the future. The scientific evaluation of ocean dumping requires a strategy of fundamental research in biological, chemical, physical and sedimentogical processes in marine systems, investigations of waste characterization, studies of toxicity mechanisms and sublethal and biological effects based on laboratory microcosm and fieldwork, determination of fates and pathways of wastes, and the development of mathematical models describing biological, chemical, and physical processes.

When possible and depending on the properties of the waste, dispersal deepwater dumpsites should be used for waste disposal instead of sites closer to shore. A program to recycle wastes will reduce the quantity of waste material produced as well as decrease the rate of depletion of those elements whose present abundance is low. Research into innovative solutions of waste disposal that minimize environmental impacts should be encouraged.

20.1. INTRODUCTION

Our complex socioeconomic civilization has created areas of dense population where large amounts of waste materials of great variety are produced. Disposal of this waste should be a high-priority problem and there are no simple solutions. In essence contemporary civilization is "fouling its own nest" and we are entering into a period of waste disposal crisis.

Economics, scarcity of land, large population densities near the ocean (Fig. 20.1), and groundwater pollution from landfills are some of the critical factors in coastal areas that make land disposal difficult, expensive, and environmentally unacceptable. The atmosphere over most urban areas is recognizably polluted, and regulations designed to clean the air are in force; in some countries land-based incineration of wastes, which aggrevates air pollution in these areas, is clearly an unacceptable option. Moreover, as land utilization, economic pressure, and sophisticated waste production increase, there will be greater demand to use the sea for waste disposal.

Industrial and sewage wastes disposed at sea are complex and highly variable mixtures. Each waste may contain hundreds of chemicals or compounds; each waste will behave differently in the ocean depending on waste characteristics, dumping practices, seasons, and physical, chemical, and biological processes in the sea; and each waste will produce different effects or impacts which will occur on varying time scales.

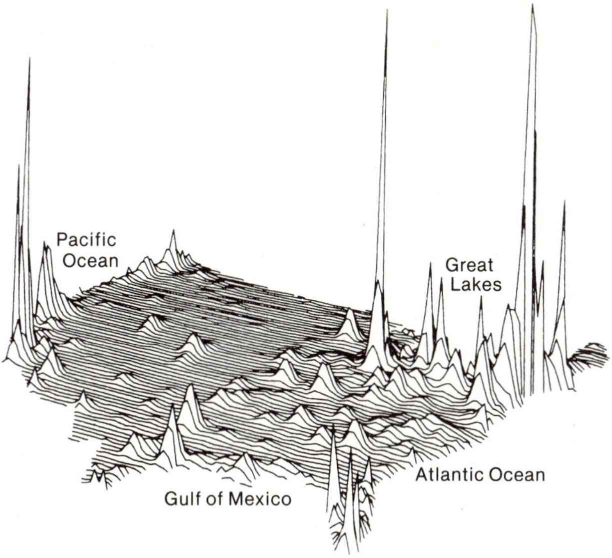

Figure 20.1. U.S. population density in 1979 (Squires, 1981) displayed on a computer-generated three-dimensional contour map. (Contour map courtesy of the Laboratory for Computer Graphics and Spatial Analysis, Harvard Graduate School of Design, Harvard University.)

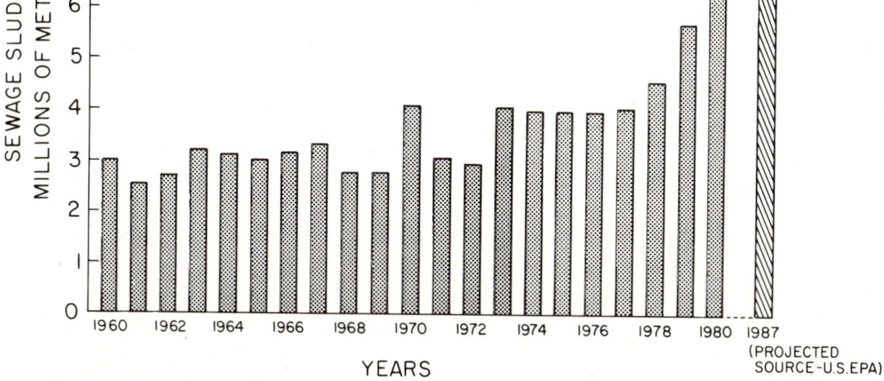

Figure 20.2. Quantities of sewage sludge in the New York Bight (Swanson et al., 1981).

Chapter 1 shows that the number of permits on a global basis for the ocean disposal of wastes increased from 1976 through 1979, the period for which data are now available. This trend will continue. The 1970s, for example, was a period in which many U.S. municipalities upgraded existing sewage treatment plants and built new ones (Chapter 2). Many plants are in service or will shortly come on line, hence sewage sludge production has been, and will be, increasing. By the year 1987 a projected 8.3 million metric tons annually of wet sludge may be ocean-dumped in the New York Bight, United States, an increase of about 30% over the 1980 quantity (Swanson et al., 1981) (Fig. 20.2). Also, now that several industrialized countries are rapidly turning to widespread use of coal to satisfy their energy needs and to produce chemicals, the future will bring large amounts of fly ash and other coal wastes which will require disposal.

The main purpose of this volume is to increase scientific understanding of the effects on the sea of ocean disposal of waste. This final chapter will emphasize a scientific strategy for industrial and sewage waste disposal in the ocean and will discuss deepwater dispersal disposal and research needs for deepwater dumpsites; we believe any disposal area should be located away from population centers and the fertile continental shelves. Finally, we discuss briefly the potential for reuse and recycling of wastes and the need to avoid bureaucratic conflicts which can beset innovative solutions to waste disposal.

20.2. STRATEGY

A critical evaluation of the problems associated with waste disposal is difficult to complete, requiring development of strategies based on scientific principles, the working together of scientists from several disciplines, and execution of large comprehensive investigations. Moreover, it will require work in agreed upon priority areas such as those listed by Goldberg (1979) and by the IMCO/FAO/UNESCO/WMO/WHO/IAEA/UN/UNEP Joint Group of Experts on the Scientific Aspects of Marine Pollution (GESAMP) (1982):

1. Types, amounts, and characteristics of wastes.
2. Toxicity of wastes to marine organisms.
3. Fates and pathways of wastes.
4. Mathematical models to predict effects of wastes.

The need to determine the ability or capacity of a body of water to assimilate pollutants or wastes without causing unacceptable biological impact is the strategy behind the selection of this particular set of priority areas (National Academy of Sciences, 1975; Goldberg, 1976, 1979). The Baltic Sea, with its very low or zero oxygen concentrations, may be cited here as an example where the capacity to assimilate organic wastes is considered exceeded sometimes (Fig. 20.3).

While this volume was not organized according to these priority areas, it has considered all of them. For example, the Inter-Governmental Maritime Consultative Organization (IMCO) has records available on the types and amounts of wastes

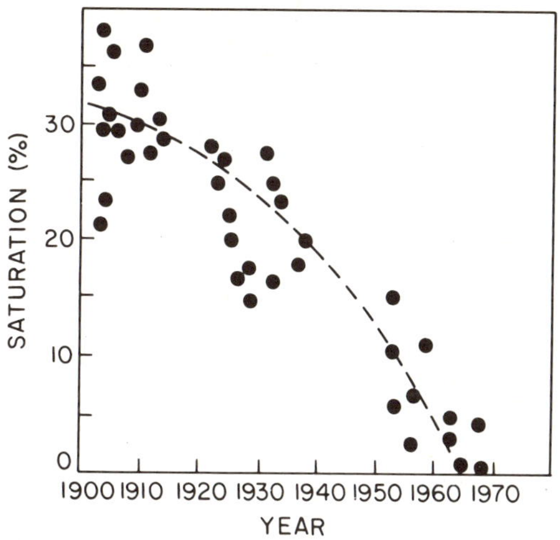

Figure 20.3. Decreasing concentrations of dissolved oxygen in the Baltic Sea (Fonselius, 1970; with permission of *Environment.*)

disposed of in the ocean by the member countries to the London Dumping Convention (Chapter 1). The compilation of these records by IMCO represents an important first step in providing a global inventory of the types and amounts of wastes dumped in the ocean. Further, these data provide a time series, albeit short, demonstrating increasing trends in ocean disposal and the location of the most active dumpsites.

Although such an inventory is useful, it is the characteristics of wastes, along with types and amounts and the conditions at the dumpsite, that determine a waste's behavior and potential hazard in the ocean. In this respect we have tried to fill partially this gap by giving in Chapter 1 some values for physical and chemical characteristics of wastes, such as density, particle size and morphology, settling velocities (of solids), composition and mineralogy, concentrations of elements and compounds, and chemical speciation. Chapters 7 through 9, 14, 18, and 19 also contain information on properties of acid-iron wastes, pharmaceutical wastes, sewage sludge, and coal ash. But in general we believe these and other wastes have not been characterized adequately. This lack of information hampers any serious effort to assign human or ecological risks associated with effects of disposal of wastes in the ocean (Goldberg, 1979; O'Connor and Stanford, 1979).

More and complete quantitative information is needed on waste properties and their fluctuations. Knowledge of this source function would permit calculations to develop hypotheses on the behavior of wastes in the sea (Chapter 3), to design appropriate field experiments, to define monitoring programs, to perform better evaluations of the biological response (human and ecological) to the potentially toxic compounds or elements present in the waste, or, under certain circumstances, to trace the movements of waste deposits in the sediment of dumpsites.

Evaluation of toxic or harmful effects of ocean-dumped wastes on marine organisms and communities will require a broad strategy of combined studies of fundamental biological and ecological processes, both in the field and laboratory, and of laboratory bioassays.

Knowledge of laboratory bioassays alone is of limited use in predicting biological effects of wastes because of the artificial conditions imposed by exposure of organisms to closed systems; in a real-world setting, waste is continually being diluted, chemical processes are modifying the waste components, time and length scales of exposure are changing, and some organisms are avoiding the wastes successfully.

But often, field experiments are difficult to control. One can, however, use microcosms such as CEPEX (Controlled Ecosystem Pollution Experiment) bags or MERL (Marine Ecosystems Research Laboratory) tanks to investigate effects of pollutants on organisms and on several components of an ecosystem. These intermediaries, which can simulate and duplicate natural seasonal cycles, are excellent research tools filling the gap between laboratory-beaker experiments and at-sea studies (Davies and Gamble, 1979; Steele, 1979). Recently, Hinga et al. (1980) were able to trace the biogeochemistry of the polycyclic aromatic hydrocarbon benzanthracene, demonstrating that 29% of the original material introduced into a MERL tank 216 days earlier appeared as respired carbon dioxide and 21 or 22% appeared in the sediment as benzanthracene and its degradation on intermediate products.

Biological effects observed in the vicinity of a dumpsite must be measured against naturally occurring changes in the community (Chapter 11). Recently, Wooster (1981) has commented critically on the problem of using biological effects of pollution monitoring in the absence of adequate data on the normally occurring biological and environmental variability. According to Wooster, "while the reaction of individuals to pollutants is relatively easy and quick to determine, it is the effects on populations that matter, and determination of these may require years." For the most part the fish consumed by humans have a relative long generation time. Consequently the effects of marine pollution on this resource may become evident only after a substantial period of time. Bacteria is one group of organisms whose changes in community structure may show a relatively quick response to effects of ocean dumping (Chapter 10). Moreover, marine bacteria play a vitally important role in regenerating nutrients and in providing a food source for benthic organisms and therefore they are very important to the marine trophic system as a whole.

Response by an organism to a waste can occur at sublethal levels which may be difficult or impossible to detect in short term (days to weeks) laboratory experiments. Responses either at the lethal level, that is causing death or serious disease or morphological changes (Chapter 12), or at sublethal levels may be a nonlinear function of concentration or doses of the waste (Fig. 20.4). And, as Waldichuk (1979) has pointed out, if one observes no response to a waste in a particular biological test we cannot assume that all possible tests will also yield no response.

It is a basic axiom of science, never to be forgotten, that *it is impossible to prove the negative*. People ask scientists to: "Prove that this waste disposal will not harm the ecosystem." Such proof is never attainable. One can only say that no effects were observed under the defined conditions and by the measurements made. However, if an effect is measured one can then calculate the limits of concentration

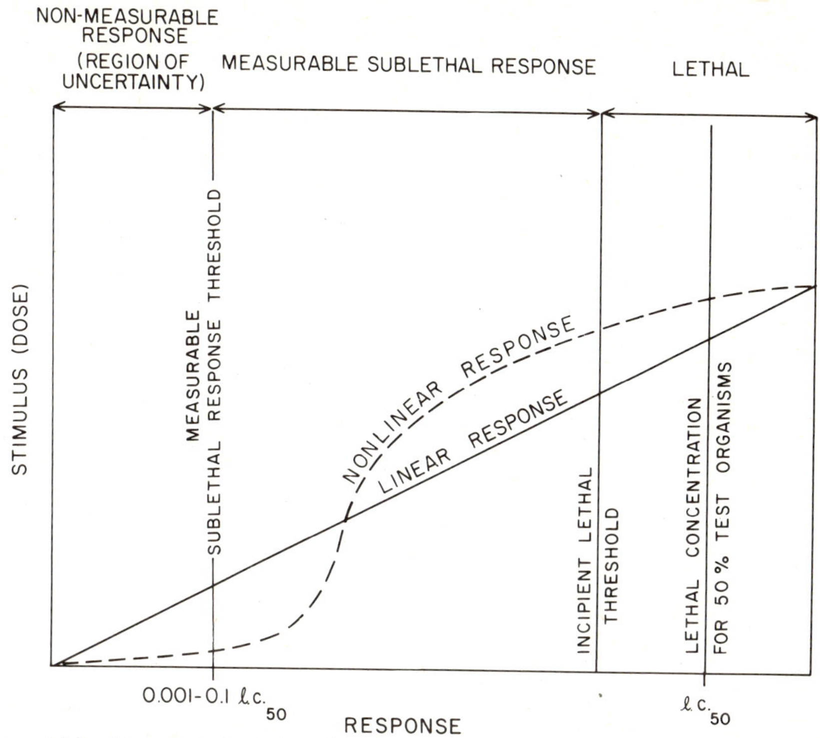

Figure 20.4. Hypothetical relation of concentration of pollutant to response of marine organisms, showing some significant points and regions on the curve (Waldichuk, 1979; with permission of The Royal Society.)

and duration of exposure which will produce the effect. Whether lesser concentrations will be relatively safe is a judgment by the scientists, and only passage of time will show whether this judgment was correct.

Chemical studies of wastes disposed in the ocean play several roles in the elucidation of fates and pathways of waste materials. Chemical analyses at disposal sites are required to verify the extent to which the marine environment is modified by the wastes. The results obtained also provide information on the rates of dispersion and displacement of the waste. This information provides the basis for estimating the biological exposure to wastes disposed in the ocean and for assessing biological impact.

Most wastes are not merely diluted passively in the marine environment. They undergo a variety of chemical and biological transformations including oxidation, incorporation into solid phases, and uptake by organisms. Many contaminants discharged to the ocean ultimately become indistinguishable from natural substances, in concentration and in chemical form, and at that point the contaminants are considered assimilated. Special care must be given to those wastes that cannot be assimilated in this sense; these include radionuclides, many synthetic organic compounds, and possibly petroleum products.

This volume includes only a small sampling of the chemical studies required for sound waste disposal management. It documents the role of seasonally variable oceanographic conditions on the dispersion of a waste (Chapter 7). It illustrates the need to devise new techniques and approaches to sampling and analysis in order to study waste disposal problems (Chapter 8). It is evident that each type of waste and each site of disposal must be evaluated individually with the present state of oceanographic knowledge. While the deepwater dumpsite in the New York Bight appears to be highly dispersive (Chapter 6), the deepwater dumpsite north of Puerto Rico shows an accumulation of wastes in the water column (Chapters 9 and 10). A bulk chemical analysis of a waste is only the first step in its characterization. It is necessary to consider the subsequent chemical processes which occur after the waste is placed in the marine environment (Chapter 19).

In dispersive dumpsites the wastes disposed in the ocean may be transported relatively long distances in relatively short periods of time (Fig. 20.5). Physical transport processes including advection, diffusion, dispersion, shear, and particle settling are responsible for the movement of wastes (Chapters 3 through 6). Know-

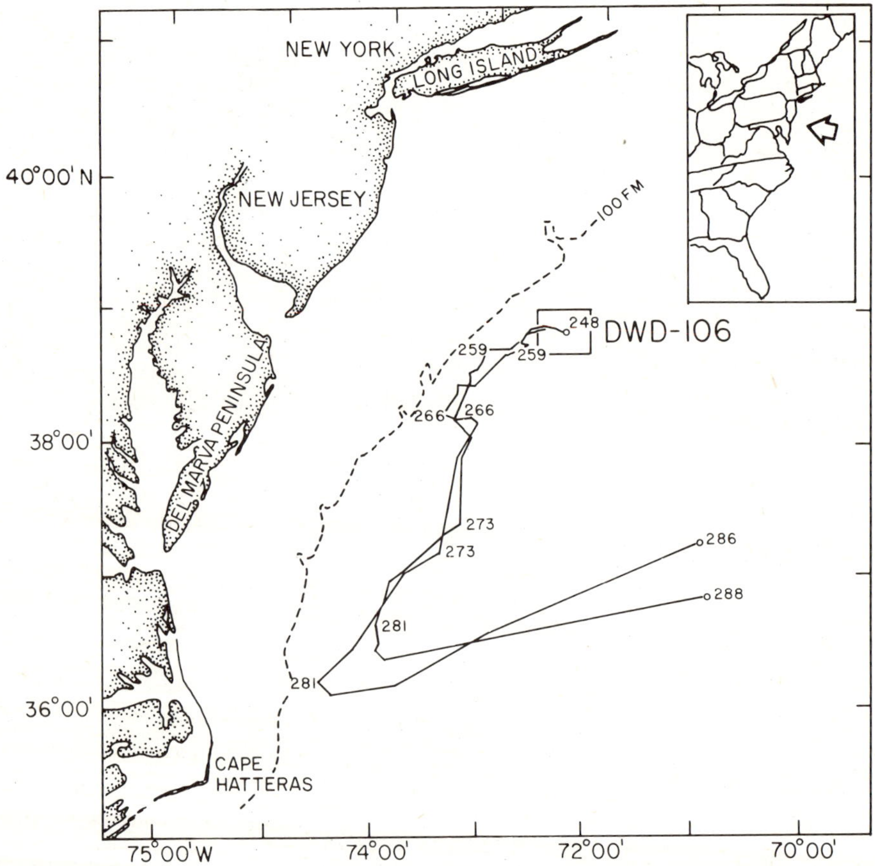

Figure 20.5. The deployment positions and trajectories of satellite traced drogued buoys from September through December 1980 (Bisagni, 1981). Numbers alongside the trajectories indicate Julian days.

ledge of these fundamental transport processes is a prerequisite to understanding fates and pathways of wastes in the ocean and also to the successful interpretation of specific ecological and chemical effects that were caused or influenced by waste dumping at sea. We must continue to develop and use mathematical models to describe the most important features of physical processes (Chapters 4 and 5).

In a more realistic sense a mathematical model to describe physical processes is a submodel of the integrated biological, chemical, and physical system. To improve predictive capability of waste effects, it is necessary to synthesize an integrated model which describes the most important biological, chemical, and physical processes. This has been attempted concisely by Flierl (Chapter 3) who treats the general problem of a waste released in a gyre at a deepwater dumpsite. The next step is to test this model.

20.3. DEEPWATER DUMPSITES

Two recent events in the United States point to a climate favoring increased ocean disposal of wastes in the future. First, the U.S. Environmental Protection Agency (EPA) was asked to revise its regulations governing the issuance of ocean dumping permits as a result of a litigation settlement with the City of New York. EPA could not prove the City's sewage sludge was harmful, and is now obtaining comments on a proposed action to designate Deepwater Dumpsite-106 (DWD-106) (Fig. 20.5) for the disposal of sewage sludge and other material (Environmental Reporter, 1981). Second, the National Advisory Committee on Oceans and Atmospheres (NACOA, 1981) recommended to the President and to members of the Congress that the U.S. policy on ocean disposal be reexamined since the scientific information available to date did not support the existing law forbidding ocean disposal of either sewage sludge after 31 December 1981 or of all industrial wastes. Two of NACOA's recommendations pertinent to the present discussion were quite specific:

> that ocean disposal of sewage sludge either by barge or through properly designed outfalls should continue to be a disposal option under appropriate management conditions and with adequate monitoring safequards in those areas where no unreasonable degradation of the environment results from sludge disposal.
>
> that ocean disposal of industrial wastes should continue at sites where evidence indicates no unreasonable environmental degradation and when human health, environmental, and economic considerations indicate that this is the preferable option.

Thus, it seems clearly evident that dispersive DWD-106 will be a strong candidate to receive existing and future industrialized and sewage wastes. At present there are no universally established procedures or guidelines to determine the suitability of a dumpsite for receiving wastes. GESAMP's (1981) discussion on biological, sedimentological, and physical considerations on dumpsite characteristics could form the basis of site selection criteria (Table 20.1); any criteria developed must also

Table 20.1. Environmental Characteristics to Be Considered on Oceanic Dumpsite Selection[a]

Physical	Sedimentology	Biological
Oceanic flow	Physical and chemical properties of wastes and sediment	Fishing grounds
Surface waves	Sorption capacity	Aquaculture sites
Wind-driven surface currents	Distribution coefficients	Breeding and nursery grounds
Interior circulation	Sedimentation	Migration routes
Turbulent diffusion	Sediment dispersion	Productivity
Shear diffusion	Bioturbation	
Vertical mixing	Sediment stability	
Modelling advection and diffusion		

[a]Based on GESAMP (1982).

consider economic factors, political or jurisdictional considerations, and resource conflicts. DWD-106, with it extensive dispersiveness, should be favored over shallower dumpsites near the shore for the disposal of wastes, especially for those wastes that do not appear to be biologically or chemically persistent in the marine environment. We do not advocate the dumping of persistent toxic wastes into the ocean. However, accidents do happen. For instance, a DDT-laden cargo ship may have to be disposed. In such instance, deep and offshore dumpsites often are less damaging than shallow and nearshore dumpsites on our society and marine ecosystems. The use of oceanic dumpsites close to shore for the disposal of wastes that either have not been characterized adequately or cannot be assimilated runs the great risk of creating short- and long-term problems for both the resident and transient biological systems in and near the dumpsite.

Although we could advocate dispersive deepwater disposal sites, we actually know little about behavior of wastes in the deepwater at these sites. Most of scientific works done at oceanic dumpsites have been confined to investigations of processes occurring in surface waters, the mixed layer, and at the thermocline. Biological communities living in the deepwater have adjusted to uniform water column properties (Sanders and Hessler, 1969). Variations in water column properties of the deepwater due to effects of dumping should be investigated with the ultimate aim of assessing impacts on pelagic biological communities.

20.4. RECYCLING VERSUS DISPOSAL

The steady and often expanding consumption of raw and processed materials used to support industry results in the production of waste. Wherever possible industry should either minimize the waste produced or reuse its waste through recycling. Recycling would decrease obviously the quantity of waste produced, thus minimizing potentially adverse environmental effects.

Recycling would have the added benefit of decreasing the rate of depletion of

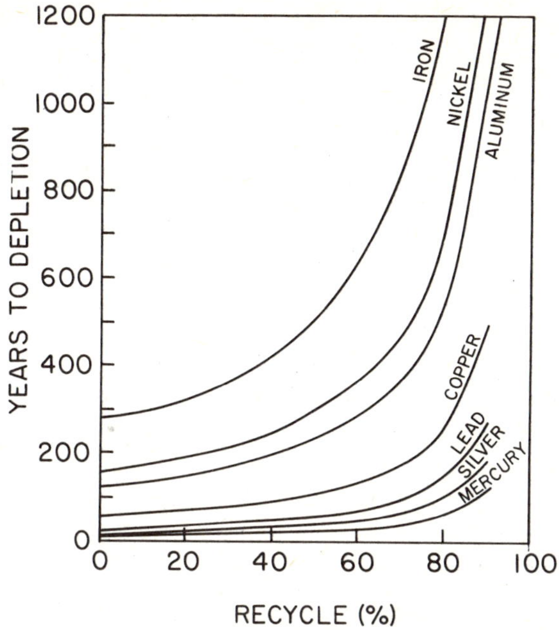

Figure 20.6. Projected hypothetical lifetimes of the present reserve of some common metals and the significance of recycling on prolonging their depletion. Figure adapted from Mann (1977).

resources. At present, for example, some of our natural resources, on a global basis, are being substantially depleted. Figure 20.6 shows the projected hypothetical lifetime of the present reserves of some commonly used metals, assuming present usage levels continue (Mann, 1977). It also shows the significance of recycling to prolong their depletion. We examined the composition of wastes in Chapter 1 and found them rich in a large number of elements. Elements such as copper, lead, silver, mercury, and others (see Mann, 1977) have relatively high rates of depletion and will be depleted from natural reserves if they are not recycled. For example, to ensure a 100-year supply, 50% of the copper, 70% of the lead, 80% of the silver, and nearly 90% of the mercury should be recycled. In addition, recycling may conserve a great quantity of energy (fuel) that is needed to refine metals such as aluminum.

The development and use of new technology should include recycling in its operation in order that waste be considered a resource. The practice of extracting silver from scrap and waste photographic supplies has recently become profitable as silver is in short supply (Fig. 20.6). Coal ash, an emerging waste that will be produced increasingly, is potentially a very large source of renewable metals (Chapter 1). The decision to extract important elements from coal waste will depend on, among other factors, the economics of the recycling procedure, market considerations, and the discovery of new materials that may be substituted for depleted elements. As we are forced to use lower grade ores, the economics of recycling wastes become more favorable. NACOA (1981) has provided other examples of recycling, but it then states that such efforts do not get very far because of Federal regulations, market structure, or public perception.

Although it is unfortunate that some societies are slow to recognize the economic and environmental benefits of recycling, it is equally comforting to know that the world's most populated country values its waste: From the book *China Men* (Kingston, 1980), we note the following passage: "He tarried so long that he had to use the outhouse, but he carefully weighed his shit on the outhouse scales so that these neighbors could return like amount to his fields."

20.5. INNOVATION

We recognize that it is impossible to eliminate waste. The waste disposal problem is ultimately governed by the second law of thermodynamics which, in simple terms, is: no real system is 100% efficient. Society can minimize the amount of waste it produces. Alternatively, it can try to develop innovative waste utilization programs.

Some innovative ideas include use of sewage effluent and sludge to improve aquaculture and to reclaim barren stripmined land, the use of polluted dredged material as a potential new material for ceramic manufacture, capping polluted dredged material (Chapters 2 and 14), and use of coal waste blocks for artificial reefs (Chapters 18 and 19). Often during the transition from waste disposal to waste utilization, a jurisdictional boundary is crossed and different environmental regulations restrict a proposed experiment or demonstration (NACOA, 1981). Thus, it becomes necessary to assess a new idea for utilization before being implemented. Answers to questions may be requested before research can be done to find the answers. Regulatory agencies should encourage the rapid development of innovative demonstration projects in waste disposal and make provisions in their regulations for such experimental projects. A case in point was experienced by one of us (I.W.D.) a few years ago during the initiation of the Coal Waste Artificial Reef Project (Parker et al., 1981), a demonstration study to determine the environmental acceptability of using coal waste blocks as artificial fishing reef substrate. If successful, coal waste reef blocks would be both an alternative solution to the disposal of coal wastes and a marine resource through underwater habitat development (Chapter 1). The project was almost stalled at the very beginning when, during the hearing to obtain approval for modification of an existing reef-building permit, answers to environmental impacts before the experiment could begin to determine them were asked to be presented.

Environmental litter provides a useful example of a waste that is derived from a nonrenewable resource, and for which innovative solutions are needed. Why do we increasingly use valuable and expensive petroleum products (plastics) for consumable containers when renewable, biodegradable, or recyclable products can serve this function? The accumulation of litter and the accelerated depletion of landfill capacity are a consequence of using long-lived materials for short-term functions. Some states in the United States have responded to this problem with the innovative solution of monetary deposits on returnable containers. Yet implementation of this approach in other states continues to be thwarted by political lobby interests using arguments that have been disproved.

The impact of marine litter is primary as aesthetic degradation of the coastal region, although litter ingestion by organisms may result in death due to blockage (Carpenter et al., 1972; Goldberg, 1976). In comparison with the potential and the real impacts of toxic substances, it may seem difficult to give much attention to the problem of marine litter. It is a real problem to the extent that we perceive it and are offended by it. One of us (D.R.K.) recently had the pleasure of an excursion to a preserved wildlife island off the coast of New Zealand. The remoteness of this island led to the development and continuation of a bird community without predators. While being impressed with the tameness of birds walking between one's legs or sitting fearlessly on the branch of a tree, there was also the impression that this remote island had not escaped the blight of marine litter. A walk along the rocky shoreline revealed the plastic artifacts of man not unlike those found on the shores of Long Island, New York. Similar observation is made along the coast of the Arctic Sea north of Canada and the United States; the plastic litters come from geophysical and oil exploration activities.

In addition to the visually evident litter along a shoreline there is evidence of persistant plastic objects distributed widely in the oceans. Carpenter and Smith (1972) reported finding plastic pellets in neuston net tows from the Sargasso Sea. The average frequency of occurrence was 3500 pellets per km^2. Carpenter et al. (1972) found polystyrene spherules in net tows of coastal New England waters. These spherules ranged from 0.1 to 2 mm in diameter. Morris (1980) found polyethylene and polypropylene pellets in neuston net tows in the South Atlantic ocean. He found 19 pellets from 10 stations and calculated an average frequency of occurrence of 2000 pellets per km^2. These and similar observations are evidence of widespread accumulation of persistent synthetic substances in the ocean resulting from casual waste disposal. The ecological impact of plastic particles in the ocean, if any, is not well known. These particles can be ingested by organisms, they can be sites for adsorption and concentration of surface active substances, and they can be colonized by hydroids and diatoms (Carpenter and Smith, 1972). The occurrence of marine litter is a sympton of inadequate waste disposal practices. It calls for innovative solutions such as greater use of recyclable and degradable materials. We must either minimize the release of these persistent wastes to the environment or we should strive to make synthetic products that are environmentally degradable by microorganisms, chemical decomposition (hydrolysis), or ultraviolet photolysis.

ACKNOWLEDGMENT

We thank the support of U.S. National Oceanic and Atmospheric Administration, Office of Marine Pollution Assessment. This work is Contribution 324 of the Marine Sciences Research Center of the State University of New York at Stony Brook.

REFERENCES

Bisagni, J. J. 1981. Near-Surface Currents at the 106-Mile Dumpsite 1979 and 1980. *In*: Report to the Congress on Ocean Dumping Monitoring and Research, January through December 1980. National Oceanic and Atmospheric Administration, Washington, D.C., pp. 2–8.

Carpenter, E. J., S. J. Anderson, G. R. Harvey, H. P. Miklas, and B. B. Peck. 1972. Polystyrene spherules in coastal waters. *Science*, **178**, 749–750.

Carpenter, E. J., and K. L. Smith, Jr. 1972. Plastics on the Sargasso Sea surface. *Science*, **175**, 1240–1241.

Davies, J. M., and J. C. Gamble. 1979. Experiments with large enclosed systems. *Philosophical Transactions of the Royal Society of London, Series B*, **286**, 523–544.

Environmental Reporter. 1981. Judge approves agreement outlining New York City, EPA ocean dumping terms. *In*: *Environmental Reporter* (27 November 1981), The Bureau of National Affairs, Inc., Washington D.C., pp. 929–930.

Fonselius, S. H. 1970. Stagnant sea. *Environment*, **12**, 2–11, 40–48.

Goldberg, E. D. 1976. The Health of the Oceans. The Unesco Press, Paris, 172 pp.

Goldberg, E. D. (Ed.). 1979. Proceedings of a Workshop on Assimilative Capacity of U.S. Coastal Waters for Pollutants, Crystal Mountain, Washington. Special Report. National Oceanic and Atmospheric Administration–Environmental Research Laboratories, Boulder, Colorado, 284 pp.

Hinga, K. R., M. E. Q. Pilson, R. F. Lee, J. W. Farrington, K. Tjessem, and A. C. Davis. 1980. Biogeochemistry of benzanthracene in an enclosed marine ecosystem. *Environmental Science and Technology*, **14**, 1136–1143.

IMCO/FAO/UNESCO/WMO/WHO/IAEA/UN/UNEP Joint Group of Experts on Scientific Aspects of Marine Pollution (GESAMP). 1982. *Annex 2*. Scientific Criteria for the Selection of Sites for the Disposal of Wastes at Sea. Rep. Stud. GESAMP Report (16), 48 pp.

Kingston, M. H. 1980. China Men. Knopf, New York, 308 pp.

Mann, A. W. 1977. Resources. *In*: Environmental Chemistry, J. O' M. Bockris (Ed.). Plenum Press, New York, pp. 121–178.

Morris, R. J. 1980. Plastic debris in the surface waters of the South Atlantic. *Marine Pollution Bulletin*, **11** (6): 164–166.

National Academy of Sciences. 1975. Assessing Potential Ocean Pollutants. National Academy of Sciences, Washington, D.C., 438 pp.

National Advisory Committee on Oceans and Atmosphere (NACOA). 1981. The Role of the Ocean in a Waste Management Strategy. Superintendent of Documents, U.S. Government Printing Office, Washington, D.C., 103 pp. + 6 appendices.

O'Connor, J. S., and H. M. Stanford. 1979. Chemical Pollutants of the New York Bight; Priorities for Research. National Oceanic and Atmospheric Administration, Environmental Research Laboratories, Boulder, Colorado, 217 pp.

Parker, J. H., P. M. J. Woodhead, and I. W. Duedall. 1981. Coal Waste Artificial Reef Program, Phase 3, Volume 2: Comprehensive Report. EPRI CS-2009, Electric Power Research Institute, Palo Alto, California, 4 Chapters (paginated separately) + appendix.

Sanders, H. L., and R. R. Hessler. 1969. Ecology of the deep-sea benthos. *Science*, **163**, 1419–1424.

Squires, D. F. 1981. The Bight of the Big Apple. New York Sea Grant Institute, State University of New York, Albany, 84 pp.

Steele, J. H. 1979. The use of experimental ecosystems. *Philosophical Transactions of The Royal Society of London, Series B*, **286**, 583–595. (Also published in: The Assessment of Sublethal Effects of Pollutants in the Sea. The Royal Society, London, 1979, pp. 185–197.)

Swanson, R. L., M. A. Champ, T. O'Connor, P. K. Park, J. O'Connor, G. Mayer, H. Stanford, E. Erdheim, and J. L. Verber. 1981. Sewage Sludge Dumping in the New York Bight Apex: A Comparison with Other Proposed Ocean Dumpsites. Third International Ocean Disposal Symposium (Abstract), Woods Hole Oceanographic Institution, Woods Hole, Massachusetts, 12–16 October 1981.

Waldichuk, M. 1979. Review of the problem. *Philosophical Transactions of The Royal Society of London, Series B*, **286**, 399–424. (Also published in: The Assessment of Sublethal Effects of Pollutants in the Sea. The Royal Society, 1979, London, pp. 1–26.)

Wooster, W. S. 1981. Review of Biological Effects of Marine Pollution and the Problems of Monitoring [Rapports et Proces-verbaux, 1980, A. D. McIntyre and J. B. Pearce (Eds.)]. *Deep-Sea Research*, **28A**, 1619–1620.

AUTHOR INDEX

Numbers in *italics* indicate pages on which full references appear.

SUBJECT INDEX

ATMOSPHERIC MOTION AND AIR POLLUTION
Richard A. Dobbins

INDUSTRIAL POLLUTION CONTROL—Volume I: Agro-Industries
E. Joe Middlebrooks

BREEDING PLANTS RESISTANT TO INSECTS
Fowden G. Maxwell and Peter Jennings, Editors

NEW TECHNOLOGY OF PEST CONTROL
Carl B. Huffaker, Editor

THE SCIENCE OF 2,4,5-T AND ASSOCIATED PHENOXY HERBICIDES
Rodney W. Bovey and Alvin L. Young

INDUSTRIAL LOCATION AND AIR QUALITY CONTROL: A Planning Approach
Jean-Michel Guldmann and Daniel Shefer

PLANT DISEASE CONTROL: RESISTANCE AND SUSCEPTIBILITY
Richard C. Staples and Gary H. Toenniessen, Editors

AQUATIC POLLUTION
Edward A. Laws

MODELING WASTEWATER RENOVATION: Land Treatment
I. K. Iskandar

AIR AND WATER POLLUTION CONTROL: A Benefit Cost Assessment
A. Myrick Freeman, III

SYSTEMS ECOLOGY: An Introduction
Howard T. Odum

INDOOR AIR POLLUTION: Characterization, Prediction, and Control
Richard A. Wadden and Peter A. Scheff

INTRODUCTION TO INSECT PEST MANAGEMENT, Second Edition
Robert L. Metcalf and William H. Luckman, Editors

WASTES IN THE OCEAN—Volume I: Industrial and Sewage Wastes in the Ocean
Iver W. Duedall, Bostwick H. Ketchum, P. Kilho Park, and Dana R. Kester, Editors

WASTES IN THE OCEAN—Volume II: Dredged Material Disposal In the Ocean
Dana R. Kester, Bostwick H. Ketchum, Iver W. Duedall and P. Kilho Park, Editors

LEAD AND LEAD POISONING IN ANTIQUITY
Jerome O. Nriagu